3D Video

3D Video

From Capture to Diffusion

Edited by
Laurent Lucas
Céline Loscos
Yannick Remion

WILEY

First published 2013 in Great Britain and the United States by ISTE Ltd and John Wiley & Sons, Inc.

ISTE Ltd
27-37 St George's Road
London SW19 4EU
UK

www.iste.co.uk

John Wiley & Sons, Inc.
111 River Street
Hoboken, NJ 07030
USA

www.wiley.com

© ISTE Ltd 2013
The rights of Laurent Lucas, Céline Loscos and Yannick Remion to be identified as the author of this work have been asserted by them in accordance with the Copyright, Designs and Patents Act 1988.

Library of Congress Control Number: 2013947317

British Library Cataloguing-in-Publication Data
A CIP record for this book is available from the British Library
ISBN: 978-1-84821-507-8

MIX
Paper from
responsible sources
FSC
www.fsc.org
FSC® C013604

Printed and bound in Great Britain by CPI Group (UK) Ltd., Croydon, Surrey CR0 4YY

Table of Contents

Chapter 7. Multi- and Stereoscopic Matching,
Stéphanie PRÉVOST, Cédric NIQUIN, Sylvie CHAMBON and
Guillaume GALES

Chapter 8. 3D Scene Reconstruction and Structuring 157
Ludovic BLACHE, Muhannad ISMAEL and Philippe SOUCHET

Foreword

The concept of giving 3D sense to flat representations (drawings, paintings, photos and films) has been progressively and deliberately re-examined and considered since the beginning of time. The rock paintings of Altamira (Spain) and Font-de-Gaume (France), for example, provide a fascinating example of the muscular systems of large herbivores. In the Lascaux cave (France), the shape of the rocks has been used to support and even accentuate the painting's form. All ancient art everywhere has, in some way or another, used depth and perspective in its representations, often awkwardly or confused, erroneous, often using more or less shared social codes, but always with the objective of understanding the real world beyond the limits of flat representation.

Formalized understanding of the mechanisms of Quattrocento perspective has largely enabled artists to move away from flat media to new, more accurate methods which have been used widely, often with competing artistic objectives and technical abilities. Complete perspective has therefore become an inseparable part of all pictures to the point of no longer even being a point of discussion: whether boring or shocking, controversial or exposing, it is no longer obvious because it is expected.

The dawn of photography, which by definition respects the canons of perspective, the undoubted problem of traditional representation, allowed artists to move away from this new norm which, over three centuries, had governed real-life representation. Artists can escape the unseen since, for example, space is no longer merely confined to perspective. Braque and Picasso, Klee and Bacon have shown us that this space is not only a matter of geometry but is also richer and holds several mysteries. Beyond perspective, it allows us to see background images and their convergence.

However, perspective, outside this small artistic field where it has somewhat faded, plays a vital role in our vision, logic and society. Unsurprisingly, the world of photography, as in painting, has quickly sought media which go beyond flat representation. Since the 19th Century, ingenious inventions have provided a third dimension to photography and then, with the dawn of cinema and its younger sibling television, it would not be long until 3D would make an impact, well before the Second World War. Binocular stereovision is the most natural input method for this mode, reliant on various separation means for optical paths, orthogonal polarizations, color decompositions, color flickers through wheels and mirrors and lens networks. Kerr's or Pockel's electromagnetic cells and liquid crystals will be examined later as part of this.

3D has not yet finished developing. Propelled by undeniable economic and social success, it has suffered from a lack of exploration followed by a new found success. The literature is evidence of this and that we are on the brink of a new dawn. However, current technologies are undeniably better than ever. Acquisition, projection, archiving and transmission technologies have come to fruition after long being suspended or in development. It has also been an opportunity and major development for production companies and commercial film distribution organizations, since virtual and augmented reality production has reached previously unseen levels of quality, performance and productivity which are indispensable for ambitious and demanding production sets. The public, expectant and demanding, desires new experiences which can be seen as evidence of the success of these new methods.

Finally, all these factors, which have made this dawn of 3D cinema possible, have played an important role in 3D television because these two fields, cinema and video, have a shared future. The opposing war between them, which has raged for 50 years, to capture an audience seen to favor one over the other, has now disappeared. We only have to think of the success of films on television or the continuation of television series through films. The public is omnivorous, consuming all kinds of images, no longer knowing whether they are from a dark room, a small screen or even a video game. This requires an abundance of pixels, bright, life-like colors and multi-sensory interaction and interactive and 3D animation, particularly when their counterparts exist in real-life but are transformed by video, as discussed in this book.

It is this which allows us to trace the progression of 3D, which has affected the entire chain of production for digital images. This book aims to examine ongoing events and describe their development, with a formal representation of theoretical tools in order to understand the approaches studied. References are provided to allow the reader to further study the developments that these numerous techniques relate to. Another aspect relates to examining all points in the technical chain which today governs 3D television. We will also examine technical tools such as cameras, screens and software. In addition , matching, detection and compression will be studied.

As a complete and complex work, *3D Video* is a welcome to the current efforts and achievements which have accompanied the the emergence of this new addition to our homes, the 3D image.

Henri MAÎTRE
September 2013

Notations

Spaces, sets

$d \in \{2, 3, \ldots\}$ space dimension

$\mathbb{R}^d$ real d-dimensional vector space

$\mathbb{Z}^d$ integer d-dimensional vector space

$][a, b][$ compact interval in $\mathbb{R}$

$\{a, \ldots, b\}$ discrete interval in $\mathbb{Z}$ or $\mathbb{N}$

$\mathbb{B} = \{false, true\}$ boolean set

$\mathbb{N}_n = \{0, \ldots, n-1\} \subset \mathbb{N}$ set of n first natural integers

$\mathbb{Z}_{a,b} = \{a, \ldots, b-1\} \subset \mathbb{Z}$ set of $b - a$ integers connecting from a to $b - 1$

Functional notations

Y^X set of applications $X \rightarrow Y$

$Y[X]$ set of tables or maps on domain X (discrete) with values in Y

$f(b)$ value of the function f in b

$t[b] \equiv t_b$ value of the sample b of the map/table t

$\lfloor x \rfloor$ integer part (the nearest inferior integer) in x

$\lceil x \rceil$ nearest superior integer to x

$\{x\} \equiv x - \lfloor x \rfloor$ fractional part of x

$[x]$ the nearest integer to x: by convention, if $x = k + 0.5$ with $k \in \mathbb{Z}$, $[x]$
$$= \begin{Bmatrix} k & \text{if } k >= 0 \\ k + 1 & \text{otherwise} \end{Bmatrix}$$

Objects

i, j, k, l, m, n	integer numbers
x, y, z	coordinates (integer or real)
t, u, v, λ, μ	real numbers
D, Δ	real lines
P, Π	real planes
$\mathbf{v}, \mathbf{w}$	vectors
$A, B, C, ...$	points in the real affine space $\mathbb{R}^2$ or $\mathbb{R}^3$
AB	bi-point vector ranging from A to B
$\mathbf{M}, \mathbf{A}, \mathbf{B}$	matrices
$\mathbf{R}, \mathbf{T}$	rotation and translation matrices
f, g, h	applications, functions
Φ, Ψ	operators on other sets $\mathbb{R}^d, \mathbb{Z}^d$
ρ, τ	rotation and translation function
G, Γ	graphs
θ, ϕ	angles
ε	threshold

Digital images

$$\Omega^{\mathbf{t}} = \prod_{i=1}^{d} \mathbb{N}_{t_i}, \quad \mathbf{t} \in \mathbb{N}^d$$

d-dimensional signal indexation domains or spaces with a size of $\mathbf{t} = (t_1, \ldots, t_d) \in \mathbb{N}^d$

$$\Omega^d \equiv \Omega^{(nc, nl[, np])}$$
$$= \mathbb{N}_{nc} \times \mathbb{N}_{nl} [\times \mathbb{N}_{np}]$$

for $d = 2[3]$ we will use the numbers of columns, lines and planes by default $(nc, nl[, np])$

$$\mathbf{p} = (x, y[, z]) \in \Omega^d$$

pixel position ($d = 2$) or voxel position ($d = 3$) within the indexation domain

$$\mathcal{C} = \mathbb{X}^{(b)}$$
$$\text{with } \mathbb{X} \in \{\mathbb{N}, \mathbb{Z}, \mathbb{R}\}$$

scalar spaces of digital values for sample components. The $b \in \{8, 10, 12, 16, 32, 64, \ldots\}$ correspond to the number of bits used for coding real values ($\mathbb{R}$) or integer values ($\mathbb{Z}$ or $\mathbb{N}$)

$$\mathcal{E} \equiv \mathcal{E}^{(\mathcal{C}_1, \ldots, \mathcal{C}_c)} = \prod_{i=1}^{c} \mathcal{C}_i$$
$$\text{with } \mathcal{C}_i = \mathbb{X}_i^{(b_i)}$$

generic vectoral space for samples' digital multi-component value

$$\mathcal{S} \in \mathcal{E}[\Omega^d]$$

2D or 3D signal (image or volume) table or map of values in $\mathcal{E}$ indexed by Ω^d

$$\mathcal{S}[\mathbf{p}] \equiv \mathcal{S}_{\mathbf{p}} \in \mathcal{E}$$

the signal sample $\mathcal{S}$ in position $\mathbf{p} \in \Omega^d$

Neighborhoods

$$\mathcal{F} = \{\mathbf{v} \in \mathbb{Z}^d\} \subset \mathbb{Z}^d$$

d-dimensional shape, set of vectors $\mathbf{v} \in \mathbb{Z}^d$ coding the positions of each point in the shape in relation to the reference point

$$\mathcal{F}_{\mathbf{p}} = \{\mathbf{p} + \mathbf{v} \mid \mathbf{v} \subset \mathcal{F}\}$$

neighborhood in the shape $\mathcal{F}$ placed in $\mathbf{p} \in \Omega^d$: translated of $\mathcal{F}$ in or by $\mathbf{p}$

$$\overline{\mathcal{F}}_{\mathbf{p}} = \mathcal{F}_{\mathbf{p}} \cap \Omega^d$$
$$= \left\{\mathbf{p} + \mathbf{v} \,\middle|\, \begin{array}{l} \mathbf{v} \in \mathcal{F} \\ \mathbf{p} + \mathbf{v} \in \Omega^d \end{array}\right\}$$

neighborhood in the shape $\mathcal{F}$ en $\mathbf{p} \in \Omega^d$, truncated by Ω^d

Multiviews

A set of N signals (known as views within the context of this book) of the same dimensions d and sizes ($nc \times nl \; (\times \; np)$) will both be considered as a table of N signals ($\in \mathcal{E}[\Omega^d][\mathbb{N}_N]$) and as a signal with a superior dimension of $d + 1$ ($\in \mathcal{E}[\Omega^d \times \mathbb{N}_N]$).

$\Upsilon^d = \Omega^d \times \mathbb{N}_N$	global indexation space of a set of N views (images or volumes) with a dimension of d
$\mathcal{M} \in \mathcal{E}[\Upsilon^d] \equiv$ $\mathcal{M} \in \mathcal{E}[\Omega^d][\mathbb{N}_N]$	multi-signal, N views indexed by Ω^d with values in $\mathcal{E}$
$\mathcal{M}[i] \equiv \mathcal{M}_i \in$ $\mathcal{E}[\Omega^d]$	digital view number $i \in \mathbb{N}_N$ in $\mathcal{M}$: d-dimensional signal
$\mathbf{r} = (\mathbf{p}, i) \in \Upsilon^d$	multiview sample index: position $\mathbf{p} \in \Omega^d$ and digital image $i \in \mathbb{N}_N$
$\mathcal{M}[\mathbf{r}] \quad \equiv \mathcal{M}_{\mathbf{r}} \quad \equiv$ $\mathcal{M}[\mathbf{p}, i] \equiv \mathcal{M}_{\mathbf{p},i} \equiv$ $\mathcal{M}[i][\mathbf{p}] \equiv \mathcal{M}[i]_{\mathbf{p}} \equiv$ $\mathcal{M}_i[\mathbf{p}] \quad \equiv \mathcal{M}_{i_{\mathbf{p}}}$	different expressions shown as equivalents to reach the value in $\mathcal{E}$ for the sample in position $\mathbf{p}$ in the view i from $\mathcal{M}$. the last (double level indice) should be avoided

Acknowledgments

We are very grateful to those who have contributed to this book through their work and research. We would like to particularly express our gratitude to Henri Maître who has overseen the compilation of this book and has generously given us his help and support. We would also like to show our recognition to ISTE Ltd. and John Wiley & Sons who have greatly assisted as throughout the production of this book. Lastly, we would like to thank all those people and organizations who have allowed us to use their data and/or illustrations within this book.

Several pieces of research data shown in this book have been made possible because of the financial support from the following organizations:

– Regional organizations from three areas: projects including CIA (CPER Nord-Pas De Calais region in France), CREATIS (CPER Champagne-Ardenne region in France) and RUBI3 (Brittany with the Image et Réseaux (Image and Networks) cluster – 2010-13).

– The Agence nationale de la recherche (ANR) (National Agency for Research, France): including projects such as SEMANTIC-3D (AAP RNRT 2002-06), CAM-RELIEF (AAP RIAM 2008-10), FAR3D (AAP CSOSG 2008-2010), COLLAVIZ (AAP COSINUS 2009-12), PERSEE (AAP Blanc 2009-13) and 3D FaceAnalyzer (AAP Blanc Inter. 2011-13).

– Competition clusters and/or Fonds unique interministériel (FUI) (Inter ministerial Funds, France): including the projects FUTURIM@GE (Image et Réseaux (Image and Networks) – 2008-10), Terra Numerica (Cap Digital – 2006-09) and 3DLive (CapDigital, Imaginove, Images et Réseaux – 2009-12).

– The Fond national pour la société numérique (FSN) (National Fund for Digital Society, France): including the RECOVER3D project (2012-2014) for future investments.

– The European Commission: including projects such as 3D Média (FEDER 2008-13), 3D-ConTourNet (ICT COST Action IC1105), HDRi (ICT COST Action IC1005), Ed-cine (FP6 IP), Apidis (FP7 Strep) and JEDI (ITEA2, co financed by DGCIS for the French Ministry of Industry).

We would like to thank these institutions for their past and ongoing support.

Lastly, we would also like to recognize those who, by their patience, understanding and encouragement, have allowed us to bring this project to completion.

Laurent Lucas, Céline Loscos and Yannick Remion
September 2013

Introduction

The extension of visual content to 3D as well as dynamically capturing scenes in 3D to generate an image on a remote site in real time has long been considered merely a part of science fiction. Today they are a reality, collectively referred to by terms such as 3D television (3DTV), free viewpoint TV (FTV) and, more generally, 3D video. This new type of image creates the illusion of a real environment, resulting from continually improving efforts in research and development over a number of years.

Numerous experts believe that 3D represents the future of media, such as television and the Internet, and will in turn improve the quality of visual experiences for the end user. The whole chain of content production must be reconsidered, beginning with recording techniques, since those designed specifically for 3D are far more numerous and varied than those used normally in conventional 2D context. The same can also be said of other aspects, such as, for example:

– the description and representation of scenes according to more or less informative structures, ranging from multiview or multiview-plus-depth videos to 3D digital reconstructed models;

– 3D reconstruction which extracts 3D models in various forms from videos acquired from multiple viewpoints, such as static or animated meshes;

– the compression of representations of scenes created by capture (stereoscopic or multiview videos) or reconstruction (3D models);

Introduction written by Laurent LUCAS, Céline LOSCOS and Yannick REMION.

– 3D display, with or without adaptation/enhancement of content and/or intermediate view synthesis.

The democratization of these technologies needs specifically designed display devices. Stereoscopic or autostereoscopic screens show a heavy tendency toward this while their use for displaying 3D content today still poses a number of problems, showing that all these techniques must yet be perfected to avoid being rejected by the end user due to reasons of poor quality and/or eyestrain.

3D videos therefore cover a multitude of aspects, collectively linking a series of recorded videos to full depth 3D visualizations, potentially using estimations of depth in video sources. The developments examined here are therefore based on methods and tools from highly varied fields, such as applied mathematics, computer imaging, computer graphics, virtual reality, signal processing as well as psychophysics and the psychology of human vision.

In this highly multidisciplinary context, the objective of this book focused on 3D video is twofold since it aims, in addition to summarizing current information about the subject, to provide:

– for students: a solid base enabling readers to carry out activities relating to this topic and to learn the underlying concepts overall;

– for researchers: as complete a reference for this subject as possible which precisely indicates current research and understanding in this field as well as future trends and perspectives.

Its organization into four parts is due to a desire to cover all phases of 3D video by bringing together formal presentations of theoretical tools and developments of more technical or technological aspects. It should be noted that all figures are also available in color at http://www.iste.co.uk/lucas/3D.zip.

The first part of this book runs through the basics of 3D video and the recording of its characterizing multiview videos. This begins with, in Chapter 1, the different fundamental aspects of this technology. Historical and mathematical aspects relating to 3D computer vision and physiology of human vision are thus presented. Chapters 2 and 3 look at technological and methodological problems in relation to capturing images, more specifically in Chapter 3 within a multiview context that characterizes 3D video. The specification of geometric elements relating to the recording and display of

3D media is then examined in Chapter 4. Chapter 5 concludes Part 1 of this book, focusing on the problems of geometric and colorimetric camera calibration.

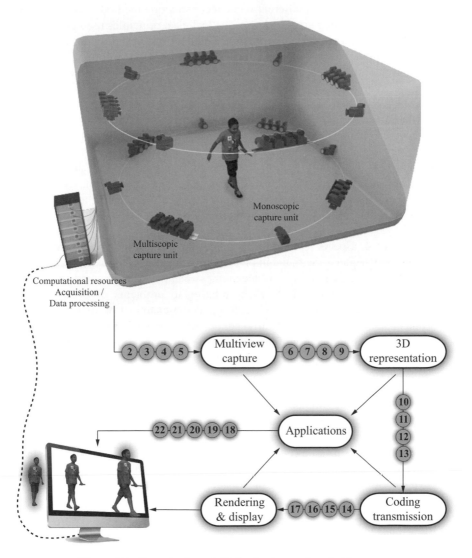

Figure I.1. *Organization of this book: the numbered chips correspond to the different chapters*

Part 2 focuses on the description and reconstruction of 3D scenes. Chapters 6 and 7 analyze the problems of local feature detection, stereo-matching and stereo-correlation through dense depth estimation. Chapter 8 then presents different scene reconstruction methods, notably using silhouettes, providing an overview of the technical principles used to structure previously constructed 3D information. Finally, Chapter 9 provides an outline of intermediate view synthesis using images with depth information. Direct and inverse projection approaches are examined alongside a description of uncovered areas filling methods.

In Part 3, the field of compression and transmission norms for 3D content is covered. Chapter 10 in particular introduces 3D video formats as well as specific techniques for stereoscopic and multiview stream coding. In Chapter 11, the multiresolution compression of meshes and mesh sequences is examined in terms of standardization and visual perception. Chapter 12 then focuses on depth video coding while Chapter 13 presents the problem of protecting stereoscopic videos by watermarking the 3D stream.

In Part 4, aspects relating to 3D rendering and display are covered. This begins, in Chapter 14 with the implementation and use of autostereoscopy and is followed, in Chapter 15, by techniques relating to augmented reality. In Chapter 16, psychophysical effects relating to problems of eyestrain and visual discomfort are discussed with a specific examination of flaws in 3D content and technologies which generate these unusual stimuli. Chapter 17 focuses on the delicate problem of 2D-to-3D conversion which remains in between technology and the arts where human intervention remains indispensable.

The practical implementation of all these technologies and their applications are considered in Part 5, the final part of this book. Aspects relating to data mining (Chapter 18), high dynamic range videos (Chapter 19), biomedical visualization (chapter 20) and sport scene reconstruction (Chapter 21) are covered. This final part is concluded by an overview of experiments in live recording and transmitting of 3D stereoscopic videos (Chapter 22).

PART 1

3D Acquisition of Scenes

Chapter 1

Foundation

1.1. Introduction

Audiovisual production has, for a number of decades, used an increasing number of ever more sophisticated technologies to play 3D and 4D real and virtual content in long takes. Grouped under the term "3D video", these technologies (motion capture (Mocap), augmented reality (AR) and free viewpoint TV (FTV) and 3DTV) complement one another and are jointly incorporated into modern productions. It is now common practice to propose AR scenes in FTV or 3DTV, either virtual or real, whether this relates to actors, sets or extras, giving virtual characters (both actors and extras) realistic movements and expressions obtained by Mocap, and even credible behavior managed by artificial intelligence.

With the success of films such as *The Matrix* in 1999 and *Avatar* in 2009 (see Figure 1.1), the acronym "3D" has become a major marketing tool for large audiovisual producers. The first, *The Matrix*, popularized a multiview sensor system containing 120 still cameras and two video cameras allowing slow motion virtual traveling, an effect known today as *bullet time*. This system has since been subject to various improvements which today not only allow the reproduction of this type of effect (FTV), but also for complete or parts of 3D reconstructions of scene content. The success of Avatar marked the renaissance of 3D cinema, a prelude to 3DTV even if it is not yet possible

Chapter written by Laurent LUCAS, Yannick REMION and Céline LOSCOS.

to free viewers from wearing 3D glasses. Glasses-free, or "autostereoscopic", 3D display is undeniably advantageous in comparison to glasses-oriented technology due to its convincing immersive 3D vision, non-invasiveness and only slightly higher production costs in relation to 2D screens. Unfortunately, the need of multiple viewpoints (generally between five and nine) to yield immersion involves a spatial mix of these multiple images which limits their individual resolution. As a result, in contrast to stereoscopy with glasses, autostereoscopic visualization is not yet available in full HD. The induced loss of detail in relation to this current standard further limits its use. The principle challenge of autostereoscopy currently concerns the conversion of the overall dedicated tool chain into full HD.

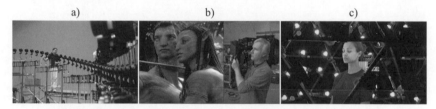

Figure 1.1. *Multiview system used to film The Matrix©Warner Bros. Entertainment Inc. a): 120 still cameras and two video cameras enabling time slicing (bullet time effect); b): stereoscopic filming; c): omnidirectional 3D capture for Avatar©20th Century Fox by James Cameron*

This profusion of technologies, a veritable 3D race, is probably the result of the rapid banalizing of effects presented to the public, despite the fact that the technologies used have not yet been fully perfected. This race therefore evidently raises further challenges. All these techniques have a point in common. They rely on multiview capture of real scenes and more or less complex processing of the resulting recorded media. They also raise a series of problems relating to the volume of data, at each stage of the media chain: capture, coding [ALA 07], storage and transmission [SMO 07], concluding with its display. It is therefore essential to be able to synthesize the characteristics of this data as systems which mark their use in order to consolidate the bases of this technological explosion.

It is this point, which is the central proposal of this book, which examines two interrelated fields of this technological domain, as summarized by Kubota *et al.* [KUB 07]:

– 3D video technologies which aim to reconstruct varying scene characteristics (geometry, lighting and movement) for various uses;

– 3DTV/FTV technologies which focus on displaying in 3D, sometimes interactively; 3D scenes with less precise reconstruction requirements but which raise more clearly the challenges of transmitting and coding 3D or multiview medias.

The aim of this chapter is to introduce the fundamental principles of 3D videos and the techniques involved in this. In the following section, we will examine an overview of the different periods of history which have marked the development and formalization of 3D. Notably, we will detail the geometric principles related to central projection (pinhole cameras) without extending these developments to stereovision, the principles of epipolar geometry [HAR 04] exposed in Chapters 3, 4 and 5. We will then examine aspects relating to the physiology of human vision before concluding, with a more taxonomic perspective, by proposing a classification of 3D visual approaches.

1.2. A short history

The term "3D images" is the name given to what was known as "perspective" during the Renaissance period. While new developments concerning 3D arose during this period, with the appearance of the first 3D drawing machine (see Figure 1.2), consciousness of this sensation, as was its corollary–3D perception is far more ancient and founded during Antiquity.

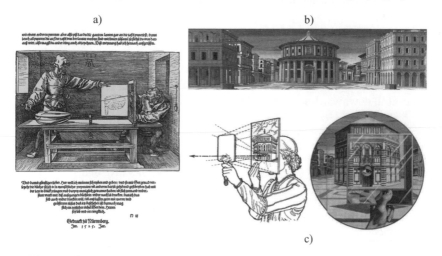

Figure 1.2. *a): the Dürer perspectograph; b): the ideal city (1475) from Piero della Francesca, c): Brunelleschi experiment*

In this section, we present a brief overview of different periods which saw the development and theorization of 3D and its extension to stereoscopy using binocular vision. These two aspects mentioned in the following sections are independent of one another for practical reasons, as they need to be examined from a more global perspective, defining our relation to imaging.

1.2.1. *The pinhole model*

The pinhole camera, or *camera obscura*, was the precursor to the modern-day camera. It is composed of a dark room with a narrow hole, from which its name is derived, by which exterior lit objects are projected, in reverse, onto the opposite internal side of the dark room.

This principle was first described by the Mohists, a pacifist Chinese sect, in a collective work [MOH 00] written around 400 B.C. under the pseudonym Mo Zi. Aristotle also referred to it in the 4th Century B.C. [ARI 36]. Its first mathematical formulation was proposed by the Persian mathematician Alhazen (Ibn Al-Haytham) [ALH 21], one of the founders of optics, notably for his descriptions of vision. In 1515, Leonardo da Vinci detailed the principle and noted that, to produce a clear image, the hole must not exceed 0.5 mm in diameter [VIN 19]. In 1556, his Italian friend Girolamo Cardano placed a convex glass lens in front of the hole which provided images with hitherto unseen clarity [CAR 56]. This added the photographic lens to his long list of scientific and technical contributions[1].

1.2.1.1. *A modern-day form of expression*

As a result, the pinhole camera is, first and foremost, a simple yet antiquated imaging device. Its principle of central projection on a plane is illustrated in Figure 1.3 that shows the object/image inversion resulting from the central downward-projection through the hole.

---

1 Among other things, we can thank Girolamo Cardano for his eponymous resolution method for quartic and cubic equations, the first use of negative and subsequently imaginary (or, in his words "fictive") numbers, previously discovered by the Hindus and then by the Fibonacci in the 13th Century, a first formulation with Raphael Bombelli of complex numbers (under the name "impossible numbers"), major, pioneering contributions to statistics, probabilities, cryptography (the Cardan grille), numerous therapeutic and diagnostic contributions to medicine, Cardan suspension and joints in mechanics, and the Baguenaudier (also known as Cardano's rings), in addition, to the photographic lens.

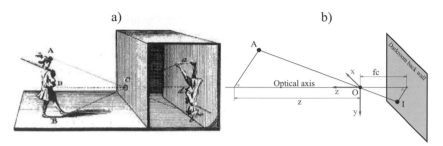

Figure 1.3. *A pinhole camera* (camera obscura)*:*
a): illustration from The Encyclopedia of Diderot & d'Alembert;
b): geometric model of the central projection involved

The geometric optical model of this device is shown in Figure 1.3. The center of projection O is the hole, located at a distance of fc from the back of the darkroom to which the optical axis is orthogonal while passing through O. It is usual to define a "viewer" orthonormal reference frame $(O, \mathbf{x}, \mathbf{y}, \mathbf{z})$, with $\mathbf{z}$ being orthogonal to the back plane of the darkroom and directed, like the implicit viewer, toward the outside of the room: $\mathbf{x}$, for example, is "horizontal", directed toward the right of the presumed viewer and $\mathbf{y} \equiv \mathbf{z} \times \mathbf{x}$.

This model gives the relation $OI = -fc/z_A.OA$ which explains the observed inversion and characterizes the projection equation in $(O, \mathbf{x}, \mathbf{y}, \mathbf{z})$ in Cartesian [1.1] as well as homogenous [1.2] coordinates:

$$
\begin{pmatrix} x_I \\ y_I \\ z_I \end{pmatrix} = -\frac{fc}{z_A} \cdot \begin{pmatrix} x_A \\ y_A \\ z_A \end{pmatrix} = -fc \cdot \begin{pmatrix} x_A/z_A \\ y_A/z_A \\ 1 \end{pmatrix} \tag{1.1}
$$

$$
\begin{pmatrix} x_I \\ y_I \\ z_I \\ 1 \end{pmatrix} = \lambda \cdot \begin{pmatrix} 1 & & & \\ & 1 & & \\ & & 1 & \\ & & -\frac{1}{fc} & 1 \end{pmatrix} \begin{pmatrix} x_A \\ y_A \\ z_A \\ 1 \end{pmatrix} \quad \text{with} \quad \lambda = -\frac{fc}{z_A} \tag{1.2}
$$

1.2.1.2. *From the pinhole to the camera*

The pinhole camera, a relatively simple design, is occasionally used today despite several disadvantages that led to the common use of its successor, the modern-day still camera:

– The hole must be narrow to maintain a clear image. The image at the back of the room of a lit point at the depth z is generated uniquely by the beams emitted by this point and passing through the hole, forming a spot of light in the same shape as the hole dilated by a factor of $1 + fc/z$.

– It cannot be too narrow to avoid too much diffraction at its edges as this may create blurring.

– The tiny surface area of the hole yields a weak illumination at the back of the room which requires a long exposure time and induces risk of motion blur.

To lessen these difficulties, according to Girolamo Cardano, the still camera replaces the hole with an objective composed of a lens or, more generally, an axial collection of lenses and a diaphragm, or iris, which controls the aperture of the admitted incidental conical beams. This camera lens improves the illumination at each point at the back of the room which facilitates the consensus between exposure time and depth of field. It solves the problems of diffraction that occur with pinhole cameras but has its own specific drawbacks:

– A depth of field controlled by the iris, yet more limited in a pinhole device because the solid angle of the conical incident and refracted beams is generally greater.

– Geometric aberrations (spherical, coma, etc.) related to thick lenses which cannot ensure perfectly precise convergence of the refraction of a conical incident beam generate a wider projection of this beam at the back of the room, even if it comes from the optimal distance.

– Chromatic aberrations related to variations in the refractive index for different wavelengths which disperse, as they exit the lens, the colored components initially carried together by incident rays.

– Radial distortions corresponding to an axial displacement of the actual optical center according to the main beam incident angle. As a result, convergences at the back of the darkroom exhibit radial barrel or pincushion deformations.

1.2.1.3. *A standard digital camera model*

These problems can be mitigated by using complex lenses (aspherical, achromatic, apochromatic, etc.) and/or digital post-processing of images with prior calibration. As a result, these improvements/corrections are generally presumed to be preprocessed when using images taken by a digital camera. This leads to the assumption that these images have been shot via central projection on a sensor placed on the back wall of the darkroom. This approximation, which ignores the impact of a non-pinhole shutter, is valid with regard to the acceptable depth of field of the digital images. It refers to the zone of depth within which a 3D point is projected at the back of the darkroom as an area smaller than the interpixel space (or pitch) of the sensor.

It should be stated that this model is somewhat of a throwback because it is exactly the model of the pinhole device, the forebear of the modern camera.

Some digital applications use non-central zones in the sensor. This is notably the case for multiview acquisition with decentered parallel geometry (see Chapter 4). Let us examine a simplified geometric model (pinhole shutter) of an ideal camera (whose residual flaws are corrected by post-processing), corresponding to a central projection through an optical center O on a decentered rectangular region of interest (ROI) in a 2D digital sensor, placed at the back wall of the darkroom. This model, which can be termed a "decentered pinhole" extends the pinhole model (centered) from which it differs only through its ability to decenter the sensor's ROI. This book uses this characteristic and this more general model. More specifically, the sensor is placed (at the back wall of the darkroom) at a distance of fc from O, has a pitch of (ph, pv) and its ROI has a size of (nc, nl) which is potentially decentered by (cx, cy) pixels in relation to the optical axis (see the downward (bottom) part of Figure 1.4).

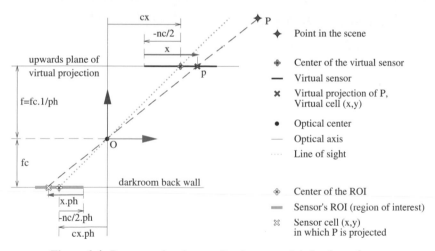

Figure 1.4. *Decentered and normalized reverse pinhole: from above (according to the **y**-axis), the projective geometries of the real sensor and its normalized virtual representation*

The inversion involved in these models, whether centered or not, between the axes of the image and that of the setting is generally countered by producing an upward rather than downward inverted projective model, known as a "reverse pinhole". The placement of the "virtual sensor" (a reverse upward avatar of the sensor's ROI) can therefore be optimized so that distances in pixels and "metric" distances can be confused, at least horizontally. It is sufficient to place the virtual projection plane, upwards, at a distance of $f = fc.1/ph$ of O. This ensures a unitary horizontal pitch for the virtual sensor whose vertical pitch is therefore equal to the anamorphosis relation $\rho = pv/ph$ of the real sensor. We refer to this as a "normalized reverse pinhole model".

The "decentered and normalized reverse pinhole model", a version decentered from its predecessor, is presented in Figure 1.4. The downward part (bottom) in the figure shows the direct decentered model of the ROI of the real sensor, while the upward part (top) presents the reverse model associated with the virtual sensor. Some specific data relating to the real sensor, its ROI and its virtual sensor includes the following:

– the sensor has a pitch of (ph, pv);

– its ROI has a size of (nc, nl) and is decentered by (cx, cy) pixels;

– its center is therefore situated at $-(cx.ph, cy.pv, fc)$ in $(O, \mathbf{x}, \mathbf{y}, \mathbf{z})$;

– a real cell (x, y) is situated at $-((x - \frac{nc}{2} + cx).ph, (y - \frac{nl}{2} + cy).pv, fc)$;

– the virtual sensor has a pitch of $(1, \rho)$;

– with a size of (nc, nl) and is decentered by (cx, cy) pixels;

– its center is therefore situated at $(cx, \rho.cy, f)$;

– a virtual cell (x, y) is situated at $(x, \rho.y, f)$.

This modeling characterizes the projection equation in the virtual sensor, in Cartesian [1.3] and homogeneous [1.4] coordinates:

$$\begin{pmatrix} x \\ y \end{pmatrix} = f. \begin{pmatrix} x_P/z_P \\ y_P/(\rho.z_P) \end{pmatrix} \tag{1.3}$$

$$\begin{pmatrix} x \\ y \\ f \\ 1 \end{pmatrix} = \lambda. \begin{pmatrix} 1 & & & \\ & \frac{1}{\rho} & & \\ & & 1 & \\ & & \frac{1}{f} & \end{pmatrix} \begin{pmatrix} x_P \\ y_P \\ z_P \\ 1 \end{pmatrix} \quad \text{with} \quad \lambda = \frac{f}{z_P} \tag{1.4}$$

We have seen that the pinhole device shares its projective model with the idealized version of its technological descendent (ideal camera with a point aperture). We have also provided a reverse, normalized and decentered version of this model which is useful, in a variety of contexts, including this book, for modeling corrected shots of digital images captured by real or virtual cameras.

1.2.2. Depth perception and binocular vision

The basic principles of 3D vision have also evolved during several periods marked by significant technological developments. As a result, in antiquity, as indicated previously, Euclid stated in his manuscript *Optics* that depth perception is "to receive in each eye the simultaneous impression of two different images of the same subject".

1.2.2.1. Pre-20th Century

It was during the Renaissance that a new stage in the development of stereoscopy came into existence. This marked one of the first explanations for the notion of parallax as the basis of understanding binocular vision, notably through the work of Jacopo Chimenti (1551–1640) of the Florentine School. It would not be until the 19th Century that the birth of photography and inventions such as Wheatstone's "stereoscope" (a stereoscopic display device, see Figure 1.5), where two reversed images are reflected by two mirrors at an angle of 90 degrees, arose. At the same time, in 1838, Brewster developed a means of reproducing 3D vision using two images. Two years later, in 1840, photography would be invented (by Daguerre in Paris and Fox Talbot in London) and the first stereoscopic photographs would be obtained. In 1844, Brewster improved his stereoscope by adding lenses to it, rendering it more compact and easier to use than models using mirrors, and described in 1849 as the first stereoscopic still camera. The distribution of the stereoscope [MIC 11] witnessed a veritable explosion, not only through the invention of devices developed primarily in Britain and France but also due to the appearance of a number of amateur clubs. One of the most popular of these models was that invented by Holmes in 1860 (see Figure 1.5). In 1896, Berthier developed the principle of *réseaux lignés* [lined networks] as a plate composed of successive black and transparent strips designed to hide from each eye the image not meant to be seen. On the basis of this principle, as a precursor to parallax barrier devices used by many current autostereoscopic screens, he also invented a chronophotographic device (see section 1.2.3), known as a *praxinographe*.

1.2.2.2. The 20th Century

The start of the 20th Century saw the democratization (mass use) of photography and the discovery of cinematography. In 1915, the Astor Theater in New York held the first public projection of a short stereoscopic film entitled *Jim, The Penman*. The stereoscopic display is provided through an enhanced anaglyphic process, a technique developed and tested during the 19th Century. During this same period, Gabriel Lippmann [LIP 08] developed a new process known as "integral photography" which creates a naturally observable 3D image. He suggested placing a grid of spherical micro-lenses upon the photographic emulsion, each acting as a mini camera. However, at this point, the process was not considered to have potential because this kind of invention was beyond the technological capabilities of the time. This method would be reexamined 30 years later and further developed by Maurice Bonnet and subsequently form the basis of the lenticular autostereoscopic devices that we know today (see Chapter 14).

a)   b)   c)

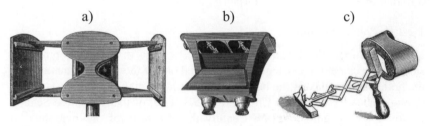

Figure 1.5. *a); The stereoscopes of Wheatstone (see [BRE 56, p. 56]);
b); Brewster (see [BRE 67, p. 67]); and c); Holmes*

In the 1950s and for two decades after, the film industry, notably Hollywood, saw the first 3D golden age. Stereoscopic techniques have since continually improved and enabled the production of several blockbusters in 3D[2]. The arrival of the 3D Imax in 1986 was also a major milestone for the industry.

1.2.2.3. *The fully digital era*

The start of the 21st Century saw the advent of "all-digital" and with it a new wave of 3D. Scientific and technological developments implied by this new kind of content today govern the whole chain of media production, from recording to display. It has opened doors to new visual experiences which will completely alter our relationship with images. We only need to look at the increasing attention given to 3D in recent (since 2010) conferences, such as the ACM SIGGRAPH conference. 3D imaging has been a strong trend in recent

---

2 *House of Wax* in 1953, http://en.wikipedia.org/wiki/House_of_Wax_(1953_film); *Creature from the Black Lagoon* in 1954, http://en.wikipedia.org/wiki/Creature_from_ the_ Black_Lagoon, etc.

years and, according to the Consumer Electronics Show, 3D television is now a reality for the audiovisual industry with 2010 being the real starting point of the industrial development of HD 3DTV.

1.2.3. *Multiview systems*

The development of photography during the 19th Century also coincided with the development of new multiview shooting devices. In this section, we will examine three systems which are today still the subject of developments. These include chronophotography, used for slow motion and video; pantascopic shooting, used for panoramic vision; and photosculpture, used for 3D modeling from several views.

1.2.3.1. *Panoramic photography*

Since the 19th Century, a number of approaches have been proposed for producing panoramic images [VAN 11]. Here, we consider the two most commonly cited [ROS 08]. First, the panoramic camera, invented by the German Friederich Von Martens in 1844, produces a 150 degree image on a curved daguerreotype plate by rotating the optical axis. Second, the pantascopic camera, patented in Britain in 1862 by John R. Johnson and John A. Harrison, is mounted on a rotating base controlled by a string-and-pulley mechanism which provides a 110 degree image by taking 24 photos successively and then recording the image on a collodion wet plate.

1.2.3.2. *High frequency movement images and the first videos*

While photography captures fixed images, shortly after its arrival, scientists were using it to capture movement using an image sequence. Two approaches were developed to do so. In 1872, Muybridge proposed a system made up of multiple cameras (ranging between 12 and 24), triggered from a distance by a wire and placed along a track to take successive images of a galloping horse (see Figure 1.6(a)). In 1878, he developed the zoopraxiscope which displayed the successive images stored on a disk. In 1882, the French scientist Jules Marey [MAN 99] developed the photographic gun and then in the same year invented "chronophotography" in order to capture the movement of animals and humans. In contrast to Muybridge's system, chronophotography involves a device with a single objective, fitted with a rotating shutter, which captures a series of movements through superposition on a single piece of film. To function, the subject must appear bright against a dark background (see Figure 1.6(b)). In 1889, this restriction was removed by allowing a transparent film to proceed jerkily, producing a sequence of up to 60 images per second.

1.2.3.3. *Multiview 3D reconstruction*

The idea of combining several images to obtain a projection of a spatial reconstruction is not new. For instance, photosculpture [BOG 81, SOR 00] proposed by François Willème (1830–1905) was inspired by two arts: photography and sculpture. The principal idea entails using photographies from several viewpoints to reconstruct a model of a portrait. The original technique positioned a system of 24 cameras placed at intervals of 15 degrees, directed toward a central point situated around 5 m away to take photographs of the model. The negatives were simultaneously produced to allow human subjects to be photographed. The images, projected successively by a lampascope on a translucent screen, were transferred via a pantograph by a potter using a clay block placed on a rotating base (see Figure 1.6(c)). The edges are then cut. The sculpture is retouched by the artist before its finalization. This technique has inspired a number of artists due to the realistic accuracy of the sculpture and the very short posing time for the subject.

a)   b)   c)

Figure 1.6. *a): Initial sequences of images with Muybridge's multiview systems; b): Marey's device superposing successive shots by a single camera; and c) the photosculpture procedure projecting 24 images on a screen connected to a pantograph*

1.3. Stereopsis and 3D physiological aspects

3D perception, visual acuity and visual field, in which details are distinguished, as well as the distance at which these details are perceived (see Figure 1.7 and Chapter 16), are important characteristics in our visual sense. Taken independently from one another, each eye can be compared to a camera whose sensory layer corresponds to the retina. Focus (visual accomodation) is carried out by a deformation of the lens and the direction toward the point being focused on by the extraocular muscles. The concept of 3D and being able to perceive distance is primarily due to binocular vision. The human visual system [LEI 06] is, therefore, evidently a complex system which uses an enormous range of indices functioning in tandem, particularly when viewing 3D. These different sources of information are normally divided into two large categories: subjective sources, which include psychophysical, graphic and dynamic indices; and objective sources, which include ocular and stereoscopic information.

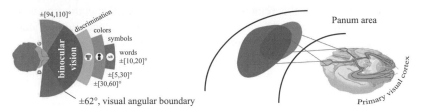

Figure 1.7. *Physiological limits and description of the human visual field; Panum's area indicates the area in which two images are fused to provide a single perception*

1.3.1. *Psychophysical indices*

According to the Gestaltist[3] theory [GUI 79, KOH 70], perception uses innate knowledge, memory and situational expectations, indicators which make perceptions that are genuinely sensed coherent. Each perceived scene is broken down into parts which are regrouped or reorganized. This theory relies on two basic principles: the distinction between foreground and background and the principles of regrouping. The brain therefore classifies, categorizes, makes sense of and regroups every tiny perception with others resembling it. The brain structures the indices in such a way that those which are small, regular or have a particular significance for us stand out against the background to create an overall structure. Each element is then perceived as a figure detached from the background, perceived as less structured and irregular. It is this foreground–background distinction that enables us to distinguish or recognize a familiar face in a crowd, as shown in Figure 1.8(a), a spiky sphere in Idesawa's figure.

1.3.2. *Monocular indices*

Perception in relation to depth within a 3D scene by monocular vision uses a coherent range of visual indices at the same time, as described below:

– occlusion: an object partially obscuring another will necessarily be in front of the masked object;

---

3 This theory takes its name from the German verb "Gestalt" which means shape.

– size and height in relation to objects: for objects with formal similarities, the observer infers their distances in relation to their size relative to the image on the retina. If they are smaller, they will be perceived as being further away;

– linear perspective: this relates to convergences toward specific points in the visual field, known as vanishing points, which appear in scenes with objects with regular edges or using motifs repeated along colinear axes;

– atmospheric diffusion: this corresponds to the decrease in contrast for distant objects. Distant objects appear more or less distinctly while closer objects are clear, giving a reinforced sensation of depth;

– shadowing: it provides information not only about the shape of an object but also its position in relation to the shadow position and size.

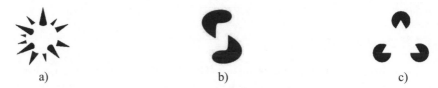

a)   b)   c)

Figure 1.8. *Gestalt and perception with: a) Idesawa's spiky sphere; b) Tse's worm; and c) the Kanizsa triangle*

To this series of static indices, we should also add dynamic indices, such as motion parallax, which provide information about visible objects' relative distances by changes in direction.

1.3.3. *Ocular indices*

These indices refer to closely related ocular movements which allow us to see from different distances. This adaptation functions using a combination of vergence movements (fusion convergence) and focus (deformation of the lens). This convergence-focus reflex is an important process in depth perception which, paradoxically, can cause significant conflicts (see Chapter 16). Indeed, as shown in Figure 1.9, the synkinetic nature of this reflex allows us to focus and converge at a single point during normal visual exploration. The problem arises, however, when we want to reproduce the sensation of depth perception using an image displayed on the surface of a 3D screen. In this case, there is a dissociation of focus and convergence movements, which may effectively induce visual discomfort.

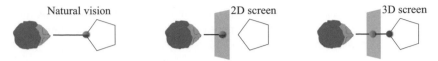

Figure 1.9. *Visual exploration using the convergence-focus reflex*
(● the focus point, ● the convergence point)

1.3.4. *Binocular indices*

Binocular or stereoscopic vision provides access to information known as retinal disparity which can be represented by the difference between the images taken from the left and right eyes (see Figure 1.10). This information, processed in the primary visual cortex, reconstructs 3D or, in other words, depth. It is this principle, also known as stereopsy, which allows us to recreate binocular vision using artificial means. This geometric model of binocular depth perception is described in further detail in Chapter 3 first within the context of stereoscopy, and then in Chapter 4 where it is extended to multistereoscopy. In both cases, problems of perceived depth distortions are examined.

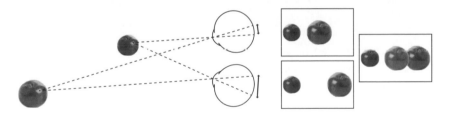

Figure 1.10. *Fusion and disparity in retinal images. Disparity accounts*
for the fact that an image is projected onto different places on the two retinas. More
than being a mere stimulus to vergence movements, the disparity between images from
the two eyes provides indications about the depth of objects in the scene

1.4. 3D computer vision

As an algorithmic representation of human vision, computer vision or artificial vision, is a discipline whose theoretical basis was first proposed during the 1960s. This processing paradigm of visual information generally operates according to two axes: ascending, related to changing sensory information into an abstract representation using a series of 3D primitives, for example, or descending, when it relates to verifying the primitives taken from the image from a series of known objects.

In 1982, one of the first formalisms of this theory related to 3D vision was proposed by D. Marr [MAR 82]. This computation model can be formally defined as follows:

– From one or several images by extracting characteristics which describe the bi-dimensional attributes of a representation known as a *primal sketch*.

– This primal sketch is the input for a number of more or less dependent processes which evaluate the local 3D properties related to the scene. This new representation, qualified by 2.5D, remains focused on the observer. These processes can often, depending on context, operate on a sequence of images if it relates to analyzing movement, on a couple of images in case of stereovision or simply a single image when, for example, it entails defining an outline on the basis of geometric, statistical, photometric or colorimetric information, etc.

– The 2.5D sketch is then compared with 3D information to construct a description of the scene in terms of objects and in relation to other objects. This is therefore a scene-focused description which no longer depends on the observer.

In 1991, E.H. Adelson and J.R. Bergen [ADE 91] proposed an alternative to modeling visual information of a real scene by applying a functional representation known as "plenoptic", noted as $\mathcal{P}(x, y, z, \phi, \theta, \lambda, t)$ (see equation [3.7] in Chapter 3) which defines at each time t and at each point p in the space with the coordinates (x, y, z), the energy at the wavelength λ traveling in any direction (θ, φ). This representation has the benefit of providing a fixed framework for problems such as capture, representing and synthesizing visual content. However, this form remains fairly difficult to use, as a result of which simplified forms of reproducing 4D light fields, or lumigraphs, have emerged. This model is also the basis for a body of work known as "image based" which is normally known as image-based modeling and rendering (IBMR) and/or computational photography.

More recently, Dai *et al.* [DAI 10] proposed another approach known as Vision Field Calculating [DAI 11], which encompasses all research related to filming and reconstructing real-life scenes.

This classification relies on a parametric space (see Figure 1.11) where 3D represents time, viewpoints and lighting. One point in this space corresponds to the conditions for capturing an image. Capture can be considered as taking a sample of the scene while analysis and synthesis are combined in what we can call its reconstruction.

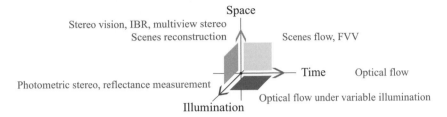

Figure 1.11. *Thematic classification in the Vision Field Calculating Space (according to [DAI 10])*

As a result, image capturing strategies in the subspace (or plane) time/viewpoints can also result in multiple viewpoint capture systems, a large number of which are examined in Chapter 3. Several acquisition solutions relate to the viewpoint/lighting planes which are also used to digitalize the optical properties of static objects' surfaces.

Alongside reconstruction, we can similarly identify classes of solutions associated with axes or planes in this space. Optical flow approaches, for example, enter into the time axis, stereovision (from two or more view points) or the light field rendering for a static scene under constant lighting in the viewpoints axis. In the time/viewpoints plane, this relates to reconstructing a dynamic scene using videos taken from several view points, such as free-viewpoint video, 3D motion capture or 4D reconstruction. The viewpoints/lighting covers problems of multi-lighting stereovision and 3D relighting in static scenes. The approaches relating to the time/lighting plane are difficult to implement because it is difficult to use multi-lighting conditions in temporal capture.

1.5. Conclusion

In this chapter, we have examined the different fundamentals of 3D video: historical, physiological in relation to human vision or mathematics and its extension to 3D computer vision. These principles are the basis for the subsequent scientific formalizations and technological developments presented in the following chapters.

Beyond this, all these subjects are treated in further detail in a number of works published in recent years, specifically the works of [CYG 11, HAR 04, JAV 09, LUK 10, MAT 12, RON 10, SCH 05, SZE 10 and WOH 13].

1.6. Bibliography

[ADE 91] ADELSON E.H., BERGEN J.R., "The plenoptic function and the elements of early vision", in LANDY M.S., MOVSHON A.J., (eds), *Computational Models of Visual Processing*, MIT Press, Cambridge, MA, pp. 3–20, 1991.

[ALA 07] ALATAN A., YEMEZ Y., GUDUKBAY U., *et al.*, "Scene representation technologies for 3DTV – a survey", *Circuits and Systems for Video Technology, IEEE Transactions on*, vol. 17, no. 11, pp. 1587–1605, 2007.

[ALH 21] ALHAZEN, *latin name of* IBN AL HAYTHAM, *Kitab al-Manazir*, in latin *De Aspectibus*, or *Opticae Thesaurus: Alhazeni Arabis*, in English *Treaty of Optics*, Cairo, Egypt, pp. 1015–1021, 1921.

[ARI 36] ARISTOTLE, *Problemata*, vol. 15, Circa-350 B.C., W.S. HETT (transl.), Harvard University Press, Cambridge, 1936.

[BOG 81] BOGART M., Photosculpture, *Art History*, vol. 4, no. 1, pp. 54–65, 1981.

[BRE 56] BREWSTER D., *The Stereoscope; its History, Theory, and Construction, with its Application to the Fine and Useful Arts and to Education: With Fifty Wood Engravings*, John Murray, 1856.

[CAR 56] CARDANO G., *De la subtilité et subtiles inventions*, L'Angelier, Paris, 1556.

[CYG 11] CYGANEK B., SIEBERT J., *An Introduction to 3D Computer Vision Techniques and Algorithms*, Wiley, 2011.

[DAI 10] DAI Q., JI X., CAO X., "Vision field capturing and its applications in 3DTV", *Picture Coding Symposium (PCS)*, IEEE, pp. 18–18, 2010.

[DAI 11] DAI QI., WU D., LIU Y.T., University (Beijing, CN), June 2011– www.freepatentsonline.com/y2011/0158507.html, Patent 20110158507.

[GUI 79] GUILLAUME P., *La psychologie de la forme*, Champ Psychologique, Flammarion, 1979.

[HAR 04] HARTLEY R., ZISSERMAN A., *Multiple View Geometry in Computer Vision*, Cambridge Books Online, Cambridge University Press, 2004.

[JAV 09] JAVIDI B., OKANO F., SON J., *Three-Dimensional Imaging, Visualization, and Display*, Signals and Communication Technology, Springer Science+Business Media, LLC, 2009.

[KOH 70] KOHLER W., *Gestalt Psychology: An Introduction to New Concepts in Modern Psychology*, Black and Gold Library, Liveright, 1970.

[KUB 07] KUBOTA A., SMOLIC A., MAGNOR M., *et al.*, "Multiview imaging and 3DTV", *Signal Processing Magazine, IEEE*, vol. 24, no. 6, pp. 10–21, 2007.

[LEI 06] LEIGH R., ZEE D., *The Neurology of Eye Movements*, Contemporary Neurology Series, Oxford University Press, 2006.

[LIP 08] LIPPMANN G., "Épreuves réversibles donnant la sensation du relief", *Journal of Theoretical and Applied Physics*, vol. 7, no. 1, pp. 821–825, 1908.

[LUK 10] LUKAC R., *Computational Photography: Methods and Applications*, Digital Imaging and Computer Vision Series, Taylor & Francis Group, 2010.

[MAN 99] MANNONI L., *Le grand art de la lumière et de l'ombre*, Nathan University, 1999.

[MAR 82] MARR D., *Vision: A Computational Investigation into the Human Representation and Processing of Visual Information*, Henry Holt and Co., Inc., New York, 1982.

[MAT 12] MATSUYAMA T., NOBUHARA S., TAKAI T., *3D Video and Its Applications*, Springer, London, 2012.

[MIC 11] MICHEL B., *La stéréoscopie numérique: Tourner, éditer, diffuser, imprimer, projeter*, Eyrolles, 2011.

[MOH 00] MOHISTS, under the pseudonym MO ZI, *Mo Jing*, China, Circa 400 B.C.

[RON 10] RONFARD R., TAUBIN G., *Image and Geometry Processing for 3D Cinematography*, Springer, 2010.

[ROS 08] ROSENBLUM N., *A World History of Photography*, 4th ed., Abbeville Press, 2008.

[SCH 05] SCHREER O., KAUFF P., SIKORA T., *3D Videocommunication: Algorithms, Concepts and Real-time Systems in Human Centred Communication*, Wiley, 2005.

[SMO 07] SMOLIC A., MUELLER K., STEFANOSKI N., *et al.*, "Coding algorithms for 3DTV – a survey", *Circuits and Systems for Video Technology, IEEE Transactions on*, vol. 17, no. 11, pp. 1606–1621, 2007.

[SOR 00] SOREL P., "Photosculpture: the fortunes of a sculptural process based on photography", in REYNAUD F., TAMBRUN C., TIMBY K. (eds), *3D: From Stereoscopy to Virtual Reality*, Paris, 2000.

[SZE 10] SZELISKI R., *Computer Vision: Algorithms and Applications*, Texts in Computer Science, Springer, 2010.

[VAN 11] VANVOLSEM M., *Chapter 1: Strip Photography and its Historical Context: A Camera Mechanism, Invention and Re-invention*, Leuven University Press, 2011.

[VIN 19] DA VINCI L., *Codex Atlanticus*, 1478–1519, set of 1119 leaflets.

[WOH 13] WOHLER C., *3D Computer Vision*, Springer, London, 2013.

Chapter 2

Digital Cameras: Definitions and Principles

2.1. Introduction

Digital cameras are a common feature of most mobile phones. In this chapter, we will outline the fundamentals of digital cameras to help users understand the differences between image features formed by sensors and optics in order to control their governing parameters more accurately. We will examine a digital camera that captures not only still images but also video, given that most modern cameras are capable of capturing both types of image data.

This chapter provides a general overview of current camera components required in three-dimensional (3D) processing and labeling, which will be examined in the remainder of this book. We will study each stage of light transport via the camera's optics, before light is captured as an image by the sensor and stored in a given format. Section 2.2 introduces the fundamentals of light transport as well as notations for wavelength and color spaces, commonly used in imaging. Section 2.3 examines how cameras have been adapted to capture and transform into digital image. This section also describes the details of different components in a camera and their influence on the final image. In particular, we will provide a brief overview of different optical components and sensors, examining their advantages and limitations.

Chapter written by Min H. KIM, Nicolas HAUTIÈRE and Céline LOSCOS.

This section also explains how these limitations can be corrected by applying postprocessing algorithms to the acquired images. Section 2.4 investigates the link between camera models and the human visual system in terms of perception, optics and color fidelity. Section 2.5 briefly explores two current camera techniques: high-dynamic-range (HDR) and hyperspectral imaging.

2.2. Capturing light: physical fundamentals

Light enables humans to see their surrounding environments and capture it as images using cameras. The image obtained by a camera represents all the light rays reaching the camera's sensors. To understand the basic concepts of this process, this chapter will examine the fundamentals of light transport. Section 2.2.1 examines generalized equations from radiometry and photometry and section 2.2.2 explains the influence of light wavelength. These aspects will be introduced due to their strong influence on digital imaging, image processing and image synthesis.

2.2.1. Radiometry and photometry

Radiometry quantitatively evaluates electromagnetic waves while photometry processes the sensation caused by this radiation for a standard observer [DEC 97]. Photometric values are obtained by combining radiometric values with the spectral response of this observer. The characteristic values of lighting and their photometric units are:

– luminous flux, denoted by Φ or F, are expressed in lumen (lm);

– luminous intensity, denoted by I, is expressed in candelas (cd);

– illuminance, denoted by E, is expressed in lux (lx);

– luminance, denoted by L, is expressed in candelas squared (cd.m^{-2});

– reflectance, denoted by f_r, is expressed in steradians sr^{-1} or in cd.m^{-2}.lx^{-1}.

Luminous flux correspond to the light energy (number of photons) per time unit. They characterize the quantity of light radiation emitted by a source. Luminous intensity characterizes the flux emitted by a light source in a specific direction per unit solid angle:

$$I = \frac{d\Phi}{d\Omega}. \qquad [2.1]$$

Luminance characterizes the luminous flux per surface unit generated by a source with an intensity I at a distance d. It decreases according to the square of the distance:

$$E = \frac{I \cos(\alpha)}{d^2},$$
[2.2]

where α is the angle between the normal of the lit surface and the direction of the source. Orders of the luminance value found in various lighting situations are presented in Table 2.1.

Light source	Luminance (lux)
Full moon, cloudless sky	0.2
Moved to 1 m	1
Lit street	20 to 30
Dawn, dusk	50
Outside, cloudy weather	5,000
Sunny	10,000–20,000
Maximum measurable	100,000

Table 2.1. *Luminance produced by different light sources*

Illuminance characterizes light intensity per unit surface in a specific direction. This is the most significant value for vision because it determines the visual sensation encountered by a user of the light being emitted, diffused or reflected according to the angle θ between the direction of emission ω_i and the angle of observation ω_o:

$$L = \frac{d^2\Phi}{d\Omega dA \cos\theta} = \frac{dI}{dA \cos\theta}.$$
[2.3]

The bidirectional reflectance distribution function (BRDF), introduced by Nicodemus [NIC 67], describes the directional reflection of light by a surface element. It is defined as the relationship between the luminance reflected by the surface and the incident lighting. The integrated luminance $L(\omega_o)$ of an environment observed from the direction ω_o is given by the radiometric transfer equation [PER 88] – L becomes radiance when considering radiometry and wavelengths:

$$L(\omega_o) = \int_\Omega f_r(\omega_o, \omega_i) E(\omega_i) \mathbf{n} \cdot \omega_i d\omega_i.$$
[2.4]

2.2.2. Wavelengths and color spaces

Electromagnetic radiation is represented as a continuous spectrum of wavelengths (see Figure 2.1). The human eye senses electromagnetic radiation with a wavelength approximately between 400 nm (upper threshold for ultraviolet light) and 700 nm (lower threshold for infrared light). White light is represented in appearance by a linear combination of three primaries, as demonstrated by Maxwell. Trivariance of a colored space was also established by Grassman. The three primary colors commonly used are red, green and blue (RGB). Metameric rays provide a colored image without having the same spectral content.

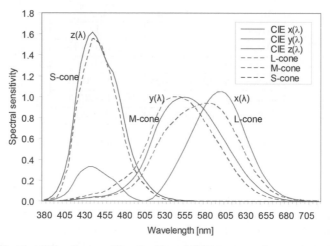

Figure 2.1. *The CIE color-matching functions (CMFs) versus the spectral sensitivities of the human cone: the solid lines represent the CIE 1931 CMFs (version: modification [VOS 78]), the dotted lines show the different physiological responses of the cones [EST 79]. In particular, the responses to the L-cone appear to be significantly different in relation to the CIE function $x(\lambda)$*

Colorimetry measures human perception of color and reduces the physical spectrum to the perceived color spectrum by correlation. It requires three essential elements: a light source (illuminantion), an object (with a standard measure of geometry) and a reference observer. In 1931, the International Commission on Illumination (Commission Internationale de l'Eclairage – CIE) carried out a series of psychophysical tests to deduce the standard colorimetric observation model that defines CIE 1931 color matching functions (CMFs) to quantify the perception of trichromatic color in human vision. In this experiment, two colors were shown to observers in normal colorimetric vision who should consolidate one of the stimuli with the appearance of the color. They used RGB lights to produce a metameric combination. The CMFs were then updated [STI 59, VOS 78]. These functions are official standards for the transformation of the visible spectrum to trichromatic color coordinates, CIEXYZ (the CIE tristimulus values). However, it was also found that the responses of the cones to long, middle and short (LMS) cones were different to physiological CMFs [EST 79, HUN 85]. A transformation of the cone's response[1] proposed by Estévez [EST 79] is widely used in modern color appearance models.

A color space is a digital model of the human chromatic vision mechanism. The first trichromatic color model, CIEXYZ, was proposed in 1931 based on the red, green and blue colors. The correspondence between the CIEXYZ model established by the CIE and the cone model set out by Estévez is illustrated in Figure 2.1. However, the CIEXYZ model, while widely used, has a number of disadvantages, the first being that it is not physiologically but perceptually plausible. Added to this is the fact that this model's representation gives negative values in relation to wavelength. These limitations have therefore led to research into other color appearance models.

Ideally, a color space is designed to represent perceived uniform colors in digital coordinates. Modern color spaces, often called color appearance models, are structured according to Zone theory [MUL 30], which describes human chromatic vision using a three-stage procedure. The calculation of color appearance includes chromatic adaptation, dynamic cone adaptation, decompositing opposite/achromatic colors and, in general, calculating perceptual color attributes. First, the input spectrum, sampled as tristimulus values XYZ, is transformed for chromatic adaptation often in relation to a specialized color space. The white adapted tristimulus values XYZ_c are then transformed in the cone color space where the cone response function is applied (often according to a power or hyperbolic function). The signal is then divided into achromatic channels A and opposite colors a and b. Finally, the perceptual correlation is based on these three channels. It is here that color perception models differ most because a large number of functions can be applied to produce these perceptual values. Popular models include CIELAB, RLAB [FAI 91], Hunt94 [HUN 94] and LLAB [LUO 96] and CIECAM97s [CIE 98], but CIECAM02 [MOR 02] has been commonly adopted as a standard appearance model.

---

1 Photoreceptors in the eye convert electromagnetic signals from light into nerve signals.

The CIELAB color space [CIE 86] is very simple and based purely on the trichromatic component values *XYZ*. It is the oldest to be taken from the psychophysical approach in 1976. Chromatic adaptation is carried out by normalizing the values *XYZ* using the white reference values XYZ_w. This form modifies von Kries' chromatic adaptation transformation [VON 70] and the cone response is modeled using a cubic root. Only lightness, chroma, hue and the color opponent coordinates (*a* and *b*) are predicted, and adaptation to different backgrounds or environments is not modeled. Despite these simplifications, the model performs well. The input parameters of the CIELAB model are the normalized CIEXYZ values (Y is equal to 100) of the test colors and the normalized values $X_n Y_n Z_n$ of the reference white.

The representation of colors in videos transmitted via television relies on the RGB model but relies on different transformation matrices to move from RGB to XYZ. This is confirmed by examples such as the National Television System Committee (NTSC) system used in the United States and Japan, the Phase Alternating Lines (PAL) system used by European countries such as the United Kingdom or Spain and the *Sequentiel Couleur A Mémoire* (SECAM – Sequential Color with Memory) system used in France, for example. Transmission takes place via signals, connected by transforming RGB components: Y, I, Q for NTSC; Y, U, V for PAL and Y, R-Y, B-Y for SECAM.

2.3. Digital camera

This section examines digital cameras and their diverse components. We will not attempt to provide an exhaustive overview of all possible configurations but purely the basic concepts necessary for understanding the choice of algorithms and the configurations of applications covered within this book.

2.3.1. *Optical components*

2.3.1.1. *Optics in cameras*

Electromagnetic radiation can be physically captured using optical mechanisms. The historical development of camera models has been examined in Chapter 1.

A camera is composed of a box and an objective lens. The objective lens contains several component lenses and a diaphragm that controls the transmission of light and is calibrated by an aperture unit (see section 2.3.3). This is represented by a number N that is defined as the focal length f of the lens divided by the diameter d of the pupil: $N = f/d$. For example, a lens with a pupil diameter of 25 mm and a focal length of 50 mm has a relative aperture of 50/25 = 2.0. The diaphragm is expressed according to the focal distance f and is described in italics, for example, by *f/4* when the relative aperture N has a value of 4. The denominator of the expression is generally indicated by the *f-number* of the lens and the relative aperture simply as its aperture or *f-stop*. If there are two apertures and aperture speed configurations, they satisfy the following relation: $\frac{t_1}{t_2} = \frac{N_1^2}{N_2^2}$ [RAY 00] because the amount of energy received by the sensor is proportional to the diaphragm's free surface and therefore the square of the aperture.

2.3.1.2. *Errors and corrections*

The mechanical and optical geometry of a camera may result in deformation and undesired effects in the image. For example, the reception of light by sensors will be much higher at the center of the image than at the edges. This effect is known as fall-off or vignetting and is illustrated in Figure 2.2(a), for an image acquired with a bigger aperture (which accentuates the effect). Also evident is the distortion of the image due to the lens (see Figure 2.2(b)). Another effect of light reflection is known as flare, which produces color effects that do not exist in real life. This is due to the multiple reflection of light onto the lenses in question. These three effects can be compensated for using automatic algorithms (see Chapter 5) precisely so that they can be measured. For example, a photograph of a white wall can be used to measure the effect of fall-off and a generic correction in cos^4 can be applied. In addition, a correction of radial distortion can be applied to the image but this creates an image that is no longer rectangular. Algorithms to counter flare have also been proposed, specifically by Ward [REI 10].

a)   b)

Figure 2.2. *Illustration of fall-off (digital effect accentuated by an improvement in contrast) and distortion in the image created by the lens used*

Filters can be directly applied to camera lenses. Neutral density filters attenuate light according to a relation producing an effect similar to a decrease in exposure time. They are used, for example, to create an effect of movement by increasing the exposure time while maintaining adapted light intensity values. Polarizing filters (circular or linear) block specific light rays. They are mobile, which means that specific desired effects can be chosen. In particular, reflective effects can be attenuated or the color contrast can be augmented.

2.3.2. *Electrical components*

The sensor in a camera has the primary function of converting light signals into electronic signals. Current cameras are generally fitted with an image sensor based on charged coupled device (CCD) technology or a complementary metal oxide semiconductor (CMOS). Either a charge coupled device (CCD) or a complementary metal oxide semiconductor (CMOS) are commonly employed as imaging sensors in current cameras. Nowadays the CCD sensors are preferred in high-end applications due to better performance with less noise, while the CMOS sensors are popularly used in commodity camera systems.

The CCD sensor system has been used in cameras for more than 20 years. Due to the smaller size of the light collecting area, light sensitivity of the CMOS is lower than that of the CCD in general. CCD sensors are, however, more costly due to the non-standard manufacturing process used to manufacture them. In addition, a phenomenon known as blooming is likely to occur when a very bright object (for example a lamp or sunlight) is present in a scene. This can produce vertical lines above and below the object.

Due to the cost of CCD sensors, CMOS sensors have been subject to a number of improvements. It is thus difficult to identify which sensor obtains the best quality. CMOS sensors reduce the overall cost of the camera because they integrate several necessary elements. They can also be used to produce small-scale cameras. One of the current limits of CMOS sensors relates to their low sensitivity to light. If the problem does not arise in normal lighting conditions, it can occur as luminance decreases. The resulting image is therefore either very dark or blurred (noise).

Two types of noise occur depending on the sensor system used: temporal noise, which varies according to each picture, and fixed noise, associated with sensor irregularities. The different types of noise are well known and there are several methods of measurement and corrections. In fact, several noises have already been compensated for by manufacturers using hardware and/or software components. The easiest noise to correct is thermal noise, which can be due to both noise on the sensors and noise due to agitated electrons, varying according to temperature. It can be measured by taking a photograph with a hood over the object in order to block light. The acquired pixels do not represent a true black while no light reaches the sensors. This noise can often be corrected by approximation by subtracting the average value obtained for this or these black image(s) from the image pixels acquired in normal conditions.

2.3.3. *Principal functions and their control*

Manual mode requires the selection of parameters such as focus, gain white balance, exposure time and the diaphragm aperture, whose values are best selected in the automatic mode.

Change of focus affects the sharpness of the image. A change in focus value will modify the position of the lens inside the objective and therefore the trajectory of light toward the sensors according to the laws of physics. The divergence or convergence of emitted light rays defines the blurred or sharp parts of the image.

As explained in section 2.3.1.1, the diaphragm's aperture is generally associated with the *f-stop* value N, corresponding to the relation between focal distance and the diaphragm's diameter. The smaller the *f-stop*, the wider the diaphragm aperture. It is characterized by values such as $f/2$ for a very wide aperture and $f/22$ for a very small aperture. The gap between *f-stops* is often calculated by dividing by a power of $\sqrt{2}$. The diaphragm aperture affects the larger or smaller passage of light. This has a strong effect on depth of field: the wider the aperture (small *f-stop* value) is, the more the depth of field is reduced. A reduced depth of field highlights a subject and blurs the background and foreground while a higher depth of field sharpens details in the image.

Exposure time t is measured in seconds, often with a relation of 2 between two exposure times such as the times $1/16$ and $1/32$ s. It refers to *the duration of the shutter's* opening. The longer the exposure time, the more light the sensors will collect. A long exposure time therefore provides a bright image while a short exposure time leads to a dark image. The exposure time and aperture *f-stop* are connected by an exposure value (EV):

$$EV = log_2(N^2/t) \qquad\qquad [2.5]$$

In photography, *f-stop* is often confused with stop, which is a relation. Stop can be used in relation to different EV values: a variation of 1 (relation of 2 on the exposure time) on an EV will give a 1 stop unit. The term "stop" is also used to size an intensity interval: n stops corresponding to a size of 2^n.

Sensor gain affects the light intensity values that can be obtained. It relates to the sensitivity of the film to light. With digital devices, this property does not exist any more since there is no longer a film. It is now related to the gain applied to the sensor signal. The higher its values are, the more the acquisition is light sensitive. For a number of photographic devices, this is accompanied by noise on the image for higher ISO values.

Lastly, the white balance reflects the color measured in Kelvins (K). The eye adapts its perception of white. If the image's color is not calibrated, the resulting image can differ, which is directly evident. A poor choice of white balance will provide an image that seems yellow (warm tones) or lacking in color (cold tones).

A number of cameras are fitted with a bracketing function, which allows photographs to be taken according to several device parameters. It is used primarily for white balance as well as exposure. It is possible to take a sequence of images in the automatic mode, which will vary one parameter while the others remain fixed. This bracketing mode does not have an equivalent in video cameras although some recording techniques have proposed alternatives (see Chapter 19).

We will examine the bracketing function applied to varying exposure time for a single aperture. This is useful for understanding Chapter 19 of this book. This bracketing mode on exposure time is based on three hypotheses: a number n (odd) images (often from three to nine depending on the camera), an aperture diaphragm with a value of N and a fixed ratio r. Generally, the camera selects the most appropriate exposure time t_0 for the content whose related EV linking N to t_0 is indicated by the symbol EV_0, producing an intermediary photograph. A series of n photographs are taken with the diaphragm aperture N with the relative EVs: $EV_{-n/2}$, $EV_{-(n-2)/2}$, ..., EV_0, ..., $EV_{(n-2)/2}$, $EV_{n/2}$. For example, the exposure time used for the image acquired in $EV_{-n/2}$ is $t_0/r(n/2)$. This provides a burst of a series of images of the same scene but representing different levels of brightness.

2.3.4. *Storage formats for images*

In low dynamic imaging, the most current format encodes photometric data using the three red, yellow and blue channels, which correspond to the dominant colors of the three ranges of wavelength that they represent. These channels are encoded using an 8-bit integer defining 256 levels for each one. However, color encoding according to the RGB addition principle cannot satisfy all applications (notably in image processing) and other color representation spaces exist to adapt it, notably the CIEXYZ format examined in section 2.2.2.

Image formats differ according to two aspects: the color representation range (which can go beyond the 8 standard bits), the compression and decompression mechanisms, and the type of compression (with or without loss). There are several types of standard, with the information loss compression formats including: Joint Photographic Experts Group (JPEG), Tagged Image File Format (TIFF), and lossless formats including: Graphics Interchange Format (GIF) and Portable Network Graphics (PNG). The majority of cameras can store images in RAW format, which is not a standard format. Each manufacturer defines its RAW format (with an appropriate extension), which must be coded differently. The RAW format encodes a multitude of different pieces of information as well as pixels' color values, in the form of metadata, which can be formalized in Exchangeable image file format (EXIF). This format, proposed initially by the Japan Electronic Industries Development Association (JEIDA), has been adopted by the majority of manufacturers. Information related to camera parameters at the point of image capture and the size and resolution of the image often constitutes part of this. Colors are represented in mosaic form using a Bayer filter (1976) or a color-filter-array such as those filtered directly by sensors. They are represented by 50% green, 25% blue and 25% red, or a format known as RGGB. When constructed, the image must be "debayerized" or subject to "demosaicing", for which there are a number of methods [LI 08]. While sometimes compressed, the RAW format can, in theory, provide access to data acquired by sensors.

2.4. Cameras, human vision and color

Human perception is examined in Chapters 1 and 16 in relation to pinhole cameras. In this chapter, however, we will examine the relationship between human perception and cameras.

2.4.1. *Adapting optics and electronics to human perception*

The human visual system is widely used to evaluate display devices [BRE 10]. However, it is rarely used to design and evaluate digital cameras, specifically their sensors, despite the fact that they are fundamental systems within an image capturing device. Equally, while the aim is that sensors must generally exceed the human eye, there is currently no means of measuring the gap between a sensor and the human visual system.

To overcome this issue, Skorka and Joseph [SKO 11] recently introduced a performance evaluation method for digital cameras in relation to the human visual system. They selected eight parameters to characterize an imaging system. The system implemented is conveyed by a quality factor that represents the gap in performance for the parameter that appears the weakest when compared to the human eye. For each of these eight parameters, an ad hoc evaluation method is proposed. The different parameters considered include energy consumption, visual field, spatial resolution, temporal resolution, the peak signal-to-noise ratio (PSNR) and the dynamic and obscurity current. Data related to the human visual system were collected from clinical or psychophysical data collected in the literature while data regarding sensors were gathered from manufacturers' own data. They applied their method to 24 modern CCD or CMOS sensors (either academic or commercial) proposing an ideal lens each time to form a complete imaging system. In the majority of cases, dynamic proved to be the most limiting factor in relation to the human visual system, followed by the dark current. Overall, the evaluation showed that modern digital cameras are not yet able to rival the human eye. The estimated gap lies between 31 and 90 dB, with an order ranging between 1.6 and 4.5.

2.4.2. *Controlling color*

2.4.2.1. *Response functions in cameras*

The image acquired by the camera differs strongly from the real scene in terms of color. The choice of constructor in the sensor response has a strong influence on the final luminance. Each camera generally has its own response function that converts the radiance values into color values to be allocated to pixels. To revert to radiance values, it is necessary to estimate the inverse response function. If we consider the irradiance E reaching the camera's sensor, there is a linear relation with the scene's irradiance, generally depending on the exposure time t. The intensity value M stored in the image corresponds to the conversion t, such that $M = f(E)$. Its inverse function

$g = f^{-1}$ is used to obtain the radiance. Generally, this process involves selecting the value of identical pixels for different exposure times in order to monitor the variation in their intensity to adjust the best curve separately for each R, G and B channel. Several methods for estimating this function exist in the literature, which generally consider f as a monotonic and increasing function, thereby ensuring that it is invertible. Additional hypotheses can be proposed in relation to its form: logarithmic [DEB 97], polynomial [MIT 99] or gamma [MAN 95]. A function can also be reconstructed using other known functions [GRO 04]. However, the procedure is very sensitive to the pixels selected and the input images. If the camera's parameters are unchanged, we can use the same curve for other images.

2.4.2.2. *Characterizing color*

If optical radiation in a reference test chart is measured and sensed simultaneously by a detection device in an image, it is possible to create a mathematical model to describe the specification of the color of an image device in coordinates with a physical meaning that is independent of the device. The device's signals or the images' colors vary according to the manufacturer's specifications or the material used. They can also vary when there are identical specifications due to the manufacturing process. The characterization of colors exceeds the variation in image devices by constructing a mathematical bridge between the device signal and the physical coordinates such that we can describe the device-dependent signals as independent signals. For example, CIEXYZ or CIELAB can be used as independent signals. To do so, it is necessary to use measuring tools for physical properties or produced desired colors in the output device. The characterization of colors requires a two-stage procedure [JOH 02]. The first is calibration, which is the setting up of a device or process so that the device gives repeatable data. The following stage is characterization, which establishes a relationship between the device's color space and the independent color space such as the CIE tristimulus values. Further information regarding this will follow in Chapter 5.

2.5. Improving current performance

Over the past 10 years, digital imaging technology has exploded. New horizons are continually opening up and research is focusing on a number of complementary domains that will converge in the near future. In this book, we can specifically cite stereoscopic photography, HDR imaging and multispectral acquisition. Here, we will concentrate on the two latter examples, given that the former will be examined in Chapter 3.

2.5.1. *HDR imaging*

Today, classic images are referred to as low-dynamic-range (LDR) images. These are images whose colors are typically represented by an interval of 8 bits. In contrast, we also have images known as HDR images, which go beyond this reduced representation. There is currently no precise consensus around the exact definition of HDR images. If we consider full HDR, this entails reconstituting the existing intensities in a given scene and storing them as an image or video. On the basis of 16 stops, all intensities are represented for a wide range of scenes. The popularity of these images is undeniable and even precedes the progression and development of the field. Photographic sensors can now cover a dynamic storable range of up to 16 or even 20 bits (the camera created by Spheron[2]) although in a 1D sequential acquisition. Video sensors are generally more limited to 12 bits and a maximum of 14 bits, with the exception of 20 bits in Spheron's experimental video camera.

Different methods have been developed to reproduce HDR content using a spectrum provided by current sensors but have faced a number of more significant obstacles with regard to producing videos. They are principally created by combining images taken with different exposure times. The bracketing mode (see section 2.3.3) is particularly useful. Nearly all approaches, software and materials rely on this approach. A model of optical reconstruction and its sensors is essential for the quality of obtained images [AGU 12]. The use of RAW format (see section 2.3.4) allows us to avoid the need of camera functions (see section 2.4.2.1) and is thus preferable. The initial HDR photograph creation approaches for static scenes examined in [BAN 11, REI 10] and applied to different commercial software can be differentiated from more recent approaches, which use dynamic contexts and video data as explained in Chapter 19.

Storing a range of HDR radiance values is subject to specific coding. Colors can be stored at 32 bits (compared to 24 bits for LDR images). Different HDR formats are generally inspired by the LDR formats (see section 2.3.4). The RGBE format [WAR 91] is an extended version of sRBG (RGB including a gamma correction normalized between 0 and 1) to which an E channel is added corresponding to the exponent to a power of 2 that is applied to the RGB values, which gives their high dynamic equivalent. From this format, we can derive an XYZE format that extends the CIEXYZ norm to HDR images. The corresponding extension of this image format is .hdr.

---

2 www.spheron.fr/.

Another available HDR format is known as LogLuv TIFF. It is inspired by the CIELUV format and encoded the L component (radiance) logarithmically [WAR 98]. However, this coding results in a greater amount of information loss than XYZE for an equivalent memory use (for a comparison of different formats, see [REI 10]). Lastly, the OpenEXR format directly encodes the RGB channels onto three floats, with 96 bits per pixel (format considered to be lossless or half-float (16 bits per component or 48 bits per pixel)) [KAI 09].

2.5.2. *Hyperspectral acquisition*

Electromagnetic radiation is generally described in terms of photon wavelength. Spectral intervals are classified according to three categories: near-ultraviolet (NUV) 300–400 nm, visible (VIS) 400–700 nm, and near-infrared (NIR) 700 nm–3.0 μm [ISO 07]. Trichromatic imaging and multispectral imaging focus on the VIS category while hyperspectral imaging (see Figure 2.3) focuses on three categories (NUV, VIS and NIR). Spectral imagery can be categorized according to different concepts, i.e., by filter of monochromatic sensor.

Several filter-based concepts have been proposed. When a complete spectrum light source shines on the surface of an object, the light reflected is captured by an extended bandwidth filter device [ATT 03, RAP 05, WAR 00]. These filters, fitted in a motorized wheel or a tunable liquid crystal filter (LCTF), are used to identify the spectrum in question. An alternative to the bandwidth filter is a spectral dispersion unit. Spectral images are reconstructed by a reverse solution, producing artifacts and a spatial resolution inferior to the bandwidth in question [KIT 10].

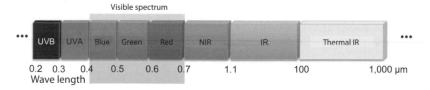

Figure 2.3. *Range represented by hyperspectral imaging*

A monochromatic sensor captures the surface of an object receiving light within a narrow bandwidth [EAS 10, FRA 10, KIM 10]. This method does not light the object with a complete spectrum light source such as a xenon light, thereby minimizing ionization damage. Because of the evolution of Light-Emitting Diode (LED) technology, the configuration of narrow bandwidth LED lights will be more effective in terms of cost than a complete spectrum light. Fluorescence (emission of light by a substance absorbing light of different wavelengths such as NUV) forms part of the reflected light and thereby interferes in measures of reflectance for each wavelength.

2.6. Conclusion

In this chapter, we have studied the fundamental of image acquisition. We have examined camera technology in relation to the transport of light since photographic sensors receive light rays that have traveled through the camera lens and diaphragm. We have also explained the image capture mechanism through various optical and electronic components and the relationship of different parameters with the final image obtained. Current technology is not free from faults, which may induce geometric or colorimetric deformations. We have explained these limitations in relation to current solutions to them. Further details on this subject are examined by Reinhard *et al.* in [REI 08], which provide a number of details regarding image acquisition and color. A number of points examined briefly in this chapter will be discussed in depth in following chapters.

2.7. Bibliography

[AGU 12] AGUERREBERE C., DELON J., GOUSSEAU Y. *et al.*, Study of the digital camera acquisition process and statistical modeling of the sensor raw data, Report Instituto de Ingenieria Eléctrica (IIE), Paris, September 2012.

[ATT 03] ATTAS M., CLOUTIS E., COLLINS C., *et al.*, "Near-infrared spectroscopic imaging in art conservation: investigation of drawing constituents", *Journal of Cultural Heritage*, vol. 4, pp. 127–136, 2003.

[BAN 11] BANTERLE F., ARTUSI A., DEBATTISTA K. *et al.*, *Advanced High Dynamic Range Imaging: Theory and Practice*, AK Peters (CRC Press), Natick, MA, 2011.

[BRE 10] BREMOND R., TAREL J.-P., DUMONT E. *et al.*, "Vision models for image quality assessment: one is not enough", *Journal of Electronic Imaging*, vol. 19, no. 4, pp. 1–14, 2010.

[CIE 86] CIE, Colorimetry, CIE Pub. no. 15.2, Commission Internationale de l'Eclairage (CIE), Vienna, 1986.

[CIE 98] CIE, The CIE 1997 interim colour appearance model (Simple Version), CIECAM97s, CIE Pub. no. 131, Commission Internationale de l'Eclairage (CIE), Vienna, 1998.

[DEB 97] DEBEVEC P.E., MALIK J., "Recovering high dynamic range radiance maps from photographs", *Proceedings of ACM Siggraph '97 (Computer Graphics)*, ACM, pp. 369–378, 1997.

[DEC 97] DECUSATIS C., (ed.), *Handbook of Applied Photometry*, AIP Press, Springer, 1997.

[EAS 10] EASTON R., KNOX K., CHRISTENS-BARRY W., *et al.*, "Standardized system for multispectral imaging of palimpsests", *Proceedings of the SPIE, Computer Vision and Image Analysis of Art*, SPIE, no. 75310D, pp. 1–11, 2010.

[EST 79] ESTÉVEZ O., On the fundamental data-base of normal and dichromatic colour vision, PhD Thesis, University of Amsterdam, 1979.

[FAI 91] FAIRCHILD M. D., "Formulation and testing of an incomplete-chromatic adaptation model", *Color Research and Application*, vol. 16, no. 4, pp. 243–250, 1991.

[FRA 10] FRANCE F. G., CHRISTENS-BARRY W., TOTH M. B. *et al.*, "Advanced image analysis for the preservation of cultural heritage", *IS&T/SPIE Electronic Imaging. International Society for Optics and Photonics*, SPIE, no. 75310E, pp. 1–11, 2010.

[GRO 04] GROSSBERG M.D., NAYAR S.K., "Modeling the space of camera response functions", *IEEE Transactions on Pattern Analysis Machine Intelligence*, vol. 26, no. 10, pp. 1272–1282, October, 2004.

[HUN 85] HUNT R.W.G., POINTER M.R., "A colour-appearance transform for the CIE 1931 standard colorimetric observer", *Color Research and Application*, vol. 10, no. 3, pp. 165–179, 1985.

[HUN 94] HUNT R.W.G., "An improved predictor of colourfulness in a model of colour vision", *Color Research and Application*, vol. 19, no. 1, pp. 23–26, 1994.

[ISO 07] ISO, ISO/20473:2007: Optics and Photonics — Spectral Bands, 2007.

[JOH 02] JOHNSON T., "Methods for characterizing colour scanners and digital cameras", in GREEN P., MACDONALD L.W., (eds), *Colour Engineering, Achieving Device, Independent Colour*, John Wiley & Sons Inc., Chichester, pp. 165–178, 2002.

[KAI 09] KAINZ F., BOGART R., "Technical introduction to OpenEXR", 2009. Available at http://www.openexr.com/index.html.

[KIM 10] KIM S.J., ZHUO S., DENG F., *et al.*, "Interactive visualization of hyperspectral images of historical documents", *IEEE Transactions on Visualization and Computer Graphics*, vol. 16, no. 6, pp. 1441–1448, 2010.

[KIT 10] KITTLE D., CHOI K., WAGADARIKAR A. *et al.*, "Multiframe image estimation for coded aperture snapshot spectral imagers", *Applied Optics Opt.*, vol. 49, no. 36, pp. 6824–6833, 2010.

[LI 08] LI X., GUNTURK B., ZHANG L., "Image demosaicing: a systematic survey" Visual Communications and Image Processing 2008, *Proceedings of the SPIE*, pp. 68221J–68221J–15, 2008.

[LUO 96] LUO M.R., LO M.-C., KUO W.-G., "The LLAB (l:c) colour model", *Color Research and Application*, vol. 21, no. 6, pp. 412–429, 1996.

[MAN 95] MANN S., PICARD R.W., "Being undigital with digital cameras: extending dynamic range by combining differently exposed pictures", *Proceedings of IST 46th Annual Conference*, Boston, Massachusetts, pp. 422–428, May 1994.

[MIT 99] MITSUNAGA T., NAYAR S. K., "Radiometric self calibration", *Proceedings of IEEE Conference on Computer Vision and Pattern Recognition*, Fort Collins, CO, pp. 374–380, June 1999.

[MOR 02] MORONEY N., FAIRCHILD M.D., HUNT R.W.G., *et al.*, "The CIECAM02 color appearance model", *Proceedings of the 10th IS&T Color Imaging Conference*, IS&T, pp. 23–27, 2002.

[MUL 30] MULLER G.E., "Über die Farbenempfindungen", *Z. Psychol.*, pp. Erganzungsbände 17 and 18, 1930.

[NIC 67] NICODEMUS F.E., "Radiometry", *Applied Optics and Optical Engineering*, vol. IV, Academic Press, New York, 1967.

[PER 88] PÉROCHE B., *La Synthèse d'Images*, Hermès Science Publications, 1988.

[RAP 05] RAPANTZIKOS K., BALAS C., "Hyperspectral imaging: potential in non-destructive analysis of palimpsests", *Proceedings of the 2005 International Conference on Image Processing, ICIP 2005*, vol. 2, pp. II, 618–21, September 2005.

[RAY 00] RAY S.F., "The Geometry of image formation", in JACOBSON R., RAY S., ATTRIDGE G.G., AXFORD N. (eds), *The Manual of Photography*, 9th ed., Focal Press, Oxford, pp. 39–71, 2000.

[REI 08] REINHARD E., KHAN E.A., AKYZ A.O., *et al.*, *Color Imaging: Fundamentals and Applications*, AK Peters, Ltd., 2008.

[REI 10] REINHARD E., WARD G., PATTANAIK S., *et al.*, *High Dynamic Range Imaging: Acquisition, Display, and Image-based Lighting*, 2nd ed. The Morgan Kaufmann series in Computer Graphics, Elsevier (Morgan Kaufmann), Burlington, MA, 2010.

[SKO 11] SKORKA O., JOSEPH D., "Toward a digital camera to rival the human eye", *Journal of Electronic Imaging*, vol. 20, no. 3, pp. 1–18, 2011.

[STI 59] STILES W.S., BURCH J.M., "NPL colour-matching investigation: final report", *Optica Acta*, vol. 6, pp. 1–26, 1959.

[VON 70] VON KRIES J., "Chromatic adaptation", in MACADAM D.L., (ed.), *Sources of Color Science*, MIT Press, Cambridge, pp. 109–119, 1970.

[VOS 78] VOS J.J., "Colorimetric and photometric properties of a 2-deg fundamental observer", *Color Research and Application*, vol. 3, pp. 125–128, 1978.

[WAR 91] WARD G., "The LogLuv encoding for full gamut, high dynamic range images", in ARVO J., Ed., *Graphics Gems II*, Academic Press Inc., pp. 80–83, 1991.

[WAR 98] WARD G., "The LogLuv encoding for full gamut, high dynamic range images", *Journal of Graphics Tools*, vol. 3, no. 1, pp. 15–31, 1998.

[WAR 00] WARE G.A., CHABRIES D.M., CHRISTIANSEN R.W., *et al.*, "Multispectral analysis of ancient Maya pigments: implications for the Naj Tunich Corpus", *IEEE Transactions*, pp. 2489–2491, 2000.

Chapter 3

Multiview Acquisition Systems

3.1. Introduction: what is a multiview acquisition system?

Multiview acquisition, the focus of this chapter, relates to the capture of synchronized video data representing different viewpoints of a single scene. In contrast to video surveillance systems, which deploy multiple cameras to visually cover a large-scale environment to be monitored with little redundancy, the materials, devices or systems used in multiview acquisition are designed to cover several perspectives of a single, often fairly restricted, physical space and use redundancy in images for specific aims:

– for three-dimensional (3D) stereoscopic or multiscopic visualization of captured videos:

– for real scene reconstruction/virtualization:

- 2.5D reconstruction of a depth map from a given viewpoint;

- textured 3D reconstruction of digital models, avatars of real objects;

- motion capture (MoCap) for realistic animation of virtual actors;

– for various and complementary adjustments in control room or during postproduction:

Chapter written by Frédéric DEVERNAY, Yves PUPULIN and Yannick REMION.

- "mosaicking" views providing a panoramic view or a high-resolution image;

- a virtual camera moving at frozen time or very slowly (bullet time);

- mixing the real/virtual (augmented reality (AR));

- view interpolation (free viewpoint TV (FTV));

- focus post-modification (refocus);

- increasing video dynamics (high dynamic range (HDR)); etc.

Depending on the final application, the number, layout and settings of cameras can fluctuate greatly. The most common configurations available today include:

– "Binocular systems" yielding two views from close-together viewpoints; these systems are compatible with 3D stereoscopic visualization (generally requiring glasses) and depth reconstruction with associated post-production methods (AR, FTV).

– Lateral or directional multiview systems[1] provide multiple views from close-together viewpoints (generally regularly spaced), each placed on the same side of a scene. These systems produce media adapted to autostereoscopic 3D visualization, "frozen time" effects within a limited range and a depth reconstruction or more robust "directional" 3D reconstruction than in the case of binocular reconstruction with the same postproduction techniques (AR, FTV). The multiplication of different perspectives also allows the use of different settings for each camera, which, with the strong redundancy in capture, renders other postproduction methods possible (refocus or HDR, for example).

– Global or omnidirectional multiview systems[1] deploy their multiple viewpoints around the target space. These systems are principally designed for bullet time in a wide angular motion, 3D reconstruction and MoCap.

Alongside these purely video-based solutions, hybrid systems adding depth sensors (Z-cams) to video sensors are also interesting. The captured depth can theoretically provide direct access to the majority of desired

---

1 Term used within this book.

postproductions. The number of video sensors as well as depth sensor resolution and spatial limitations can, however, restrict some of these postproduction processes. These hybrid systems, however, will not be examined within this book.

All these materials share the need to synchronize and calibrate (often even with geometric and/or colorimetric corrections) information captured by different cameras or Z-cams, and often have different accompanying capabilities regarding:

– recording signals from all sensors without loss of data;

– processing all data in real time, which demands a significant computation infrastructure (often using distributed calculating).

This chapter introduces the main configurations mentioned above in a purely video multiview capture context, using notable practical examples and their use. We will also propose links to databases providing access to media produced by devices within each category.

3.2. Binocular systems

3.2.1. *Technical description*

Capturing binocular video, also known as stereoscopy or, more recently "3D stereoscopy" (3DS), requires the use of two cameras[2] connected by a rigid or articulated mechanical device known as a "stereoscopic rig". The images taken can be projected either on a stereoscopic display device (such as a cinema screen or a 3D television, most commonly) [DEV 10], or used to extract the scene's 3D geometry, in the form of a depth map, using stereo correspondence algorithms.

3.2.1.1. *The shooting geometry*

Filming is carried out using two cameras with the same optical parameters (focal length, focus distance, exposure time, etc.), pointing roughly in the same direction, orthogonal to the line connecting their optical centers (which is known as the *baseline*). The optical axes can be parallel or convergent.

---

2 In photography, where the scene is fixed, we only need a single device that is moved along a slider between the left and right views.

Ideally, to simplify stereoscopic correspondence, the two optical axes must be strictly parallel, orthogonal to the baseline, and the two image planes must be identical. In this situation, the corresponding points have the same y-coordinate in both images. However, if the cameras are convergent (i.e. the optical axes converge at a finite distance) or if the alignment is approximate, the images taken by the camera can be rectified (see section 5.4) to get back to the ideal situation. Rectification is therefore an important postproduction phase for stereoscopic films (see section 3.2.2.1).

The main geometric parameters for stereoscopic recording and stereoscopic visualization are shown in Figure 3.1. b, W and H are the parameters of the stereoscopic camera and Z is the distance from a 3D point to the plane passing through the stereoscopic baseline and parallel to the image planes. The triangles $\mathbf{M_l P M_r}$ and $\mathbf{C_l P C_r}$ are homothetic. As a result: $(Z - H)/Z = dW/b$. This allows us to simply express the relations between the stereoscopic disparity d, expressed as a fraction of the image's width W and the distance Z, similar to that shown in Chapter 7:

$$d = \frac{b}{W}\frac{Z - H}{Z}, \quad \text{or} \quad Z = \frac{H}{1 - dW/b} \qquad [3.1]$$

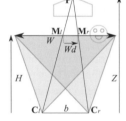

Symbol	Camera	Display
$\mathbf{C}_l, \mathbf{C}_r$	Optical center	Optical center (eyes)
$\mathbf{P}$	Point in the scene	Perceived 3D point
$\mathbf{M}_l, \mathbf{M}_r$	Points in the image in $\mathbf{P}$	Points on the screen
b	Interocular	Interocular (eyes)
H	Convergence distance	Distance from the screen
W	Width of the convergence plane	Size of the screen
Z	Real depth	Perceived depth
d	Right-left disparity (as a fraction of W)	

Figure 3.1. *Geometry of the stereoscopic shooting device and that of the stereoscopic display device can be described by the same low number of parameters*

3.2.1.2. *Perceived geometric distortions*

If stereoscopic video is designed to be projected onto a stereoscopic display device whose parameters are b', W' and H', the depth Z' perceived by stereoscopy[3] can be calculated according to the disparity d

---

3 Stereoscopy is combined with a number of other monocular indices to create the 3D perception of the scene [LIP 82]: light and shade, relative size, interposition, texture gradient, aerial perspective, perspective, flow, etc.

(equation [3.2]). By eliminating the disparity d from [3.1] and [3.2], in [3.3] we obtain the relation between the real depth Z and the perceived depth Z', which will be applied to the multiscopic example in Chapter 4:

$$Z' = \frac{H'}{1 - dW'/b'} \tag{3.2}$$

$$Z' = \frac{H'}{1 - \frac{W'}{b'}\left(\frac{b}{W}\frac{Z-H}{Z}\right)} \quad \text{or} \quad Z = \frac{H}{1 - \frac{W}{b}\left(\frac{b'}{W'}\frac{Z'-H'}{Z'}\right)} \tag{3.3}$$

There is ocular divergence when $Z' < 0$ ($d' > \frac{b'}{W'}$), i.e. when the on screen binocular disparity is larger than the viewer's interocular. In general, real objects that are very far away ($Z \rightarrow +\infty$) are perceived at a finite distance or create divergence, depending on whether $\frac{W'}{b'}\frac{b}{W}$ is smaller or greater than 1. We then consider that an ocular divergence in the order of $0.5a$ is acceptable for short durations, and that this trick is used by stereographers to artificially augment the depth available behind the movie screen.

In the case of 3D television, the disparity limits due to the conflict between convergence and accommodation [EMO 05, UKA 07, YAN 04] render large (either positive or negative) disparities uncomfortable. The depth of focus of the human eye is in the order of approximately $0.3\ \delta$ (diopters) in normal situations[4], which, on a screen placed 3 m away, gives a depth of focus ranging from $1/(\frac{1}{3} + 0.3) \approx 1.6$ m to $1/(\frac{1}{3} - 0.3) = 30$ m. In practice, TV production rules are much stricter. 3DTV programs are produced with disparities ranging from -1% to $+2\%$ of the screen width[5] to remain in this comfort zone[6], with disparities temporarily ranging from -2.5% to $+4\%$, which completely prevents reaching the divergence limit on private projection devices.

We can see also that the situation where the perceived depth is strictly identical to the real depth ($Z' = Z$) can only be obtained if all parameters are equal, which is known as the "orthostereoscopic" configuration (this

---

4 More precise studies [MAR 99] have shown that this also depends on parameters such as pupil diameter, wavelength and spectral composition.

5 Negative disparities correspond to points closer to the screen and positive disparities correspond to disparities further away.

6 See, for example, the production guidelines of Sky 3D in the UK: www.sky.com/shop/tv/3d/producing3d.

configuration is often used for IMAX 3D films since the geometry of the projection device is known beforehand).

For a 3D fronto-parallel plane placed at a distance Z, we can calculate the scale factor s between the distances measured within this frame and the distances in the convergence plane: $s = H/Z$. We can also calculate the image scale factor σ', which explains the extent to which an object placed at a depth of Z or the disparity d is perceived as being enlarged ($\sigma' > 1$) or reduced ($\sigma' < 1$) in the directions X and Y with respect to objects in the convergence plane ($Z = H$):

$$\sigma' = \frac{s'}{s} = \frac{H'}{Z'}\frac{Z}{H} = \frac{1 - dW'/b'}{1 - dW/b} \qquad [3.4]$$

Of course, for objects in the screen plane ($d = 0$), we have $\sigma' = 1$. The relation between Z and Z' is linear if, and only if, $W/b = W'/b'$, in which case $\sigma' = 1$ and $Z' = ZH'/H$. We refer to this configuration as "orthoplastic" configuration (an orthostereoscopic configuration is, above all, orthoplastic).

A small object with a width of ∂X and a depth of ∂Z, placed at Z, is perceived as an object with the dimensions $\partial X' \times \partial Z'$ at a depth of Z', and the *roundness factor* ρ measures how much the object's proportions are modified:

$$\rho = \frac{\partial Z'}{\partial Z}\bigg/\frac{\partial X'}{\partial X} = \frac{\partial Z'}{\partial Z}\bigg/\frac{W'/s'}{W/s} = \sigma'\frac{W}{W'}\frac{\partial Z'}{\partial Z} \qquad [3.5]$$

In the screen's frame ($Z = H$ and $Z' = H'$), the roundness factor can be simplified as:

$$\rho_{\text{screen}} = \frac{W}{W'}\frac{\partial Z'}{\partial Z}_{(Z=H)} = \frac{b}{H}\frac{H'}{b'} \qquad [3.6]$$

A roundness factor equal to 1 indicates that a sphere is perceived exactly as a sphere, a smaller roundness factor indicates that it is perceived as a sphere flattened in the depth direction and a larger roundness factor indicates that it is perceived as an ellipsoid stretched in the depth direction. The roundness of an object in the screen plane is equal to 1 if, and only if, $b'/b = H'/H$. In order for this to be the case in the whole space, it is necessary that $b'/b = W'/W = H'/H$. As a result, the only geometric configurations that preserve roundness everywhere are identical to the display configuration up to a scale factor; these are "orthoplastic" configurations. Even if the geometry of the display device is known during filming, this imposes strict constraints on how the film is shot, which can be very difficult to follow in different situations (i.e. when filming sports events or wildlife documentaries). On the other hand, since the viewer's interocular b' is fixed, this indicates that a film can only be projected on a screen of a given size W' placed at a given distance H', which is in contradiction with the large variability of projection devices and movie theaters. We therefore refer to "hyperplastic" or "hypoplastic" configurations when the roundness is larger or smaller than 1, respectively. The roundness in the screen plane also increases when we move away from the screen and it is independent of screen size, which is counterintuitive; the majority of viewers expect to perceive "more 3D" when approaching a large screen.

Another important point to make is that a film, shot to have a specific roundness for a cinema screen positioned 15 m away on average, will see its roundness divided by 5 once projected on a 3DTV screen placed 3 m away, which, in part, explains the current dissatisfaction of 3DTV viewers. This effect can be counter balanced by specific post production for media designed for private viewing (home cinema), e.g. for 3D Blu-ray, although there are few titles that benefit from this treatment. Of course, this reduction in roundness is, in part, compensated by monoscopic depth cues. Besides, the roundness used in 3D cinema films is, in reality, between 0.3 and 0.6, depending on the desired dramatic effect [MEN 09], in order to favor the viewer's visual comfort.

3.2.2. *Principal uses*

3.2.2.1. *Cinema and 3D television*

Cinema and television rigs are, for the most part, heavy systems that often use a semi-reflective mirror to obtain interocular distances for the camera shorter than the diameter of the lens [MEN 11] (see Figure 3.2 (a)). Today a number of manufacturers produce compact semi-professional integrated stereoscopic cameras but their field of use is reduced, notably due to the fact that the interocular of these cameras is generally fixed while stereoscopic filming requires an adequate tuning of all stereoscopic parameters; merely adding a second camera alongside the first is not enough for 3DS filming.

3.2.2.1.1. Stereoscopy, a new and different art

2D cinema, in order to exist, has (1) to study the function of the brain in order to trick it into believing that a series of fixed images are really showing movement, (2) to survey, through experience gained from photography, the techniques that enable this illusion and develop a complete cinematographic chain and (3) to invent the parameters of a new art, which is the role of artists involved in the production of films, followed by engineers producing tools enabling these new artistic practices.

Stereoscopy is both a continuous evolution and a turning point in cinematography due to the fact that, as with photography, it must use current techniques and develop others. To do so, it is essential to:

– restudy the brain and the visual system and examine how to trick it, not only temporally but also spatially by recreating the illusion of a 3D space while, in reality, there are only two 2D images;

– improve recording and postproduction stereoscopy tools in the cinematographic chain and produce new tools based on cerebral observations in order to ensure that this new illusion is comfortable;

– enable the invention of a filming technique based on these different parameters that contribute to creating this illusion.

The cinematographic parameters on which traditional filming relies are well known. However, the rules that govern the stereoscopic parameters in order to create this new illusion have not yet been established. Based on the way the human visual system works, they should simulate (1) how convergence is, in general, coupled with accommodation, and (2) 3D vision resulting from the distance between both eyes, a parameter that varies slightly throughout the lifespan of each individual and between individuals.

However, simply shooting with an interocular equal to the average interocular of a population sample cannot, contrary to some ophthalmological studies, be considered sufficient. Indeed, stereoscopy uses these two parameters (interocular and convergence) to create emotion and feeling, exactly as the lenses used on a camera do not try to reproduce human perspective vision but reform it depending on the medium used. If we push these variations in distance to the extreme, on the one hand, we have the value 0, which corresponds to two identical 2D images and, on the other hand, interaxial distances without any relationship with the geometry of the human visual system. NASA, for example, has produced stereoscopic images of Earth with a distance of almost 70 m between the two viewpoints.

To create a rig, the interocular distance must be able to vary from 0 to the greatest usable value for some kind of scene. In general, for a standard configuration for comedy, a variation from a few millimeters to several centimeters corresponds to 90% of needs for fiction-based filming. As a result, rigs used for close-ups have interocular ranges between 0 and 100 mm.

Lastly, for long-distance shots of clouds, for example, the distance between the two cameras may even extend to several meters and the side-by-side rigs are often adapted to the specific needs of a given shot.

3.2.2.1.2. Computer-assisted production

While the rules for recreating a universe in 3D have been known since the 19th Century, the possibility of stereoscopic filming using rigs is much more recent and involves the use of a computer to analyze video streams and correct any potential faults. Given the fact that no mechanical, optical or electronic device is perfect, it is imperative to correct the recorded images as precisely as possible with a 3D corrector, in real time for television and in postproduction for cinema. This was enabled by the invention of digital images, which can correct each pixel individually.

3.2.2.1.3. Robotized rigs

A rig must use synchronized cameras and lenses with perfectly synchronized and calibrated zoom, point and diaphragm movements. The rig itself is robotized and contains motors that adjust distance and convergence in real time, as well as yaw/pitch/roll adjusting plates used to converge the two optical axes (the optical axes must be concurrent). In some cases, rigs have been used with more than two cameras, as was the case for the French language film *La France entre ciel et mer* [France between sky and sea], which was filmed by Binocle with four cameras on a helicopter (see Figure 3.2). In this case, the matching of four zooms and adjusting plates with four cameras demanded a huge degree of expertise since all optical centers had to be aligned as closely as possible.

Examples of materials used to pilot the rig, and to directly control the geometric and photometric quality and faults include TaggerLive and TaggerMovie by Binocle[7], *Stereoscopic Analyzer* (STAN) by Fraunhofer HHI, *Stereoscopic Image Processor* (SIP) by 3ality Technica[8], the real-time correction processor MPES-3D01 – often referred to as "3DBox" – by Sony and Pure by Stereolabs[9].

3.2.2.1.4. Stereoscopic postproduction

Postproduction tools have also been adapted to 3D cinema and algorithms specific to stereoscopy have been integrated into this software such as

---

7 www.binocle.com.

8 www.3alitytechnica.com/3D-rigs/SIP.php.

9 www.stereolabs.tv/products/pure/.

rectification, viewpoint interpolation and depth modifications, 2D to 3D conversion, color balancing of two streams and production of a depth map for 3D scene compositing. These tools include the Ocula plugins suite for Nuke (The Foundry)[10], DisparityKiller (Binocle), and Mistika Post (SGO)[11].

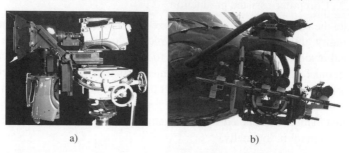

a)   b)

Figure 3.2. *Examples of rigs: a) Binocle Brigger III in a studio configuration, a robotized rig for 3DTV; b) a heliborne rig with four cameras used by Binocle for the film La France entre ciel et mer*

3.2.2.2. *Depth reconstruction*

Binocular systems designed to produce a stereoscopic reconstruction of "partial" 3D data[12] are generally much simpler than those used for cinema or television. These are most often lightweight systems that are small, consume little energy and can be used by a vehicle or mobile robot, for example, and they almost always have a fixed interocular distance in order to simplify their calibration.

The majority of these systems use monochrome cameras, since brightness alone is sufficient for stereoscopic correspondence, but color may bring additional functions such as the possibility of using color for segmentation tasks (such as skin color) or object recognition. Cameras used in this kind of system generally use a single sensor, since the use of color (by the way of a Bayer matrix filter) results in a loss of spatial resolution in images and therefore affects the precision of reconstructed depth.

The choice of the optimal interocular distance value for reconstruction is a disputed subject but a simple rule of thumb can predict the final precision.

---

10 www.thefoundry.co.uk/products/ocula/.

11 www.sgo.es/mistika-post/.

12 In the sense that they only contain the 3D information about the scene as seen from the stereo rig viewpoint.

The precision of the disparity d obtained by the stereoscopic correspondence algorithm can be presumed constant in the image (let us say 0.5 pixels). The error in the reconstructed depth Z is obtained by deriving equation [3.1]: $\partial Z/\partial d = bHW/(b - dW)^2$, and $\partial Z/\partial d = Z^2W/(bH)$. The error increases with the square of the distance and theoretically decreases with the interocular distance b, so that theoretically the larger the interocular distance, the better the precision in depth reconstruction. However, when we increase the distance, stereoscopic matching between the images is more difficult and the precision of disparity d is strongly degraded when the b/H value increases. Experience shows that, as a rule of thumb, a b/H value between 0.1 and 0.3 represents a reasonable compromise between ease of stereoscopic correspondence and precision in depth reconstruction.

Any pair of rigidly linked and synchronized cameras can be used[13] to reconstruct depth using stereoscopic correspondence algorithms (the OpenCV software library provides calibration functions, stereoscopic correspondence and simple 3D reconstruction algorithms).

Commercial off-the-shelf systems are also available. They have the advantage of being solidly constructed, precalibrated or easy to calibrate, and sometimes propose optimized stereoscopic correspondence algorithms, using the CPU or a dedicated FPGA. Point Grey has developed the Bumblebee system[14] using two or three cameras with different sensors or focal length options and a Software Development Kit (SDK) for calulating depth maps on the CPU. The Tyzx DeepSea stereo vision system[15], proposed with several interocular distance options, uses a FPGA and a PowerPC CPU to compute disparity, and transmits the 3D data via ethernet.

Focus Robotics has developed nDepth[16], with a fixed interocular distance of 6 cm, and a factory-calibrated monochrome sensor. Videre Design[17] has created stereo vision systems with fixed or variable interocular distances, with disparity computation carried out by the Small Vision System software (developed by SRI) or by a special chip (Stereo On Chip (STOC)). Surveyor

---

13 Synchronization is carried out either by a specific master–slave trigger connection between cameras or by the image transfer bus (for example, the majority of cameras manufactured by Point Grey are automatically synchronized when they are on the same "firewire" bus).

14 www.ptgrey.com/products/sterco.asp.

15 www.tyzx.com.

16 www.focusrobotics.com/.

17 http://users.rcn.com/mclaughl.dnai/.

Corporation[18] sells the *Stereo Vision System* (SVS), which is a low-cost solution for stereo with options such as embedded image capture, motorization and Wi-fi transmission, based on an open-source firmware.

3.2.3. *Related databases*

The European QUALINET project[19] has collated and classified a number of multimedia databases with a specific section dedicated to 3D Visual Content Databases directing users toward databases of fixed images or multiview stereoscopic video. The MOBILE-3DTV project[20] also contains a number of reference stereoscopic sequences. Other high-quality databases are also made available because of IEEE-3D *Quality Assesment Standard Group*[21] and the Sigmedia team at Trinity College Dublin[22].

3.3. Lateral or directional multiview systems

3.3.1. *Technical description*

This section examines systems and devices with close-together (relative to the scene being filmed) multiview sensors, often distributed evenly along a curve (whether rectilinear or not) or on a grid (flat or not). Thus, there are systems designed by mechanical assembly (linear or matricial) and synchronization of usual cameras as well as devices constructed by integrating optoelectronic components situated in order to provide the desired layout of viewpoints and then synchronized using specifically designed electronics. Lastly, these capture tools differ by the target use of the multiview media they capture (direct multiscopic visualization, FTV, reconstruction, refocus, etc.), which has a direct impact on the compromise between the number of views and their resolutions to maintain an acceptable volume of pixels to be captured, transmitted and stored.

These close multiview capture tools (either assembled or integrated) are often known as "camera arrays" (grids or linear layouts of cameras or viewpoints) and "plenoptic" systems or cameras. Camera arrays are generally

---

18 www.surveyor.com/.

19 www.qualinet.eu, dbq.multimediatech.cz.

20 www.focusrobotics.com/.

21 http://grouper.ieee.org/groups/3dhf, ftp://165.132.126.47.

22 www.tchpc.tcd.ie/stereo_database/.

focused on capturing multiple images with significant resolution for the depth reconstruction and 3D and/ or interactive visualization (FTV), while plenoptic systems generally aim to capture the "light field", and are more balanced in terms of the number of views and resolution to extract interpolated views (FTV) or variable focus images (refocus) as well as, sometimes, depth reconstructions. This classification is more nuanced than it seems because the similarity of their shooting geometries and improvements in shooting and pixel processing volumetric capabilities tend to bring closer those ratios number of views/number of pixels per view ratios and therefore mean that intended applications are accessible by both types of system. This classification could, however, soon be a historical artefact related to the appearance in successive waves of these technologies as well as their original objectives.

Undeniably, the first devices proposed fell within the class of linear viewpoint arrangements. Initially limited to capturing static scenes (in terms of composition as well as lighting), the very first systems achieved multiple perspective captures by controlling sequential positions of a still camera, as developed by Stanford University [LEV 96]. They were quickly overtaken by multisensor devices taking images of the same dynamic scene simultaneously, such as that proposed by Dayton Taylor in 1996 [TAY 96], and/or in low-level and controlled desynchronization, such as the system developed by Manex Entertainment for the film *The Matrix*. The majority of these devices were often designed and build specifically for their desired function: the MERL 3DTV project by Mitsubishi [MAT 04] positioned 16 cameras on a rail to produce multiscopic content designed for their *ad hoc* autostereoscopic screens while the University of California in San Diego, with Mitsubishi [JOS 06], used a rail with eight cameras for an automatic video matting application. Several prototypes of integrated devices have also been proposed for specific applications. We can, for example, cite the cameras with eight viewpoints developed in Reims, France [PRE 10], which are illustrated in Figure 3.3, and which were specifically designed to produce multiscopic content with controlled distortion (see Chapter 4) for autostereoscopic screens on the market.

These linear layouts have, in addition, also been extended by several laboratories to more complex systems of 2D grids of cameras. The most well known is probably that created by Stanford University[23] [WIL 05], which has been used for multiple applications, notably aimed at FTV and refocus. It is composed of a variable number of cameras (usually more than 100) organized

---

23 http://graphics.stanford.edu/projects/array/.

according to various configurations in planar or piecewise planar 2D grids. Another 2D grid, albeit irregular, has been developed by the Carnegie Mellon University [ZHA 04] with 48 cameras in individual horizontal and vertical positions controlled to optimize the calculation of depth in order to generate the desired perspective (FTV). We can also cite Sony in partnership with Columbia University [NOM 07], which have proposed flexible and stretchable 1D and 2D grids, composed of elastic supports on which 20 cameras are fixed in regular positions (at rest state). The deformation of the support therefore modifies the system's configuration to adapt to the situation and the desired requirements (more or less panoramic mosaicking in [NOM 07]).

The emergence of grids has also enabled research dealing with ray-space associated with plenoptic function, notably summarized by [ADE 91]. This plenoptic function (an aggregation of the Latin *plenus* – complete – and optics) is the function that gives the light intensity of all the rays in a scene. Yielding real values, it is defined for seven real variables; three for the position of a point of the ray, two for its 3D direction of propagation, one for the wavelength from which we measure intensity and the last for the point in time of this measure:

$$\mathcal{P} \quad \mathbb{R}^3 \times \mathbb{R}/2\pi\mathbb{Z} \times \mathbb{R}/\pi\mathbb{Z} \times \mathbb{R}^+ \times \mathbb{R} \longmapsto \mathbb{R}^+$$

$$((x,y,z),(\phi,\theta),\lambda,t) \longrightarrow \mathcal{P}(x,y,z,\phi,\theta,\lambda,t)$$

[3.7]

Usually, this function is reduced to five variables by externalizing the wavelength in the result that becomes a spectrum and by considering the intensity to be constant at the time of measure along the whole length of the ray[24]. According to this hypothesis, all the points in the ray deliver almost the same spectrum at the time studied and we can therefore reduce this redundancy by suppressing one of the space variables. In practice, we commonly select coplanar points by no longer "managing" the rays parallel to this ray-capturing plane. This gives:

$$\mathcal{P} \quad \mathbb{R}^2 \times \mathbb{R}/2\pi\mathbb{Z} \times \mathbb{R}/\pi\mathbb{Z} \times \mathbb{R} \longmapsto \mathbb{R}^{+\mathbb{R}^+}$$

$$((x,y),(\phi,\theta),t) \longrightarrow \mathcal{P}(x,y,\phi,\theta,t) \equiv \text{spectrum } \mathcal{S}(\lambda)$$

[3.8]

---

24 Given that we temporally sample time at a step dt, and then that the light intensity if transported to the speed of light c yielding $\mathcal{I}(x,t) = \mathcal{I}(x0, t - (x - x0)/c)$, this hypothesis is reasonable if the maximum width of the scene is slightly less than the distance traveled by a photon between two time steps, namely $299{,}792{,}458.dt$ m $\approx$ 12,491 km at 24 Hz, 2,998 km at 1 kHz or even 300 m at 1 MHz.

The domain's dimension can again be reduced to four by fixing the time of study or by transferring it in the result that becomes a temporal spectrum:

$$\mathcal{P} \quad \mathbb{R}^2 \times \mathbb{R}/2\pi\mathbb{Z} \times \mathbb{R}/\pi\mathbb{Z} \longmapsto \mathbb{R}^{+^{\mathbb{R}^+ \times \mathbb{R}}}$$

$$((x,y),(\phi,\theta)) \longrightarrow \mathcal{P}(x,y,\phi,\theta) \equiv \text{temporal spectrum } \mathcal{S}(\lambda,t)$$

[3.9]

Digitalizing the reduced plenoptic function involves spatial, angular, spectral and temporal windowing and sampling operations followed by quantification of the intensities that limit the domain as well as the value space. These operations create a temporal series of 4D digital signals indexed by the indices i, j (connected to x, y) from the capture points arranged in a grid and the coordinates s, t of the image pixel captured (in i, j), representative of the direction ϕ, θ of the ray measured in i, j, s, t. For each sample, they contain a set of intensities quantified for a discrete number of spectral bands (generally three – red, green, blue (RGB)). These light fields can be easily obtained from the data captured by a camera array by simply stacking up the views captured according to the grid's layout:

$$\mathcal{LF}[s,t,i,j] \equiv Quantify\,(\mathcal{P}(x(i,j),y(i,j),\phi(i,j,s,t),\theta(i,j,s,t)))$$

[3.10]

The growing attraction for this multiview capture representation and, specifically for its resulting models and applications (FTV, refocus, to name but a few), has led to the arrival of dedicated optics, such as that proposed by Todor Georgiev from Adobe-Qualcomm[25], and integrated solutions, such as the "plenoptic cameras" proposed in recent years by companies such as Raytrix[26] or Lytro[27] (see Figure 3.3). These cameras generally include a microlens grid in front or behind the lens in order to separately capture, after deviation, the light rays that are combined in a standard camera (see Figure 3.4 for an illustration with a lenticular array at the back wall of the darkroom). If the object is captured in the focus plane (example B in Figure 3.4), instead of a clear pixel, we obtain a homogenous microimage that is synonymous with the object's position being in the focus plane. Otherwise (examples A and B), we obtain, instead of a blurred pixel, a local sampling of the object that, coupled with those of the neighboring capture positions,

---

25 www.wired.com/gadgetlab/2007/10/adobe-shows-off/.

26 www.raytrix.de/index.php/Cameras.html.

27 https://www.lytro.com/camera.

allows reconstructing the points outside the focus plane. Other approaches, notably that by Mitsubishi [VEE 07][28], replace the lenticular array with a printed mask similar to parallax barriers. As a result, the debate between masks and microlenses, well known with autostereoscopic displays, also applies to plenoptic cameras.

a)   b)

Figure 3.3. *Examples of integrated cameras: a) a Cam-Box prototype camera with eight integrated perspectives developed by 3DTV Solutions and the University of Reims and b) the Lytro plenoptic camera*

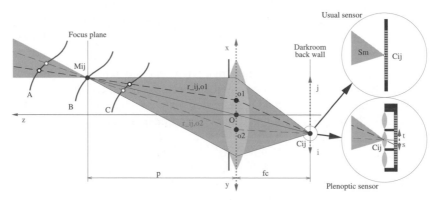

Figure 3.4. *Differences between standard and plenoptic cameras: from above (axes x, j, s) or the side (axes y, j, t) the rays converging as a single point at the back wall of the darkroom are summed in the first and differentiated by refraction and sampling in the second*

There has also been a recent tendency to miniaturize small grids within new integrated components, designed specifically for mobile terminals. The

---

28 http://web.media.mit.edu/~raskar//Mask/.

Californian company Pelican Imaging has produced a 5×5 microgrid component, which is the size of a current monoview sensor[29].

3.3.2. *Principal uses*

Linear layouts of different viewpoints allow, by simple selection (or even interpolation) of a specific viewpoint, the effect of camera movement around a frozen or slow-motion scene. These technologies, known as bullet time, were largely brought to the fore in 1999 by the film *The Matrix*. It has since been used by a number of companies using more or less integrated proprietary systems that can be used with varied and occasionally surprising applications such as surfing[30].

With the emergence of multiscopic visualization devices (see Chapter 14), the question of creating adapted content using real capture has been developed, notably leading to several improvements in camera arrays. Linear layouts have also focused on autostereoscopic devices with a simple horizontal parallax. Similarly, grids have also been used for double parallax devices, known as "integral imaging displays" in reference to its precursor, "integral photography" proposed [LIP 08b] and then experimentally demonstrated [LIP 08a] in 1908 by Gabriel Lippmann.

The generation of intermediary viewpoints (FTV, "image-based rendering" (IBR)) also had a strong influence on the emergence of different camera arrays. This technology is somewhat an extension of the frozen time virtual camera technique using camera position interpolation. Its implementation is, however, different and relies either on a depth reconstruction to project the available views on the virtual camera (see Chapter 9) or on a planar section of the light field (with the real, coordinates i, j fixed), yielding a digital signal that samples the reduced plenoptic function according to equation [3.10].

The strong redundancy of close-together multiple perspective captures in a single scene can provide a depth reconstruction with increased reliability. As the quality of both depth maps (or disparity maps with parallel geometry capturing) and occlusion detection is essential in related applications (such as FTV and AR), a number of teams have studied the opportunity to use these strong redundancies which imply additional new challenges. Multiple

---

29 www.pelicanimaging.com/index.htm.

30 www.core77.com/blog/technology/rip_curl_time-slice_camera_array_collaboration_ lets_ you_perceive_surfing_as_never_before_20925.asp.

solutions have been proposed, seeking coherence between multiple binocular matches or directly examining multiocular matches across all views. Regardless of the approach, managing occlusions, which is accessible in multiocular vision, is an opportunity that remains difficult to manage. Chapter 7 provides a more detailed description of this area.

Similarly, the availability of strongly redundant views allowing for a global matching process has been used (see Chapter 19) to create HDR capturing devices by postprocessing views captured with moderate but varied dynamic from different viewpoints. The allocation of different dynamic ranges to viewpoints is obtained by neutral filters of different densities or by distinct exposure time settings.

To conclude, let us present an example of application of multiview capture, either by grids or plenoptic cameras, which is surprising since the notion of depth of field, a crucial aspect of photography, seemed definitely set at shooting. The numerous multiview captures as well as ray-space modeling have given rise to a flurry of activity relating to a new opportunity with highly promising possibilities: the choice to refocus postcapture. This includes, for example:

– the selection of the focus plane (by averaging pixels from several perspectives corresponding to the rays geometrically issued from the same points in this plane);

– the choice of aperture and therefore depth of field (by selecting the neighboring viewpoints from which the averaged pixels are taken);

– the possibility of selecting an "all-in-focus" infinite depth of field (by selecting non-averaged pixels, which corresponds to a pinhole camera);

– removing the foreground from some images, to show the partially hidden background, if it is far enough away to be visible from several other viewpoints.

3.3.3. *Related databases*

Without attempting to provide an exhaustive list, there are a number of databases created using the devices discussed in this section. The University of California in San Diego and Mitsubishi[31] deliver some captures in a linear layout with eight-view videos and a series of 120–500 still images. The Light

---

31 http://graphics.ucsd.edu/datasets/lfarchive/lfs.shtml.

Fields library at the University of Stanford[32] is full of highly varied multiple scenes captured in high resolution, often from several hundred viewpoints, created by moving the camera on robotized arms or the Stanford grid. This information is available as raw or modified data with calibration information and the possibility of interacting online with their light field form by selecting a perspective and handling refocus (choice of shutter and focus plane). This library completes and surpasses its predecessor[33], which proposed less complex series, both in terms of the number of views as well as their resolution. A simpler example is also available on Todor Georgiev's site[34], which contains a number of plenoptic images with several tens of millions of rays. Lastly, the University of Heidelberg maintains a library[35] of several synthesized light fields, accompanied by genuine depth information, as well as real scene captures by the Raytrix plenoptic cameras using a 9×9 grid.

3.4. Global or omnidirectional multiview systems

3.4.1. *Technical description*

In this section, we will examine multicamera systems with spaced out and approximately convergent layout in order to "cover", with enough redundancy, a scene volume large enough to encompass evolving objects and/or actors. The first systems of this kind have been used for bullet time or MoCap techniques. "Global systems" used for frozen time are generally composed of a rail forming a curve representing the desired trajectory for the virtual camera (i.e. closed or not, not always planar or circular, etc.) often hosting a significant number of cameras with a viewing direction set according to that desired for the virtual camera at this place, and with controlled synchronization depending on the desired effect (frozen time or more or less slow-motion). In MoCap using video markers, for the most part, we use fewer synchronized infrared cameras freely positioned, and a geometric calibration obtained by moving a target object bearing fixed markers.

The fairly intensive use of these techniques by the film and video games industries (whose business-model makes it profitable) has raised a marked interest in a more advanced technology using markerless multiview capture with more varied results: 3D video. Proposed in 1997 [KAN 97, MOE 97]

---

32 http://lightfield.stanford.edu/.

33 http://graphics.stanford.edu/software/lightpack/lifs.html.

34 www.tgeorgiev.net/Gallery/.

35 http://hci.iwr.uni-heidelberg.de/HCI/Research/LightField/lf_archive.php.

and intensively studied and developed since then [MAT 12], it allows the reconstruction within an entire sequence of the geometry as well as the texture of the object or actor being filmed to create an animated digital avatar of sufficient quality that it can be reused by synthesizing the image from loosely restricted angles.

This requires a synchronized multiview capture system with numerous viewpoints distributed around the scene space, characterized as the intersection of camera fields of view (see left of Figure 3.5). The compromise between the number of cameras (completion) and the gap between cameras (precision of reconstruction) has been suggested by Kanade *et al.* [KAN 97] to be between 9 and 16 for a circular and regular layout placed at mid-height of the scene space with converging axis at the circle center (see top left of Figure 3.5 for an example with 12 cameras). More complete solutions have also been proposed to reconstruct the top of objects by adding cameras overlooking the scene from above and then selecting layouts sampling the directions of capture more evenly (several circles at different heights with aerial cameras[36], domes[37,38], in more *ad hoc* studio or outside layouts [KIM 12][39]) with the number of cameras fluctuating depending on the applicative context from a few units (University of Surrey[39], *Max Planck Institute* [DE 08] or the "GrImage" project[40]) to several hundreds (1,000 for the "Virtualized reality" project[41]).

These complex systems must also have networking, storage and calculatory capabilities in order to manage generated video streams and very precise geometric and colorimetrics calibration technologies. Lastly, controlling lighting conditions and simplifying objects outlining facilitates image processing. This renders these systems complex, delicate and costly and explains their normal use in dedicated rooms known as "3D video studios".

---

36 Recover3D, a project, 2012–2014, run by XD Productions, see far right and bottom of Figure 3.5.

37 www.cs.cmu.edu/ virtualized-reality/page_History.html.

38 The 3D-COFORM FP7 project 2007-2013, www.vcc-3d.eu/multiview and www.3dcoform.eu, digitalizing heritage for small objects exhibiting complex light/matter interactions.

39 www.surrey.ac.uk/cvssp/research/3d_video/index.htm.

40 www.inrialpes.fr/grimage/.

41 www.cs.cmu.edu/~virtualized-reality/.

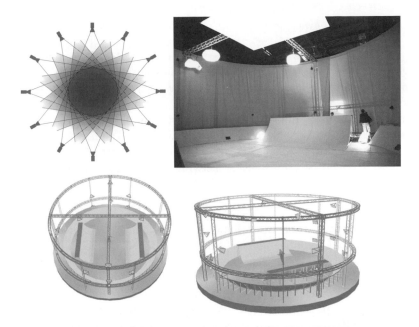

Figure 3.5. *Examples of 3D video studios: from top left, circular arrangement of 12 cameras showing the scenic space used as an intersection of camera field depth zones (in light gray); top right and below, the studio of the Recover3d project*[36]

The "bullet time" market is principally structured around service providers[42] that operate proprietary systems while MoCap also concerns several companies[43] that distribute off-the-shelf solutions. With regard to 3D video, the service has developed with specialized production companies with 3D studios[44] while the commercialization of these systems is just beginning[45].

---

42 Such as Reel EFX www.reelefx.com/ and Time Slice www.timeslicefilms.com/#1.

43 Such as Vicon (www.vicon.com/), Animazoo (www.animazoo.com/) and Moven (www.moven.com/).

44 For example, XD Productions (www.xdprod.com/) and 4D View Solutions (www.4dviews.com/).

45 4D View Solutions (www.4dviews.com/) has also been marketing solutions for some time.

3.4.2. *Principal uses*

In this section, we will not discuss frozen time or MoCap technologies at length as their fairly specific capturing systems position them at the edge of the scope of this book. Hence, the main use of "global multiview systems" concerns 3D video, which has witnessed a boom both in research and production, as noted in [MAT 12] who focuses entirely on this technique. 3D video relies on complex systems including a number of cameras synchronized, distributed and calibrated in terms of geometry and colorimetry within a video stream transfer network with significant storage and calculation capabilities.

The extraction of avatars' geometry from multiple video streams initially requires a precise geometric calibration of all cameras. This reconstruction can be operated according to three techniques classed as "model based" or, in contrast, free methods. The first class corresponds to searching the configuration of a predefined model that optimizes the geometric model's degrees of freedom so that its projections correspond to the images captured as closely as possible. The second contains two competing techniques; multiview stereo, which aims to reconstruct 3D points by triangulation using supposedly homologous pixels in different images, and "silhouette-based" methods, which reconstruct the visual hull of the avatar by intersecting generalized cones supported by the outlines of its projections in all images. However, searching a predefined model configuration has shown a fairly fatal flaw in its construction; it lacks adaptability although it can, nevertheless, guide a silhouette-based reconstruction using fewer cameras ([DE 08], the "Free Viewpoint Video of Human Actors" project[46] [CAR 03]). Stereovision methods are sensitive to errors in colorimetric calibration and to specular reflections, and are generally very costly in terms of computation time but can provide geometric information in concave zones where the visual hull is naturally convex. In contrast, visual hulls are easier to obtain, can be calculated efficiently and are more reliable although these envelopes provide, by their very nature, only rough results in concave zones of the objects. The model-based techniques are often employed to digitize human actors. Among free methods (non-model based), even when applied to humans, "Visual Hull" techniques (examined in Chapter 8), are often used in production due to their reliability, although their limitations have restricted their progression so far. It is for this reason that the complimentary combination of multiview stereo and silhouettes has inspired projects based on creating hybrids of them such as

---

46 www.mpi-inf.mpg.de/ theobalt/FreeViewpointVideo/.

Recover3d[46] in which monoscopic and multiscopic cameras are distributed around the scene space to produce a robust geometric model (by integrating it into the visual hull), which is more detailed (through multiview stero reconstruction), notably in concave areas.

Once the geometric model has been reconstructed at each time step, it has to be given a temporally coherent visual content (texture) taken from the captured images. One may apply geometric models' temporal tracking solutions (see Chapter 8) to create semantic coherence between texture hooks, followed by video texturing techniques that involve locally mixing photometric information reprojected onto the geometric model from the images where this local zone is not hidden. Difficulties here relate to what decision to make when there are gaps between retro-projected data. These gaps can originate in geometric reconstruction faults, colorimetric calibration faults, as well as characteristics related to the scene itself such as reflections, or other specular phenomena. These complex visual phenomena are the basis of further study, such as the Light Stages series[47], which examines systems dedicated to capturing complex optical properties in a camera array context with lightning conditions modulation or, more recently, the 3D-COFORM project[48], which focuses on the high-quality digitalization of heritage and cultural objects through capturing static objects in multiple lighting conditions (151 sources) from 151 viewpoints and different exposures to create HDR views (one per source/viewpoint pair), thereby enabling mapping of optical properties in the form of bidirectional function textures (BFTs).

3D video capture is more costly than MoCap because it is more complex. However, its results are far more versatile. Indeed, the producer and his/her graphics technicians can, in postproduction, easily select the angles of view with few spatial limitations while editing the animated avatars acquired in these scenes (spatiotemporal movement/deformation, duplication, transposition into other scenes, relighting[48]). These possibilities make these acquired avatars more reusable and profitable, thereby reducing production costs. This creates a kind of technology that is both open to creativity and cheaper and, as a result, is more accessible for televisual production. The digitalization of animated avatars is also of interest for other applicative domains such as culture[48], sport [KIM 12] and collaborative telepresence [PET 10].

---

47 http://gl.ict.usc.edu/LightStages/.

48 A number of illustrations of this can be found on the XD Productions Website www.xdprod.com/Xd Productions_RD.swf.

Lastly, a recent tendency, outside the scope of this chapter, extrapolates the 3D video capabilities described previously, by targeting 3D reconstruction using non-calibrated collective sources (such as Web-found amateur captures) in the form of photos [GOE 07, SNA 09] or videos ([BAL 10], the "Virtual Video Camera" project[49]).

3.4.3. *Related databases*

Several academic sites offer multiview sequences captured by their systems. The University of Surrey gives eight-view captures in a circular layout   (www.ee.surrey.ac.uk/cvssp/visualmedia/visual-contentproduction/ projects/surfcap), MIT proposes a number of complete data sets (images, exposure, results, etc.) that have been captured and processed according to [VLA 08] (http://people.csail.mit.edu/drdaniel/mesh_animation/) and Inria Grenoble-Rhône-Alpes has made public its "4D repository" of several tens of data sets captured by their GrImage system (http://4drepository.inrialpes.fr/).

3.5. Conclusion

This chapter has shown that multiview capture entails the use of varied and highly complex technologies. These technologies have opened up new perspectives on more creative postproduction processes, which could revolutionize audiovisual production while offering further potential for qualitative editing of recorded media postfilming. They also provide an increasingly rich means of digitalizing our environment, as well as a number of other applicative fields requiring 3D reconstruction and/or motion recognition. While these technologies are currently being developed as laboratory prototypes mainly, as *ad hoc* systems for service providers or limited batch production devices, the importance of these applications will enable their commercial development, as shown by the arrival of plenoptic cameras and microgrids for mobile devices.

3.6. Bibliography

[ADE 91] ADELSON E.H., BERGEN J.R., "The plenoptic function and the elements of early vision", in LANDY M.S., MOVSHON A.J., (eds), *Computational Models of Visual Processing*, MIT Press, Cambridge, MA, pp. 3–20, 1991.

---

49 http://graphics.tu-bs.de/projects/vvc/.

[BAL 10] BALLAN L., BROSTOW G.J., PUWEIN J., *et al.*, "Unstructured video-based rendering: interactive exploration of casually captured videos", *ACM SIGGRAPH Papers, SIGGRAPH'10 2010, ACM*, New York, NY, pp. 87:1–87:11, 2010.

[CAR 03] CARRANZA J., THEOBALT C., MAGNOR M.A., *et al.*, "Free-viewpoint video of human actors", *ACM SIGGRAPH 2003 Papers, SIGGRAPH'03*, ACM, New York, NY, pp. 569–577, 2003.

[DE 08] DE AGUIAR E., STOLL C., THEOBALT C., *et al.*, "Performance capture from sparse multi-view video", *ACM Transitions on Graphics*, vol. 27, no. 3, pp. 98:1–98:10, August 2008.

[DEV 10] DEVERNAY F., BEARDSLEY P., "Stereoscopic cinema", in RONFARD R., TAUBIN G. (eds), *Image and Geometry Processing for 3-D Cinematography*, vol. 5 of *Geometry and Computing*, Chapter 2, Springer, Berlin, Heidelberg, pp. 11–51, 2010.

[EMO 05] EMOTO M., NIIDA T., OKANO F., "Repeated vergence adaptation causes the decline of visual functions in watching stereoscopic television", *Journal of Display Technology*, vol. 1, no. 2, pp. 328–340, December 2005.

[GOE 07] GOESELE M., SNAVELY N., CURLESS B., *et al.*, "Multi-view stereo for community photo collections", *Proceedings ICCV, IEEE International Conference on Computer Vision*, Rio de Janeiro, Brasil, pp. 1–8, October 2007.

[JOS 06] JOSHI N., MATUSIK W., AVIDAN S., "Natural video matting using camera arrays", *ACM SIGGRAPH 2006 Papers, SIGGRAPH '06*, vol. 25, ACM, 2006.

[KAN 97] KANADE T., RANDER P., NARAYANAN P.J., "Virtualized reality: constructing virtual worlds from real scenes", *IEEE MultiMedia*, vol. 4, no. 1, pp. 34–47, January 1997.

[KIM 12] KIM H., GUILLEMAUT J.-Y., TAKAI T., *et al.*, "Outdoor dynamic 3-D scene reconstruction", *IEEE Transactions on Circuits and Systems for Video Technology*, vol. 22, no. 11, pp. 1611–1622, November 2012.

[LEV 96] LEVOY M., HANRAHAN P., "Light field rendering", *ACM SIGGRAPH 1996 Papers, SIGGRAPH '96*, ACM, pp. 31–42, 1996.

[LIP 82] LIPTON L., *Foundations of the Stereoscopic Cinema*,   Van Nostrand Reinhold, 1982.

[LIP 08a] LIPPMANN M.G., "Epreuves réversibles donnant la sensation du relief", *Journal of Physics*, vol. 7, pp. 821–825, November 1908.

[LIP 08b] LIPPMANN M.G., "Epreuves réversibles. photographies intégrales", *Comptes Rendus de l'Académie des Sciences*, vol. 146, no. 9, pp. 446–451, March 1908.

[MAR 99] MARCOS S., MORENO E., NAVARRO R., "The depth-of-field of the human eye from objective and subjective measurements", *Vision Research*, vol. 39, no. 12, pp. 2039–2049, June 1999.

[MAT 04] MATUSIK W., PFISTER H., "3D TV: a scalable system for real-time acquisition, transmission, and autostereoscopic display of dynamic scenes", *ACM SIGGRAPH 2004 Papers, SIGGRAPH '04*, vol. 24, ACM, 2004.

[MAT 12] MATSUYAMA T., NOBUHARA S., TAKAI T., *3D Video and its Applications*, SpringerLink: Bücher, Springer, London, 2012.

[MEN 09] MENDIBURU B., *3D Movie Making: Stereoscopic Digital Cinema from Script to Screen*, Focal Press, 2009.

[MEN 11] MENDIBURU B., *3D TV and 3D Cinema: Tools and Processes for Creative Stereoscopy*, 1st ed., Focal Press, 2011.

[MOE 97] MOEZZI S., TAI L.-C., GERARD P., "Virtual view generation for 3D digital video", *IEEE MultiMedia*, vol. 4, no. 1, pp. 18–26, January 1997.

[NOM 07] NOMURA Y., ZHANG L., NAYAR S., "Scene collages and flexible camera arrays", *Proceedings of Eurographics Symposium on Rendering*, Eurographics Association, June 2007.

[PET 10] PETIT B., DUPEUX T., BOSSAVIT B., *et al.*, "A 3d data intensive tele-immersive grid", *Proceedings of the International Conference on Multimedia, MM '10*, ACM, New York, NY, pp. 1315–1318, 2010.

[PRE 10] PREVOTEAU J., CHALENÇON-PIOTIN S., DEBONS D., *et al.*, "Multi-view shooting geometry for multiscopic rendering with controlled distortion", *International Journal of Digital Multimedia Broadcasting (IJDMB), special issue Advances in 3DTV: Theory and Practice*, vol. 2010, pp. 1–11, March 2010.

[SNA 09] SNAVELY K.N., Scene reconstruction and visualization from internet photo collections, PhD Thesis, University of Washington, Seattle, WA, 2009.

[TAY 96] TAYLOR D., "Virtual camera movement: the way of the future?", *American Cinematographer*, vol. 77, no. 9, pp. 93–100, 1996.

[UKA 07] UKAI K., HOWARTH P.A., "Visual fatigue caused by viewing stereoscopic motion images: background, theories, and observations", *Displays*, vol. 29, no. 2, pp. 106–116, March 2007.

[VEE 07] VEERARAGHAVAN A., RASKAR R., AGRAWAL A., *et al.*, "Dappled photography: mask enhanced cameras for heterodyned light fields and coded aperture refocusing", *ACM Transactions on Graphics*, vol. 26, no. 3, pp. 69-1–69-12, July 2007.

[VLA 08] VLASIC D., BARAN I., MATUSIK W., *et al.*, "Articulated mesh animation from multi-view silhouettes", *ACM Transitions on Graphics*, vol. 27, no. 3, pp. 97:1–97:9, August 2008.

[WIL 05] WILBURN B., JOSHI N., VAISH V., *et al.*, "High performance imaging using large camera arrays", *ACM SIGGRAPH 2005 Papers, SIGGRAPH '05*, ACM, New York, pp. 765–776, 2005.

[YAN 04] YANO S., EMOTO M., MITSUHASHI T., "Two factors in visual fatigue caused by stereoscopic HDTV images", *Displays*, vol. 25, no. 4, pp. 141–150, November 2004.

[ZHA 04] ZHANG C., CHEN T., "A self-reconfigurable camera array", in KELLER A., JENSEN H.W. (eds), *Proceedings of the 15th Eurographics Workshop on Rendering Techniques*, Eurographics Association, pp. 243–254, 21–23 June 2004.

Chapter 4

Shooting and Viewing Geometries in 3DTV

4.1. Introduction

A three-dimensional (3D) perception induced by a 3D display, operated according to various modalities (optics, colorimetrics and alternately shutters) through spatial and/or temporal, generally planar, mixing of colocalized 2D images in front of viewers, is essentially only an illusion. These mixed images are separated before being received by viewers' eyes so that, through stereopsis, their minds are tricked into seeing a deceptive 3D scene instead of two superposed flat images. This generic viewing geometry must be taken into account when capturing media for 3D television (3DTV) because the relationship between shooting and viewing geometries directly affects the quality of the viewer's experience, as well as depth distortion of the perceived scenes.

In this chapter, we will describe and characterize the viewing geometry and then present compatible shooting geometries. We will then study the potential distortions in perceived scenes that a combination of these shooting and viewing geometries may cause. The relations between these distortions and the parameters of the geometries used will allow us to propose a specification methodology for the shooting geometry, which will ensure that scenes are perceived with a set of arbitrarily selected possible distortion on

Chapter written by Jessica PRÉVOTEAU, Laurent LUCAS and Yannick REMION.

the 3DTV device used. Lastly, we will also provide practical details on how to use this methodology in order to place and configure virtual cameras when calculating synthetic content for 3DTV.

4.2. The geometry of 3D viewing

4.2.1. *Description*

In this section, we focus on display devices delivering deceptive 3D scene perception using multiview colocalized planar mixing. All these systems are based on spatial, optical, colorimetric and/or temporal mixing, within a single region of interest (ROI, an area occupied by the image shown on the display), of the $n \times m$ initial images of the scene, shot from different points of view. These devices temporally and/or physically separate the images reaching the eyes of one or more viewers.

In the case of stereoscopic systems, this separation of images can be achieved within a single optical beam[1] (see Figure 4.1) regardless of the position of the viewer within this beam [DUB 01, PEI 09, SAN 03]. Therefore, the device only uses two images ($n = 2$, $m = 1$), which are transported by this same optical beam and then physically (polarization, color, etc.) or temporally (shutter) separated by the viewer's glasses.

In contrast, autostereoscopic systems, where separation is carried out by the display device, deliver images as distinct optical beams (see Figure 4.1(b)), structured, for example, as a horizontal "fan" of n images (in this case, $n \geq 2$ and $m = 1$) [DOD 02, PER 00]. Optical beams could also be organized as both horizontal and vertical "fans". However, today only integral imaging delivers vertical disparity, although this will surely change in coming years. We then have an array of $n \times m$ optical beams ($n \geq 2$ for horizontal distribution and $m \geq 2$ for vertical distribution), each transporting a distinct image.

As such, all known systems provide alternating or simultaneous $n \times m$ images ($n \geq 2$ and $m \geq 1$) within one or several beams so that each eye of the viewer, correctly positioned in relation to the device, receives a coherent image (i.e. one of the initial images and not a combination of them), which is different to that received by the other eye. The viewer's brain therefore reconstructs the

---

1 An optical beam is a series of light rays emanating from a single wide source, which may be a screen, projector or another source. However, only its restriction to a single point source on the display is shown in Figure 4.1(a).

scene depth by stereopsis if the received images form a sound stereoscopic pair [HIL 53].

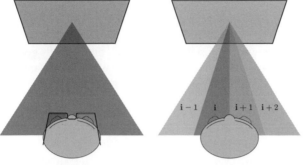

a) A stereoscopic system: a single optical beam transporting two images that are physically separated by glasses

b) An autostereoscopic system: several distinct optical beams, each transporting an image

Figure 4.1. *Image transport according to the technology used: glasses-based stereoscopy and autostereoscopy*

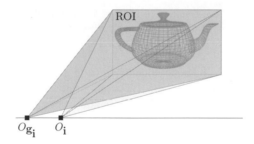

Figure 4.2. *Image viewing pyramids: the shared base is the device's ROI and the apices are the users' eyes O_i and O_{g_i}*

Multiscopic flat displays involve the planar colocalized mixing of images of the same scene taken from distinct points of view. The displayed images are rarely orthogonal to each viewer target axis (the axes between each eye and the center of the system's ROI). The viewing of images generally involves pyramids whose shared base is the system's ROI and whose apices are the users' eyes. Since the target axes are generally not orthogonal to the plane of the observed image, the viewing of these images creates trapezoid distortions if the "skew" of these viewings is not taken into consideration when these images are shot. If these trapezoid distortions are not coherent for the two images received by the viewer, stereoscopic pairing by the brain is more complex, and even impossible, which reduces or removes perception of 3D.

4.2.2. *Setting the parametric model*

Figure 4.3 shows a possible model of the shared geometry of 3D multiscopic displays (multiview colocalized planar mixing) and provides a set of parameters that completely characterize this geometry.

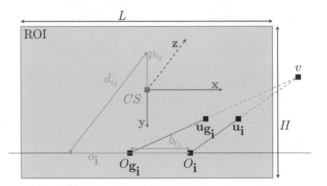

Figure 4.3. *Characterization of geometry of the 3D multiscopic display using colocalized planar mixing*

Our analysis of the characteristics of viewing geometry relies on a global reference frame defined in relation to the display device $r = (CS, \mathbf{x}, \mathbf{y}, \mathbf{z} \equiv \mathbf{x} \wedge \mathbf{y})$, selected at the center CS of the ROI with the axis $\mathbf{x}$, parallel to the ROI's lines and directed toward the viewer(s) right side, and the axis $\mathbf{y}$, parallel to the ROI columns and directed toward the bottom.

The 3D display system mixes $n \times m$ images within its ROI with the dimensions L (width) and H (height). Each of these images (denoted by $\mathbf{i} = (i_1, i_2) \in \mathbb{N}_n \times \mathbb{N}_m$) is presumed to be "correctly" visible (without being mixed with other images), at least from the preferred selected position $O_{\mathbf{i}}$. These positions are arranged as m lines parallel to the ROI lines situated at a distance of d_{i_2} from the system's ROI. Preferential positions are placed on these lines to ensure that the viewer, whose binocular gap is b_{i_2}, with the eyes parallel to the lines on the display, will have his/her right eye at $O_{\mathbf{i}}$ and his/her left eye at $O_{\mathbf{g_i}}$. The parameter b_{i_2} is often identical to the average human binocular gap of 65 mm but it is possible to select a different gap depending on the target audience, i.e. children. The right eye at $O_{\mathbf{i}}$ will see image number $\mathbf{i}$ while the left eye $O_{\mathbf{g_i}}$ will see image number $\mathbf{g_i}$, knowing that $\mathbf{g_i} = \mathbf{i} - (q_{i2}, 0)$ where q_{i_2} represents the gap between image numbers composing coherent stereoscopic couples that are visible with a binocular gap of b_{i_2} with a distance of d_{i_2}. As such, by combining the preferential positions of both the left and right eyes, we have: $O_i = O_{g_i} + b_{i_2}\mathbf{x}$ and $o_i = o_{g_i} + b_{i_2}$.

We also place the preferential position lines on the vertical axis by p_{i_2}, which represents the drop, i.e. the vertical gap between line i_2 of preferential positions and the center CS of the ROI. When $m = 1$, the device does not create any vertical separation and any drop is acceptable *a priori*. However, not all drops create the same perception and it is therefore necessary to know the average effective drop of target viewers during the design stage. If we do not know this expected drop p_{i_2}, we use the drop of an average size viewer.

Supposing that the pixels $\mathbf{u_i}$ and $\mathbf{u_{g_i}}$ are stereoscopic homologues in the images $\mathbf{i}$ and $\mathbf{g_i}$, their perception by the right and left eye at O_i and O_{g_i} leads the viewer's brain to perceive the 3D point v by stereopsis.

4.3. The geometry of 3D shooting

4.3.1. *Choosing a convenient geometry*

Providing 3D content to selected display systems requires sets of $n \times m$ images of a scene obtained using well-selected distinct points of view according to adapted projective geometries. A correctly positioned viewer will therefore receive two distinct images that form a stereoscopic couple that allows his brain to perceive the scene depth. Each eye receives an image that physically originates from the same area (which is normally rectangular), which corresponds to the display's ROI. What differs is that each eye is evidently positioned differently and therefore views the device's ROI according to different target axes. Depending on the desired application, three types of multiview shooting geometries are used primarily: convergent geometry, parallel geometry and decentered parallel geometry (see Figure 4.4).

Convergent shooting geometry (see Figure 4.4(a)) relates to cameras whose optical axes, equivalent to the target axes[2], converge at a single point without a shared base for shooting pyramids. Solutions for this type of system have been proposed [SON 07, YAM 97]. Since images have different trapezoid distortions, it is necessary to apply a systematic trapezoid correction

---

2 The optical axis is the line orthogonal to the sensor passing by the optical center while the target axis is the line passing through the optical center and the center of the sensor's ROI.

to enable the perception of 3D. However, this is not necessarily desirable due to the fact that it slows the chain of production and deteriorates the quality of images.

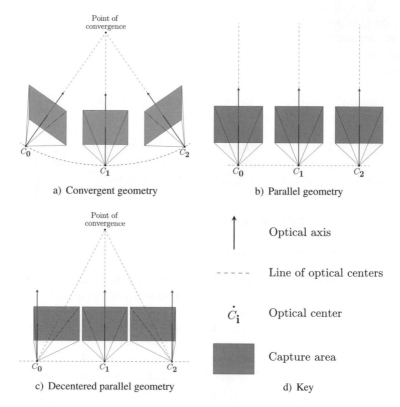

Figure 4.4. *The different shooting geometries represented in reversed pinhole model*

Another standard geometry, known as parallel (see Figure 4.4(b)), involves optical axes, equivalent to the target axes, parallel with each other, passing by optical centers aligned on m "optical center" straight lines parallel to the sensors' lines. It can be considered as a specific example of convergent geometry (with an infinite distance of convergence), as well as a specific decentered parallel geometry (with null decentering). If it does not require any prior correction in the images to enable 3D perception, this configuration is not entirely the best suited. Indeed, the perceived scene only appears to be protruding from the display since all the captured points are in front of the point of convergence at infinity, which is reproduced at the center of the display's ROI.

Lastly, decentered parallel shooting geometry (see Figure 4.4(c)) shares features with parallel geometry (parallel optical axes, optical centers aligned on m straight lines parallel to the sensors' lines) but separates the optical axes that converge at infinity and target axes that converge at a single point in the scene. As a result, the convergence distance of the target axes is no longer necessarily infinite. This is achieved by decentering the actually-used zone on each sensor (capture area) so that its center is aligned with the optical center and the chosen point of convergence. The visualization pyramids are therefore decentered and share a rectangular base, a projection of their capture area via their optical center on the plane parallel to the sensors passing by the point of convergence. Since their apices (optical centers) are distributed along a straight line parallel to the lines of this shared base (see Figure 4.5), these image shooting pyramids correspond qualitatively to those of the target display devices. With the shared base being displayed on the display's ROI, it is possible to render a scene both protruding (points in front of the point of convergence, perceived in front of the ROI) and hollow (points behind the point of convergence, perceived behind the ROI). Dodgson *et al.* have used this scheme for their autostereoscopic camera system with temporal multiplexing [DOD 97].

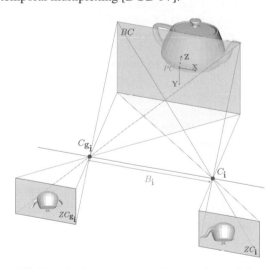

Figure 4.5. *Generic description of a decentered parallel geometry*

As we have seen, display on a flat multiscopic system involves selecting, for the two images aimed at the same viewer, capture pyramids sharing a rectangular base in the scene with apices placed on a straight line parallel to the lines in this shared base. For contexts of collective viewing of a single scene (autostereoscopic systems), which share views between several potential observation positions within one or several "chains" of key positions, this shared base should be applied to all captures destined for this single chain and even all chains if we want coherence between viewing of these different chains. The target axes are therefore all necessarily convergent at the center of this shared base and the apices of the pyramids must pairwise form straight lines parallel to the shared base's lines. Each "chain" of images must therefore be captured from positions located on a straight line parallel with the lines of the shared base. As such, so that the capture areas yielding to these pyramids and therefore the images that they capture remain rectangular, they must be parallel to this shared base. We must thus use a decentered parallel system (see Figure 4.4(c)), as Yamanoue and Woods have shown [WOO 93, YAM 06].

4.3.2. *Setting the parametric model*

Figure 4.6 provides a perspective representation of a decentered parallel shooting geometry. This figure shows the plane of the capture areas (ZC_i) and the optical centers (C_i) and specifies a set of parameters that completely characterize the shooting geometry. Figures 4.6(b) and (c) show the view from above and the front view of this geometry.

Our analysis of shooting geometry relies on the global shooting reference frame $R = (PC, \mathbf{X}, \mathbf{Y}, \mathbf{Z} \equiv \mathbf{X} \wedge \mathbf{Y})$, which is centered at the desired point of convergence PC (which is also the center of the shared base BC in the scene) and is directed so that the first referential vectors are colinear to the axes of the shared base BC in the scene and are therefore colinear with the axes in the capture areas. In addition, the first axis is presumed to be parallel to the lines in the capture areas and the second axis is parallel to the columns in these areas. The size of the shared base BC has the dimensions Lb and Hb. This reference frame defines the position and direction of all the projection pyramids representing the capture areas by specifying the direction of observation $\mathbf{Z}$ and the m alignment lines of the optical centers.

In line with these principles, the $n \times m$ shooting pyramids are specified by:

– optical axes in the direction $\mathbf{Z}$;

– the optical centers C_i aligned on one or several (m) straight lines parallel to the lines in the shared base and therefore the direction $\mathbf{X}$;

– rectangular capture areas ZC_i.

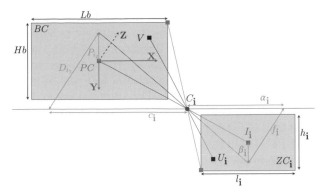

a) Perspective view

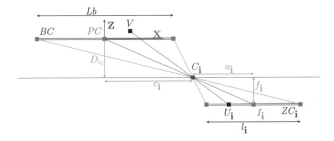

b) View from above

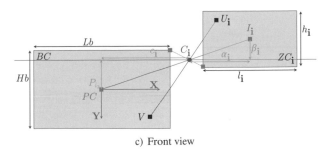

c) Front view

Figure 4.6. *Characterization of the decentered parallel shooting geometry*

The capture areas must be orthogonal to $\mathbf{Z}$ and therefore parallel to each other and the shared base BC as well as the straight lines holding the optical centers C_i (which are defined by their distances to PC, D_{i_2} in relation to $\mathbf{Z}$, P_{i_2} in relation to $\mathbf{Y}$ and c_i in relation to $\mathbf{X}$). These capture areas are placed at distances of f_i in relation to $\mathbf{Z}$, β_i in relation to $\mathbf{Y}$ and α_i in relation to $\mathbf{X}$ from their respective optical centers C_i. Their dimensions are l_i and h_i. They are decentered in relation to their respective optical axes in the points I_i such that the straight lines $(I_i C_i)$ define the target axes intersecting at the fixed point of convergence PC. The centers C_i and C_{g_i} must be on the same "center line" and spaced from B_i in relation to $\mathbf{X}$ ($C_i = C_{g_i} + B_i\mathbf{X}$ and $c_i = c_{g_i} + B_i$).

This kind of shooting configuration ensures a depth perception on a multiscopic system with colocalized planar mixing with the possibility of a protruding as well as hollow image effect. However, this does not ensure that the perceived scene will not be distorted in relation to the original scene. The absence of distortion implies that the viewing pyramids are perfect homologues of the shooting pyramids, i.e. they have exactly the same opening and deviation angles in both horizontal and vertical directions. Any flaw in this shooting and viewing pyramids' homology involves a potentially complex distortion of the 3D image perceived in relation to the captured scene. In some cases, however, this is desirable when creating special effects among other things. This implies that the shooting and viewing configurations must be specified as a set, which must ensure the desired distortion (or non-distortion) effect.

We will now model these distortion effects that are potentially implied by the combination of shooting and viewing geometries.

4.4. Geometric impact of the 3D workflow

4.4.1. *Rendered-to-shot space mapping*

In this section, we will use perfect lenses and sensors without distortion.

According to our analyses of viewing and shooting geometries, it is possible to connect the coordinates (X, Y, Z), in the reference frame R, from the point V in the scene captured by the previously identified cameras with the coordinates (x_i, y_i, z_i) in the reference frame r of its homologue v_i perceived by an observer of the display device, placed in a preferential position (the right eye at O_i). Supposing that the point V in the scene is visible in image number **i**, its projection U_i verifies:

$$U_i - C_i = \frac{f_i}{Z + D_{i_2}} \cdot (C_i - V) = \frac{f_i}{Z + D_{i_2}} \begin{bmatrix} c_i - X \\ -P_{i_2} - Y \\ -D_{i_2} - Z \end{bmatrix}_R \qquad [4.1]$$

Knowing that I_i, the center of ZC_i, is the projection on the image **i** of PC, the center of R, we can calculate $I_i - C_i$. By subtracting this result from equation [4.1], we obtain the positions of the projections of the point V in the different images:

$$U_i - I_i = \frac{-f_i}{Z + D_{i_2}} \begin{bmatrix} X + Z\frac{c_i}{D_{i_2}} \\ Y - Z\frac{P_{i_2}}{D_{i_2}} \\ 0 \end{bmatrix}_R \qquad [4.2]$$

Since the images are captured downward the optical centers, the images' implicit axes are the opposite of those in the global shooting reference frame R. In addition, the images are resized for display according to the display device's ROI. This places the projections U_i of V at their positions u_i on the ROI:

$$u_i|_r = (u_i - CS)|_r = - \begin{bmatrix} \frac{L}{l_i} \\ \frac{H}{h_i} \\ 1 \end{bmatrix} (U_i - I_i)_{|R} \quad \forall i \qquad [4.3]$$

Thalès' theorem used in Figure 4.6 yields $f_i L_b = D_{i_2} l_i$ and $f_i H_b = D_{i_2} h_i$. Therefore, we express u_i in the reference frame r by:

$$u_i|_r = \frac{D_{i_2}}{Z + D_{i_2}} \begin{bmatrix} \left(X + Z\frac{c_i}{D_{i_2}} \right) \frac{L}{L_b} \\ \left(Y - Z\frac{P_{i_2}}{D_{i_2}} \right) \frac{H}{H_b} \\ 0 \end{bmatrix} \quad \forall i \qquad [4.4]$$

Let us remark that the image g_i comes from the sensor associated with the optical center C_{g_i}, which is on the same "centers' line" that the optical centers as C_i (same secondary index i_2) and is spaced from B_i in relation to **X** ($C_i = C_{g_i} + B_i X$ and $c_i = c_{g_i} + B_i$) . Then, supposing that V is visible in the two images g_i and **i**, we can see that u_{g_i} and u_i are situated on the same line of the ROI. This responds to the epipolar constraint and therefore enables the stereopsis reconstruction of $v_i = [x_i, y_i, z_i]_r^t$ from O_{g_i} and O_i. In Figure 4.3, Thalès' theorem gives us:

$$(u_i - u_{g_i})|_r = \begin{bmatrix} \frac{z_i}{z_i + d_{i_2}} b_{i_2} \\ 0 \\ 0 \end{bmatrix} , \text{ which gives } z_i$$

[4.5]

By inverse projection, we find v_i:

$$v_i - O_i = \frac{z_i + d_{i_2}}{d_{i_2}} (u_i - O_i) , \text{ which then gives } x_i, y_i$$

[4.6]

Therefore, the relation between the 3D coordinates of points at the scene and those of their images perceived by the viewer at position number **i** can be characterized by:

$$a_i \begin{bmatrix} x_i \\ y_i \\ z_i \end{bmatrix} = k_{i_2} \begin{bmatrix} \mu_i & \gamma_i \\ \rho\mu_i & \delta_i \\ & 1 \end{bmatrix} * \begin{bmatrix} X \\ Y \\ Z \end{bmatrix}$$

[4.7]

Since a_i affinely depends on Z, we progress onto homogeneous 4D coordinates:

$$a_i \begin{bmatrix} x_i \\ y_i \\ z_i \\ 1 \end{bmatrix} = \begin{bmatrix} k_{i_2} \begin{vmatrix} \mu_i & \gamma_i \\ \rho\mu_i & \delta_i \\ & 1 \end{vmatrix} & \begin{matrix} 0 \\ 0 \\ 0 \end{matrix} \\ 0 \quad 0 \quad \frac{k_{i_2}(\varepsilon_i - 1)}{d_{i_2}} & \varepsilon_i \end{bmatrix} * \begin{bmatrix} X \\ Y \\ Z \\ 1 \end{bmatrix}$$

[4.8]

Equation [4.8], corresponds to the transformation matrix given by Jones *et al.* [JON 01], with a characterization of distortion parameters depending on the shooting and viewing parameters in the analytical distortion model for a viewer at position **i**.

4.4.2. *3D space distortion model*

This model (see equation [4.8]) clearly highlights all the distortions that could be obtained during a multiscopic viewing experience using planar colocalized mixing systems, regardless of the number of views or the nature of the images (whether real or virtual). It also underlines, in addition to a_i

(leftover calculation with no other significance), new parameters that quantify their distortions. Homogeneous matrices therefore define the transformations between the initial space in the scene and the viewing space for each favored observation position numbered as **i**. These parameters can be analytically expressed using geometric parameters from shooting and viewing systems. Their relations with the geometric parameters are presented in Table 4.1 and their impacts on distortion are described below:

– k_{i_2}: the global magnification factor, which does not really distort the scene.

– $\varepsilon_{\mathbf{i}}$: the control parameter of the potential nonlinear distortion, which transforms a cube in a truncated pyramid of the axis **Z**. Depending on the rate of global reduction $a_{\mathbf{i}} = \varepsilon_{\mathbf{i}} + k_{i_2} (\varepsilon_{\mathbf{i}} - 1) \frac{Z}{d_{i_2}}$, which can vary according to Z, if $\varepsilon_i \neq 1$, this creates a distortion in the displayed volume as a "truncated pyramid" of the axis **Z**.

– $\mu_{\mathbf{i}}$: the rate of width magnification in relation to the depth. When $\mu_{\mathbf{i}} \neq 1$, a horizontal/depth anamorphosis producing unequal dilations in **X** in relation to **Z** is applied.

– ρ: the rate of height magnification in relation to width. When $\rho \neq 1$, a vertical/horizontal anamorphosis producing unequal dilations in **Y** in relation to **X** is applied.

– $\gamma_{\mathbf{i}}$: the horizontal/depth skew of the perceived scene. When $\gamma_{\mathbf{i}} \neq 0$, a horizontal drift according to the depth is applied.

– $\delta_{\mathbf{i}}$: the rate of vertical/depth skew of the scene perceived by a viewer of a drop conforms to expectations. When $\delta_{\mathbf{i}} \neq 0$ and/or when the viewer's real drop differs from the optimal drop, a vertical drift according to depth is applied.

$k_{i_2}= \dfrac{d_{i_2}}{D_{i_2}}$	$\varepsilon_{\mathbf{i}} = \dfrac{b_{i_2}}{B_{\mathbf{i}}} \dfrac{L_b}{L}$	$\gamma_{\mathbf{i}} = \dfrac{c_{\mathbf{i}} b_{i_2} - o_{\mathbf{i}} B_{\mathbf{i}}}{d_{i_2} B_{\mathbf{i}}}$
$\rho = \dfrac{L_b}{H_b} \dfrac{H}{L}$	$\mu_{\mathbf{i}} = \dfrac{b_{i_2}}{k_{i_2} B_{\mathbf{i}}}$	$\delta_{\mathbf{i}} = \dfrac{p_{i_2} B_{\mathbf{i}} - P_{i_2} b_{i_2} \rho}{d_{i_2} B_{\mathbf{i}}}$

Table 4.1. *Expression of parameters quantifying the distortions in relation to shooting and viewing geometric parameters*

This defines all the depth distortion possibilities using the previously established shooting and viewing geometries. In addition, this model quantifies these distortions for any pair of settings (shooting and viewing) by a simple calculation based on their geometric parameters.

4.5. Specification methodology for multiscopic shooting

In this section, we propose a methodology that specifies the shooting configuration required to obtain the desired 3D distortion on a given display system [PRE 10]: a perfect 3D effect or selected distortions. Indeed, knowing how the distortion, shooting and display parameters are related, it is possible to adapt the shooting to the choice for distortions and viewing. We describe two shooting schemes using the distortion model: a generic scheme that gives precise control of each distortion parameter, and a second more specific, but very significant, scheme since it is involved in creating accurate depth perception (i.e. without any distortion).

4.5.1. *Controlling depth distortion*

It is possible to calculate the ideal shooting configuration for a desired distortion and fixed display. To do so, we have a range of distortion parameters that can be defined according to the desired result. By adjusting the magnification factor k_{i_2} and by fixing the distortion parameters ε_i (and therefore $a_i = \varepsilon_i + k_{i_2} \left(\varepsilon_i - 1 \right) / d_{i_2}$), μ_i, ρ, γ_i and δ_i, we obtain the desired 3D distortion. The last condition on δ_i is more difficult when $m = 1$ because this depends on the height of the viewer, which inevitably affects the real viewing drop (height of the eyes in relation to the center of the device) in relation to the display system. The selected vertical skew δ_i can therefore only be obtained in this case for a viewer whose real drop corresponds to that defined in the viewing rules for this viewing position.

Using the desired viewing and distortion parameters, it is possible to calculate the shooting parameters, which ensure that the desired distortion is obtained effectively by displaying on the chosen 3D device. The equations used, obtained by inverting those which define the distortion parameters, are given in Table 4.2.

This control scheme for depth distortion obtains shooting parameters that produce the desired content for any colocalized planar mixing multiscopic display system and any combination of depth distortions according to the model in equation [4.8]. This can prove highly useful within the cinema and the special effects' industry where we may want to accentuate 3D effect in certain parts of the scene to draw the viewer's attention.

	Controlling depth distortion	Accurate 3D effect
Global parameters	$P_{i_2} = (p_{i_2} - \delta_i d_{i_2}) / (k_{i_2}\rho\mu_i)$ $D_{i_2} = d_{i_2}/k_{i_2}$ $L_b = L\varepsilon_i/(k_{i_2}\mu_i)$ $H_b = H\varepsilon_i/(k_{i_2}\rho\mu_i)$ $c_i = (o_i + \gamma_i d_{i_2})/(k_{i_2}\mu_i)$	$P_{i_2} = p_{i_2}/k_{i_2}$ $D_{i_2} = d_{i_2}/k_{i_2}$ $L_b = L/k_{i_2}$ $H_b = H/k_{i_2}$ $c_i = o_i/k_{i_2}$
Local parameters	$l_i = L_b f_i/D_{i_2} = L f_i \varepsilon_i/(\mu_i d_{i_2})$ $h_i = H_b f_i/D_{i_2} = H f_i \varepsilon_i/(\mu_i d_{i_2})$ $\alpha_i = c_i f_i/D_{i_2} = f_i(o_i + \gamma_i d_{i_2})/(\mu_i d_{i_2})$ $\beta_i = P_{i_2} f_i/D_{i_2} = f_i(p_{i_2} - \delta_i d_{i_2})/(\mu_i \rho d_{i_2})$	$l_i = L_b f_i/D_{i_2} = L f_i/d_{i_2}$ $h_i = H_b f_i/D_{i_2} = H f_i/d_{i_2}$ $\alpha_i = c_i f_i/D_{i_2} = o_i f_i/d_{i_2}$ $\beta_i = P_{i_2} f_i/D_{i_2} = p_{i_2} f_i/d_{i_2}$

Table 4.2. *Shooting parameters in the case of controlling depth distortion and in the case of accurate depth. The darkroom depth f_i can be imposed or chosen (globally – by center line, $\forall i_2 \in \mathbb{N}_m$ – or individually, $\forall i \in \mathbb{N}_n \times \mathbb{N}_m$)*

4.5.2. *Accurate depth effect*

A specific example of depth distortion is perfect or accurate depth (a perception of the scene without any distortion in comparison to the real scene). To do so, it is necessary that the shooting of images, the display device's configuration (its ROI) and the conditions of use (favored positions) are jointly defined according to the desired distortion parameters. To create an accurate depth display (with a global magnification factor k_{i_2}), we need to configure the shooting in order to prevent potential distortions. This is achieved by ensuring that the distortion parameters verify $\varepsilon_i = 1$, $\mu_i = 1$, $\rho = 1$, $\gamma_i = 0$ and $\delta_i = 0$. This last condition, $\delta_i = 0$, is more difficult if $m = 1$ because it depends on the height of the viewer, which inevitably affects the effective drop in relation to the device. It can therefore only be guaranteed for viewers with a real drop equal to that defined in the display parameters for this viewing position. The shooting configuration parameters can therefore be calculated by replacing the distortion parameters fixed previously. The equations used are provided in Table 4.2.

This specific case is highly interesting due to its realism and has a number of potential applications. This is particularly the case, for example, with medical visualization software where accurate display is essential for a correct diagnosis or computer aided design (CAD) software that eliminates interpretation errors when designing a mechanical part. It could also be used to convince investors or planning committees by giving a 3D impression of the real size of a building.

4.6. OpenGL implementation

The shooting specification method examined previously can be used to develop OpenGL applications that can render 3D media. This method places virtual cameras around a standard monocular camera depending on the chosen display device and the desired depth effect. The advantage of this virtual scheme is that virtual cameras are perfect and there is therefore no problem with potential distortions due to sensors or lenses. The camera arrangement in OpenGL is explained below.

The geometry of a single camera is normally defined by its position, direction and vision pyramid. However, in stereoscopic environments, it is necessary to have two virtual cameras, one for each eye, whereas multiview autostereoscopic displays require up to N virtual cameras. Each virtual camera is laterally shifted and has its own decentered pyramid of vision such that the center of its image is located on the line linking its optical center and the center of convergence lying on the optical axis of the reference monocular camera. The optical axes are parallel to each other. The distance of the "depth-clipping" planes, i.e. the planes perpendicular to the optical axes between which the objects are viewed, remains unchanged. The convergence distance (depth of the convergence point) must be manually defined. This determines whether the objects appear "in front of" or "behind" the display, thereby creating (or not) a protruding effect.

Algorithm 4.1, identifies the six parameters l, r, b, t, n, f required by the OpenGL function $glFrustum$: l and r are the coordinates (left and right) of the vertical "clipping" planes, b and t are the coordinates (top and bottom) of the horizontal "clipping" planes and n and f are the distances of the depth "clipping" planes.

We then apply the OpenGL command chain to correctly place the camera and therefore obtain the desired view. As such, we can create the N views desired by varying c from 0 to $N - 1$. The use of these views depends on the autostereoscopic display device chosen and is examined in Chapter 14.

Algorithm 4.1. Obtaining N views

Data: The selected distortion parameters and the focal.

begin

$\quad l_\mathbf{i} = (L\varepsilon f_\mathbf{i})/\mu d_{i_2}$

$\quad ratio = f_\mathbf{i}/(d_{i_2}/k_{i_2})$

$\quad h_\mathbf{i} = (H\varepsilon f_\mathbf{i})/(\mu\rho d_{i_2})$

$\quad P = -(\delta d_{i_2})/(k_{i_2}\rho\mu)$

$\quad b = -(h_\mathbf{i}/2) - P * ratio$

$\quad t = (h_\mathbf{i}/2) - P * ratio$

$\quad n = f_\mathbf{i}$

$\quad f = focal$

$\quad$**foreach** $c \in [0 - N[$ **do**

$\quad\quad horizontal = (b_{i_2}/k_{i_2}\mu)(c - (N - 1)/2.0) + (\gamma d_{i_2}/b_{i_2})$

$\quad\quad l = -(l_\mathbf{i}/2.0f) - horizontal * ratio$

$\quad\quad r = (l_\mathbf{i}/2.0f) - horizontal * ratio$

$\quad\quad glMatrixMode(GL\_PROJECTION)$

$\quad\quad glPushMatrix()$

$\quad\quad glLoadIdentity()$

$\quad\quad glFrustum(l, r, b, t, n, f)$

$\quad\quad glMatrixMode(GL\_MODELVIEW)$

$\quad\quad glPushMatrix()$

$\quad\quad glLoadIdentity()$

$\quad\quad glTranslatef(-horizontal, -P, -d_{i_2}/k_{i_2})$

$\quad\quad$// **Scene rendering**

$\quad\quad glMatrixMode(GL\_MODELVIEW)$

$\quad\quad glPopMatrix()$

$\quad\quad glMatrixMode(GL\_PROJECTION)$

$\quad\quad glPopMatrix()$

$\quad$**end**

end

4.7. Conclusion

In this chapter, we have examined a methodology that can be used to describe and qualify the geometric relations linking a 3D shooting device with a display device mixing 3D colocalized planar images. These relations, defined by a restricted set of parameters, provide global control of distortions and can be used to specify the shooting geometry, which will guarantee, for both a real and a virtual scene, that the images will be accurately displayed on a given 3DTV device. Principally aimed at autostereoscopy, this chapter can be applied to a number of areas in 3D visualization in line with, for example, biomedical imaging, audiovisual production and multimedia production.

4.8. Bibliography

[DOD 97] DODGSON N.A., MOORE J.R., LANG S.R., "Time-multiplexed autostereoscopic camera system", *Proceedings of Stereoscopic Displays and Virtual Reality Systems IV*, SPIE, vol. 3012, 1997.

[DOD 02] DODGSON N.A., "Analysis of the viewing zone of multi-view autostereoscopic displays", *Proceedings of Stereoscopic Displays and Applications XIII*, SPIE, vol. 4660, pp. 254–265, 2002.

[DUB 01] DUBOIS E., "A projection method to generate anaglyph stereo images", *Proceedings of the IEEE International Conference on Acoustics, Speech, and Signal Processing*, IEEE Computer Society Press, pp. 1661–1664, 2001.

[HIL 53] HILL A.J., "A mathematical and experimental foundation for stereoscopic photography", *Journal of SMPTE*, vol. 61, pp. 461–486, 1953.

[JON 01] JONES G., LEE D., HOLLIMAN N., *et al.*, "Controlling perceived depth in stereoscopic images", *Proceedings of Stereoscopic Displays and Virtual Reality Systems VIII*, vol. 4297, pp. 42–53, June 2001.

[PEI 09] PEINSIPP-BYMA E., REHFELD N., ECK R., "Evaluation of stereoscopic 3D displays for image analysis tasks", in WOODS A.J., HOLLIMAN N.S., MERRITT J.O. (eds), *Stereoscopic Displays and Applications XX*, SPIE, pp. 72370L–72370L–12, 2009.

[PER 00] PERLIN K., PAXIA S., KOLLIN J.S., "An autostereoscopic display", *ACM SIGGRAPH 2000 Conference Proceedings*, vol. 33, pp. 319–326, 2000.

[PRE 10] PRÉVOTEAU J., CHALENÇON-PIOTIN S., DEBONS D., *et al.*, "Multi-view shooting geometry for multiscopic rendering with controlled distortion", *International Journal of Digital Multimedia Broadcasting (IJDMB), special issue Advances in 3DTV: Theory and Practice*, vol. 2010, pp. 1–11, March 2010.

[SAN 03] SANDERS W.R., MCALLISTER D.F., "Producing anaglyphs from synthetic images", in WOODS A.J., BOLAS M.T., MERRITT J.O., BENTON S.A. (eds), *Stereoscopic Displays and Virtual Reality Systems X*, SPIE, pp. 348–358, 2003.

[SON 07] SON J.-Y., GRUTS Y.N., KWACK K.-D., *et al.*, "Stereoscopic image distortion in radial camera and projector configurations", *Journal of the Optical Society of America A*, vol. 24, pp. 643–650, 2007.

[WOO 93] WOODS A., DOCHERTY T., KOCH R., "Image distortions in stereoscopic video systems", *Proceedings of SPIE: Stereoscopic Dispalys and Applications IV*, vol. 1915, pp. 36–48, 1993.

[YAM 97] YAMANOUE H., "The relation between size distortion and shooting conditions for stereoscopic images", *SMPTE Journal*, vol. 106, pp. 225–232, 1997.

[YAM 06] YAMANOUE H., "The differences between toed-in camera configurations and parallel camera configurations in shooting stereoscopic images", *IEEE International Conference on Multimedia and Expo*, vol. 0, IEEE Computer Society, pp. 1701–1704, 2006.

Chapter 5

Camera Calibration: Geometric and Colorimetric Correction

5.1. Introduction

This chapter analyzes camera calibration from a geometric and colorimetric perspective. The first part of this chapter introduces the mathematical model that describes a camera (position, orientation, focal, etc.) as well as its applications (e.g. 3D reconstructions). This chapter also analyzes different geometric processes for stereoscopic images such as corrections of radial distortion as well as image rectification. The second part of this chapter will focus on colorimetric models relating to digital acquisition systems. We will demonstrate how to characterize a colorimetric camera and then analyze different elements involved in color correction for images taken from a system of cameras.

5.2. Camera calibration

5.2.1. *Introduction*

The camera model used in computer vision is an idealized representation of a pinhole camera (see Chapter 1). This model, as well as problems such as projective geometry calibration, is analyzed by Hartley and Zisserman

Chapter written by Vincent Nozick and Jean-Baptiste Thomas.

in [HAR 04], a key reference in the field of 3D vision as it is essential for all those interested in problems of this kind and the following sections are largely based on this work.

The objective of the matricial camera model is to represent this camera and its geometric relation to the scene. By extension, it also creates geometric relations between several cameras from which it is possible to create 3D reconstructions of the scene or locate an object.

The camera calibration process entails calculating its projection matrix and then splitting this up into several sub-matrices relating to the focal, position and direction parameters. Calibrating a camera is achieved either using 3D $\leftrightarrow$ 2D correspondence points or using 2D $\leftrightarrow$ 2D correspondence points on several successive images.

5.2.2. *Camera model*

A camera is modeled according to a pinhole representation using a projection matrix P. A point in the space $\mathbf{X} = (X, Y, Z, W)^\top \in \mathbb{P}^3$ is projected onto the pixel $\mathbf{x} = (x, y, w)^\top \in \mathbb{P}^2$, so that $\mathbf{x} = P\mathbf{X}$. The matrix P includes both the parameters relating to the camera's lenses (intrinsic parameters) and the position and direction of the camera (extrinsic parameters). It is composed as follows:

$$P_{3\times4} = K_{3\times3}\left[R_{3\times3} \mid - R_{3\times3}c_{3\times1}\right]$$

where the operator [A|B] represents the matricial concatenation of A and B and K is the matrix of intrinsic parameters representing the camera's lens. This expresses the perspective of an image (the further away an object, the smaller it appears) via the lens's focal. It is a 3×3 triangular matrix in the form:

$$K = \begin{bmatrix} f_x & s & u_x \\ 0 & f_y & u_y \\ 0 & 0 & 1 \end{bmatrix}$$

with

– f_x and f_y: being the focal distance of the camera expressed in pixel units. If the sensor's photosites are square (as is generally the case), we have $f_x = f_y$;

– s: represents the non-orthogonal character of the sensor's photosite grid. In general, the photosites are square, which can be simplified as $s = 0$;

– $(u_x, u_y)^\top$: coordinates in pixels of the principal point that correspond to the intersection of the optical axis with the image. This point is generally fairly close to the center of the image but generally does not perfectly coincide.

The matrix R is a 3×3 rotation matrix that represents the direction of the camera in the scene's referential. Finally, the vector $\mathbf{c} = (c_x, c_y, c_z, c_w)^\top$ represents the position of the camera's center of projection in the scene's referential.

If we know the projection matrix P, we can break it down into $P = K\,[R| - Rc]$ with a few calculations. The vecotr $\mathbf{p}_i$ corresponds to the ith column of P such that $P = [\mathbf{p}_1\mathbf{p}_2\mathbf{p}_3\mathbf{p}_4]$. The aim is to extract the camera's parameters from this matrix P. The breakdown of P can also be written as:

$$P = K\,[R| - Rc] = [KR| - KRc] = [M|\mathbf{p}_4]$$

where $M = [\mathbf{p}_1\mathbf{p}_2\mathbf{p}_3]$. The intrinsic parameter matrix K and the rotation matrix R can be calculated by a QR decomposition of the matrix M. Therefore, we obtain the decomposition $KR = M$ (ensuring that f_x and f_y are positive). Finally, the position of the camera can be calculated by:

$$
\begin{aligned}
c_x &= \det([\mathbf{p}_2\mathbf{p}_3\mathbf{p}_4]) \\
c_y &= \det([\mathbf{p}_1\mathbf{p}_3\mathbf{p}_4]) \\
c_z &= \det([\mathbf{p}_1\mathbf{p}_2\mathbf{p}_4]) \\
c_w &= \det([\mathbf{p}_1\mathbf{p}_2\mathbf{p}_3])
\end{aligned}
$$

As is always the case in projective geometry, the equivalent point in the Euclidean space is $\mathbf{c} = (c_x/c_w, c_y/c_w, c_z/c_w)^\top$.

5.2.3. *Calibration using a sight*

A direct but highly accurate means of calibrating a camera entails defining the 3D ↔ 2D correspondence points, i.e. finding the pixels $\mathbf{x}_i$ on the image corresponding to points $\mathbf{X}_i$ known in the 3D scene. It is therefore necessary to solve the following system:

$$
\begin{bmatrix}
\mathbf{0}_4^\top & -w_1\mathbf{X}_1^\top & y_1\mathbf{X}_1^\top \\
w_1\mathbf{X}_1^\top & \mathbf{0}_4^\top & x_1\mathbf{X}_1^\top \\
\mathbf{0}_4^\top & -w_2\mathbf{X}_2^\top & y_2\mathbf{X}_2^\top \\
w_2\mathbf{X}_2^\top & \mathbf{0}_4^\top & x_2\mathbf{X}_2^\top \\
\vdots & \vdots & \vdots \\
\mathbf{0}_4^\top & -w_i\mathbf{X}_i^\top & y_i\mathbf{X}_i^\top \\
w_i\mathbf{X}_i^\top & \mathbf{0}_4^\top & x_i\mathbf{X}_i^\top \\
\vdots & \vdots & \vdots
\end{bmatrix}
\begin{pmatrix}
p_{11} \\ p_{12} \\ p_{13} \\ p_{14} \\ p_{21} \\ p_{22} \\ p_{23} \\ p_{24} \\ p_{31} \\ p_{32} \\ p_{33} \\ p_{34}
\end{pmatrix}
=
\begin{pmatrix}
0 \\ 0 \\ 0 \\ 0 \\ \vdots \\ 0 \\ 0 \\ \vdots
\end{pmatrix}
$$

where the p_{ij} corresponds to the elements (i,j) in the matrix P. To solve this system, there need to be at least six correspondences $\mathbf{X}_i \leftrightarrow \mathbf{x}_i$, and the more there are, the more robust the solution. This solution is achieved by calculating the right null space of the previous matrix using a singular value decomposition (SVD). Once the p_{ij} is located and the matrix P has been reconstructed, it is broken down into $P = K[R| - Rc]$ using the method described in section 5.2.2 in order to extract the parameters.

Another simpler, and fairly common, approach involves calibrating a camera using photographs of a chessboard from several angles to highlight the vanishing points generated by the camera's lens (see Figure 5.1). All of these images can be used to calculate the matrix K of the intrinsic parameters. Finally, one of the images is selected to serve as the scene's referential for the direction R and position c of the camera. The details of this method are analyzed by Zhang [ZHA 00].

5.2.4. *Automatic methods*

There are methods that can be used to calibrate cameras using correspondence points between images as well as an accurate estimation of cameras' intrinsic parameters. If the points are automatically detected (see Chapter 6), the method is completely automatic.

The first stage involves estimating the respective positions of the cameras using correspondence points. It is possible to use epipolar geometry and, in particular, the essential matrix ([HAR 04], p. 258), or even Nister's *5-points algorithm* [NIS 04]. The latter is fairly complex to implement, although there are several versions available on the Internet.

Figure 5.1. *Calibration with a chessboard photographed from several angles to show the leak points generated by the camera's perspective. The chessboard is automatically detected by the OpenCV library*

This estimation of camera calibration is generally not very precise but can still be refined. To do so, we apply a bundle adjustment [TRI 00] to the data. First, this involves triangulating the correspondence points on each image in order to define an estimation of the 3D points in question. The bundle adjustment entails minimizing the reprojection error of each 3D point on the correspondence points in question (i.e. the distance between a point x detected on an image and the projected image x′ on the same image taken from the 3D point calculated by triangulating x). The parameters of this nonlinear minimization method are the positions and orientation of the cameras as well as their intrinsic parameters and the 3D points' positions. In general, the nonlinear method used is a "sparse" version of the Levenberg–Marquardt method. In the end, we obtain a series of relatively well-calibrated cameras as well as a 3D reconstruction of correspondence points. It should be noted that this method is highly sensitive to the initial conditions, i.e. the first estimation of the position and orientation of cameras. Without a strong estimation, the bundle adjustment has a tendency to diverge or even converge toward a local minimum and the solution obtained will be unusable.

5.3. Radial distortion

5.3.1. *Introduction*

As specified in section 5.2.2, the camera model used in computer vision is an ideal representation of the pinhole camera that does not include optical camera imperfections. There are two main types of geometric distortions: tangential distortion and radial distortion (see Figure 5.2). Tangential distortion is often less important in contrast to radial distortion that often requires correction.

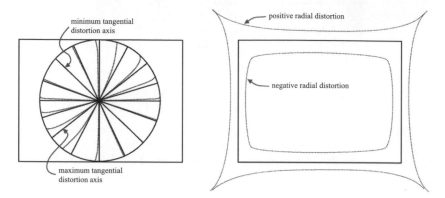

minimum tangential distortion axis

maximum tangential distortion axis

positive radial distortion

negative radial distortion

Figure 5.2. *Tangential and radial distortion*

Radial distortion is a transformation of the image, which creates the straight lines in the scene that appear curved in the image. We call this radial distortion because its amplitude depends on the distance between the optical center in the image and the affected pixel. This distortion is due to light rays that pass through the lens and that do not satisfy Gauss's estimation, i.e. the rays must be parallel to the optical axis and pass through the lens's optical center. To control these rays, the majority of lenses are fitted with a diaphragm that blocks undesirable rays. Ideally, it is preferable to have two diaphragms placed around the lens. In practice, there is only a single diaphragm that allows distortions around the edge of the image. Depending on whether it is in front of or behind the lens, radial distortion is positive (pincushion) or negative (barrel shaped).

5.3.2. *When to correct distortion?*

The matricial camera model introduced in section 5.2.2 is a linear model that does not take into consideration potential radial distortion in images, particularly for those taken with a wide-angle lens. Correction of radial distortion (see Figure 5.3) is therefore necessary in order for the camera model to remain valid. This process is particularly important for 3D reconstructions and for calculating the disparity map.

Figure 5.3. *Correction of radial distortion*

In the case of stereoscopic rendering, radial distortion can generate vertical parallax between two images, notably around the upper and lower edges of the image, i.e. where distortion is not the same for a point in the image and for its counterpart in the other image. Correction of radial distortion is therefore necessary in order to obtain stereoscopic images that are comfortable to look at.

5.3.3. *Radial distortion correction models*

The most commonly used model in the computer vision literature is a polynomial model that calculates the corrected position $(\hat{x}, \hat{y})$ of a pixel (x, y) according to the following equation:

$$\begin{cases} r^2 = (x - x_c)^2 + (y - y_c)^2 \\ \hat{x} = x + (x - x_c)(k_1 r^2 + k_2 r^4) \\ \hat{y} = y + (y - y_c)(k_1 r^2 + k_2 r^4) \end{cases}$$

where k_1 and k_2 are the first correction parameters of radial distortion and where (x_c, y_c) represents the center of radial distortion. This center is generally associated with the principal point in the image, i.e. the intersection of the optical axis and the image.

It is possible to use a polynomial of higher degree while keeping only the paired powers of the ray r. While commonly used, this method has the major disadvantage of being non-invertible. In effect, the parameters used to correct image distortion are not the same as those used to redistort the corrected image.

A highly effective way of applying this method, known as plumbline, entails finding the points of a curve presumed to be straight in the image and finding k_1 and k_2 that minimize the "dis-alignment" of these points. This method, introduced by Devernay and Faugeras [DEV 01], is effective but nonlinear.

There is a more recent but far simpler method, developed by Strand and Hayman [STR 05] who have created a non-polynomial, linear and invertible model that only uses a single parameter λ:

$$\begin{cases} \hat{x} = x + (x - x_c)/(1 + \lambda r^2) \\ \hat{y} = y + (y - y_c)/(1 + \lambda r^2) \end{cases}$$

This model estimates that a curve, which should normally be straight in the image, can be considered as a circle. The plumbline method is largely simplified since it is only necessary to find the ray and the center of the circle to calculate the correction coefficient λ. Since the equation of a circle is $x^2 + y^2 + ax + by + c = 0$, we find a series of points (x_i, y_i) on the supposedly straight line and solve the following system in terms of least squares:

$$\begin{bmatrix} x_i & y_i & 1 \\ \vdots & \vdots & \vdots \end{bmatrix} \begin{pmatrix} a \\ b \\ c \end{pmatrix} = \begin{pmatrix} -x_i^2 - y_i^2 \\ \vdots \end{pmatrix}$$

A simple means of solving an overdetermined system $\mathsf{A}\mathbf{x} = \mathbf{b}$ is to calculate the pseudo-inverse matrix $\mathsf{A}^+ = (\mathsf{A}^\top \mathsf{A})^{-1} \mathsf{A}^\top$. Therefore, the solution is $\mathbf{x} = \mathsf{A}^+ \mathbf{b}$. Once $(a, b, c)^\top$ is found, the center (C_x, C_y) and the ray R of the circle are:

$$\begin{cases} \left(C_x, C_y \right)^\top = \left(-\dfrac{a}{2}, -\dfrac{b}{2} \right)^\top \\ R = \sqrt{-c + C_x^2 + C_y^2} \end{cases}$$

Finally, the radial distortion correction parameter is calculated by:

$$\lambda = \frac{-1}{R^2 - C_x^2 - C_y^2}$$

5.4. Image rectification

5.4.1. Introduction

Image rectification involves transforming two images from the same scene so that each pixel from the first image has its corresponding pixel in the other, horizontally aligned in the same image (see Figure 5.4).

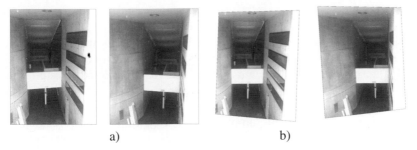

Figure 5.4. *Rectification of two images. a), The original images. b), The corrected images. Each pixel in an image is aligned horizontally with its counterpart in the other*

More precisely, the rectification process involves transforming the images so that they seem to have been taken using cameras with the same image plane (see Figure 5.5). In terms of epipolar geometry, rectifying a pair of images is a linear transformation of each image that places the epipoles at infinity, in the direction given by the straight line passing through the centers of the two cameras. Image rectification is generally calculated using correspondence points between the images but can also be calculated using projection matrices from the two cameras associated with each image. Finally, image rectification can be applied to several images under certain conditions.

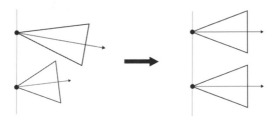

Figure 5.5. *The rectification process entails transforming the images so that they appear to be from two cameras sharing the same image plane*

Image rectification is mainly used to calculate disparity maps or to search for the corresponding point in an image which is done using the same line in the other image. It is also used within the context of stereoscopic image production in order to limit vertical parallax between the two images, thereby creating a more comfortable stereoscopic image to look at.

5.4.2. *Problems*

The principal difficulty in image rectification relates to the fact that there are infinite solutions to the problem. In effect, two perfectly corrected images with a horizontal magnification or skew will remain corrected. This problem is illustrated in Figure 5.6. The aim of image rectification methods is to find a solution that does not alter the original images too much. There are two approaches for solving this problem: image-based methods, focused on explicitly minimizing deformations in the two images, and camera-based methods, which generate corrected images that are compatible with the pinhole camera model. As a result, this second approach is slightly better adapted to correcting images specifically within pinhole cameras.

Figure 5.6. *There are infinite solutions to the image correction problem*

Finally, a more recent problem relates to how to correct more than two images simultaneously. We will see in section 5.4.5 that this process is possible if the cameras' centers are aligned. However, if this is not the case, correction will be imperfect or even impossible.

5.4.3. *Image-based methods*

Since the solution to the rectification problem is not unique, the main aim of rectification methods is to find a solution that does not alter the original images too drastically. The initial motivation of rectification is to generate disparity maps and the scientific community has therefore developed a number of means of minimizing the deformation of processed images without considering the compatibility of the images transformed using camera models. These models are generally linear, i.e. very easy to calculate, and the obtained results are suitable for calculating disparity.

One of the most commonly used methods is that introduced by Hartley [HAR 99]. This method is used within the context of epipolar geometry and involves finding transformations in the two images that place the epipoles at infinity.

This method is fast, easy to implement and available in a large number of computer vision libraries. Its main drawback, however, is its dependence in relation to the distribution of correspondence points on the images which is critical in the last harmonization phase for the second image in relation to the first.

There are also other rectification methods of which the following are the most well known. Robert *et al.*'s method [ROB 95] seeks transformation that best preserves orthogonality in the image marker as centered coordinates. Loop and Zhang [LOO 99] have proposed a nonlinear model that breaks down the rectification process into an affine transformation followed by a perspective transformation. Finally, Gluckman *et al.* have researched a transformation that best maintains the bijective character of pixel correspondence from one image to another [GLU 01].

5.4.4. *Camera-based methods*

Camera-based methods involve finding transformations in images corresponding to those that may occur if we modify the cameras so that they share the same image plane. These methods are nonlinear, i.e. they take longer to calculate than linear methods, but the generated images are equivalent to those that may have been created using decentered lenses combined with an appropriate zoom. In practice, transformations generated in each image are a combination of rotations and a scale, as analyzed in Figure 5.5.

From a more formal perspective, we will consider the two cameras $P_1 = K_1 [R_1 | - R_1 c_1]$ and $P_2 = K_2 [R_2 | - R_2 c_2]$, where K is the matrix of the camera's intrinsic parameters, which includes focal length information, R is a rotation matrix specifying the camera's direction and c is the camera's position. The aim of rectification is to find new projection matrices for the cameras $P_1' = K_1' [R_1' | - R_1' c_1]$ and $P_2' = K_2' [R_2' | - R_2' c_2]$ so that the cameras share the same image plane. The corresponding transformations in the images are the homographies H_1 and H_2 such that:

$$H = P'P^{-1} = K'R' (KR)^{-1} = K'R'R^{-1}K^{-1} = K'\widehat{R}K^{-1}$$

The aim of this method is to find the homographies H_1 and H_2 so that the transformed images are rectified. Even an approximate estimation of the matrices K_1 and K_2 will be sufficient to correct the images. The details of this method are further analyzed in [NOZ 11].

5.4.5. *Correcting more than two images simultaneously*

It is possible to correct more than two images on the condition that the cameras' centers of projection are aligned. Moreover, epipolar geometry is only defined between two images and is therefore not suitable for correcting more than two images at the same time. The vast majority of rectification methods use epipolar geometry and are therefore not appropriate for correcting several images simultaneously.

The camera-based method [NOZ 11] presented in section 5.4.4 can be directly adapted to rectify more than two views. Therefore, the problem becomes one of finding the homographies $H_i = K'_i \widehat{R}_i K_i^{-1}$, minimizing the vertical disparity of each correspondence point, no longer between the two images but between all images. Figure 5.7 illustrates a correction of several images.

Figure 5.7. *Correction of several images*

5.5. Colorimetric considerations in cameras

This section explores the problem of color information raised when filming a scene using a trichromatic red, green, blue (RGB) color camera. Indeed, the relationship between the color perceived when the scene is viewed directly and the digital value acquired by the camera is not straightforward. In addition, we will investigate the concept of color image correction in order to overcome the differences created by the scene's geometry, given that the scene's perception is dependent on lighting and angle of view.

To explain these problems, we first analyze several elements of color science and colorimetry that can be used to understand how the digital value of a camera is related to the sensation of perceived color. These elements will also enable us to understand the metrics used to estimate quality as well as cost functions of affiliated inverse problems. Second, we will analyze colorimetric characterization of cameras, i.e. how to connect the sense of color with a digital value. To conclude, we will introduce color corrections in pairs of 3D images.

5.5.1. *Elements of applied colorimetry*

If we think of the eye as a camera that contains three types of receptors whose spectral sensitivity is known, we can deduce three values that characterize a single point generating the sensation of color. This base simulates a triplet in the large, medium, small (LMS) space, which represents the spectral sensitivities of the photosensitive cells in the eye (see Figure 5.8).

It is important to relate this information to the notion of luminance as defined by the International Commission on Illumination (*Commission Internationale de l'Eclairage* – CIE) and it is also useful to use the virtual spectral sensitivities $\bar{x}$, $\bar{y}$, $\bar{z}$ defined by the CIE for a given angle of view, i.e. $2°$ or $10°$ (Figure 5.8). These data define the space XYZ, characterizing the color of an object in a scene, according to its reflection properties $R(\lambda)$ (the quantity of energy reflected by the object in relation to wavelength) and a given light level $I(\lambda)$ (the quantity of energy reaching an object in relation to wavelength).

To evaluate the difference between two similar colors, the CIE recommends working within a more perceptual space where the Euclidean distance is less representative of the perceived difference than the physical projection space XYZ, i.e. $CIELAB$. The distinctive feature of a typical pseudo-perceptual space is a nonlinear transformation (pseudo-logarithmic, see Weber–Fechner's law) of the values XYZ weighted by a so-called reference white (X_n, Y_n, Z_n), representing the illumination in the scene. Note that a colorimetric system, which organizes colors in relation to each other, such as the "Munsell" system, is also pseudo-perceptual but is not a colorimetric space in itself.

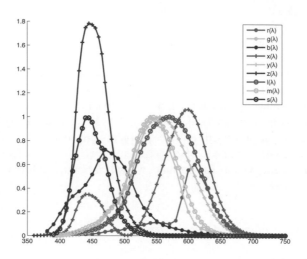

Figure 5.8. *Spectral sensitivities of cones (LMS, normalized by M), of primaries defined by the International Commission on Illumination (CIE), $\bar{x}$, $\bar{y}$, $\bar{z}$ and of a camera r, g, b (Sinarback 54 Camera [DAY 03])*

In $CIELAB$, the perceptual difference between two similar colors C_1 and C_2 can be expressed by using the Euclidean distance, denoted as ΔE^*_{76}. The subscript 76 refers to the formulation's year of acceptance, and similarly we will use 94, CMC, 00 among others for distinction between metrics. Euclidean distance being an approximation, several improvements based on experimental data have been proposed, giving rise to non-Euclidean metrics that better fit our own perception and whose principal feature is that they weight the Euclidean distance for strongly chromatic colors, where the perceived difference is less sensitive to small variations.

In particular, ΔE^*_{94} incorporates a weight function for hue and chromaticity while ΔE^*_{00}, in addition to modifying specific parameters present in ΔE^*_{94}, adds an intermediary *crossing* term between chroma and hue as well as a weight on the axis a^*, which affects similar colors within the chromatic axis where very small signal variations can be perceived.

We will not analyze the specification of a color in terms of luminance/brightness, hue and saturation, or the appearance of color or the different systems such as Atlas or Espace that predict the appearance of color or a physical signal. However, there is a broad literature on this subject, particularly in relation to CIE standards for a more extensive formulation of the material analyzed previously [CIE 01, CIE 04], Wyszecki and Stiles' comprehensive study of modern colorimetry [WYS 00] or Fairchild's excellent investigation of the appearance of color [FAI 98], among others.

The connection with capturing a scene via an RGB color camera becomes apparent when we analyze spectral sensitivities in cameras (see Figure 5.8). If the spectral sensitivities of a camera can be defined by a linear combination of the eye's spectral sensitivities, our camera is said to be colorimetric (Luther-Yves condition). However, physical and material constraints mean that this condition is often not respected, and the three RGB values provided by a camera are not directly representative of the perceived color. The quality index of a set of filters defined by Vora and Trusell [VOR 93] evaluates the difference in sensitivity between the camera and the human eye via the angle between the space defined by the camera's filter and the space formed by the sensitivity of the human visual system. Different quality indicators can be used and have been compared by Hung [HUN 00]. To conclude this brief overview, we should highlight two fundamental implications. On the one hand, the integration of the same spectral signal by different systems will not give the same result (the value representing color and the corresponding digital value after acquisition will be different) and, on the other hand, the integration of two different signals may give similar values between two systems and therefore create confusion (difference of metamerism). The other significant aspect of this section is that we have defined a space and metric to evaluate the difference between two colors. In the next section, we will explain colorimetric characterization/calibration of a color camera.

5.5.2. *Colorimetric calibration in cameras*

Colorimetric characterization of RGB trichromatic cameras is designed to establish the relationship between the digital value acquired and a value representing color. The calculation of digital values from a color value is called "model" and an "inverse model" relates to a color value taken from digital values. Different methods have been proposed in the literature. However, we will focus on a purely conceptual study. There is an extensive body of literature available on this subject for further information.

The characterization of a camera is generally carried out in two stages: a linearization stage and a colorimetric transformation stage.

The following discussion is primarily inspired by the excellent article by Barnard and Funt [BAR 02] as well as that by Cheung *et al.* [CHE 05]. We also highly recommend Gaurav Sharma's book [SHA 97] for a more encompassing view of colorimetric characterization as well as other problems related to color imaging. First, we will define the acquisition model and use this to define the spectral sensitivities in a camera, i.e., the characterization in the proper sense of the term. Second, we will analyze the practical aspect by focusing on application, i.e. how to obtain a colorimetric value from digital values given by the camera. We will use the model previously defined for the human eye and apply it to cameras.

We suppose that the sensor's nonlinearity is independent of wavelength. The effects of "vignetting" and other geometric effects are not included, but can be integrated into the model either in $F^{(k)}$ or in $S^{(k)}$, if we consider them as independent of wavelength (see the following equations). Equally, the effects of diaphragm opening, focal and acquisition time are fixed and appropriate. In general, a relative characterization is enough to guarantee color information, but it is also possible to find the absolute values by following the propositions of Martinez-Verdu *et al.* [MAR 03] or Debevec and Malik [DEB 97].

Given $L(\lambda)$, the radiance of the scene is the convolution over a set of wavelenghts (λ) of the light source illuminationg the scene, $I(\lambda)$ and of the reflectance properties of the objects composing the scene, $R(\lambda)$. Given, the spectral sensitivities of the sensor $S^{(k)}(\lambda)$, and given $\nu^{(k)}$, the value given by the sensor in answer to $L(\lambda)$ for a channel k, typically $k \in R, G, B$, we have:

$$\rho^{(k)} = F^{(k)}(\nu^{(k)}) = \int_{\Lambda} L(\lambda) R^{(k)}(\lambda) d\lambda \qquad [5.1]$$

where $F^{(k)}$ is a linearization function and $\rho^{(k)}$ is the linearized value (the camera's response). $F^{(k)}$ is typically a correction of the camera's gamma and offset.

The measure and the representation of this information is usually discrete, and $L(\lambda)$ and $S(\lambda)$ become the vectors $\mathbf{L}$ and $\mathbf{S}$. Therefore, equation [5.1] becomes:

$$\rho^{(k)} = F^{(k)}(\nu^{(k)}) = \mathbf{L}.\mathbf{S}^{(\mathbf{k})} \qquad [5.2]$$

The characterization of a camera therefore requires estimating $F^{(k)}$ and $\mathbf{S}^{(\mathbf{k})}$.

5.5.2.1. *Estimation of $F^{(k)}$ and $\mathbf{S}^{(\mathbf{k})}$*

If we suppose that $F^{(k)}$ is independent of wavelength, we can estimate it by stimulating the camera with a light source, which has its intensity varied (with different density filters or even by taking the light source further away). The data can be interpolated or approximated using a smooth function and a "look-up table" (LUT) can be created. In general, the camera is considered to be linear throughout most of the domain but for the weakest intensity values we often see a nonlinearity, which must be accounted for by a more precise characterization. Regardless of this, however, this hypothesis must be verified in order for it to be reliable. Often a gamma correction is applied by the constructor, by default, to the camera data to bring the camera's RGB values closer to the sRGB standard and the inverse of this correction must be applied.

In order to obtain $\mathbf{S}^{(\mathbf{k})}$, the simplest method to use conceptually is a monochromator, i.e. to stimulate the camera using signals in a very narrow bandwidth. This approach requires expensive material and is not commonly used in practice outside of laboratories.

Other methods have also been developed in order to overcome this material constraint. The general approach consists of evaluating $F^{(k)}$, and then measuring a finite number of spectra (generally of a smaller size than the discretization required to reconstruct a spectrum) and the corresponding values after acquisition by the camera. If we consider $\mathbf{r}^{(\mathbf{k})}$, a vector whose elements are the responses of the linearized camera, L a matrix whose lines are samples of known spectrums, we can write using equation [5.2]:

$$r^{(k)} = L.\mathbf{S}^{(\mathbf{k})} \tag{5.3}$$

We therefore find $\mathbf{S}^{(\mathbf{k})}$ by finding a solution to the linear system above. However, the problem is said to be hill-posed, which means that the amount of data is often too small to strictly solve the equation. When L contains enough information, we have a situation similar to that found with a monochromator. By directly calculating the pseudo-inverse of L using a reduced number of measurements, the result is highly unstable, particularly because it is the noise that is mostly adjusted. Therefore, the estimation of the sensor's sensitivity curves shows a number of peaks and negative values.

Various methods providing acceptable results include finding an approximate solution to the problem rather than an exact solution. This solution is subject to a number of constraints and uses a number of different methods and tools. The constraints typically include the positivity of values, ensuring that the curves are smooth, analyzing their modality (the sensors' responses are often uni- or bimodal, not in the strict mathematical sense but in an overall sense), limiting the maximum error tolerated, etc. Various methods may include second derivative analysis, Wiener methods, using a combination of Fourier series or the addition of regularization terms. For further information on the advantages, disadvantages and implementation of such methods, see [BAR 02, FIN 98, JOH 96, SHA 96, SHA 97].

5.5.2.2. *In practice*

Beyond the physical aspects, in practice, it may be sufficient to have a mapping of a data set taken from a camera (in an RGB space specific to the camera) to a space representative of color perception, such as XYZ or $CIELAB$. If we reuse a linear model, we have:

$$v = \mathbf{M}.\mathbf{c} \qquad\qquad [5.4]$$

where v is a vector containing colorimetric values, c is a vector containing data from the camera and $\mathbf{M}$ is a transformation matrix. If the Luther-Yves conditions are respected, M is a matrix with a size of 3×3. If this is not the case, matrix $\mathbf{M}$ must be of a greater size ($3 \times n$) and c must contain values of cross-components, typically $m = 10$, including the terms R, G, B, RG, RB, BG, R^2, G^2, B^2 and an offset term. The matrix $\mathbf{M}$ is typically adjusted using a least squares method based on a training data set (consisting of known values obtained from a color chart with known reflectances), such as the X-rite Colorchecker Classic, depending on the function to be minimized. This function is based on the perceived difference between the reference values and the values calculated by the model (ΔE_{76}^* or similar). Similar to adjusting polynomials, this transformation can be achieved using networks of neurons or different interpolation techniques. In particular, good results have been obtained using *radial basis functions* (RBF) with polyharmonic kernels [THO 11]. There is an abundance of literature on this subject, beyond the scope of this chapter [CHE 02, CHE 04, CHE 05, HON 01, JOH 96, SHA 00].

These methods are evaluated based on a statistically significant set of known reflectances, independent from the data used to establish the model. The difference between the actual colors and the evaluated colors is measured using one of the metrics mentioned above. Significant values may include average error, variance and maximum error. The 90th or 95th percentiles also provide good indications. It is often useful to know where in the color space the model shows the least amount of precision, and visualizations can therefore provide a good indication of the model's behavior.

To conclude, while assuming the response in intensity to be linear or corrected for its nonlinearity, there are three main possibilities to consider:

– the Luther-Yves conditions are respected and a simple linear transformation of the camera's values is sufficient;

– illumination is consistent but the Luther-Yves conditions are not respected: the transformation becomes less linear, the transformation may include more terms and can be adjusted by a regression process based on a collection of known values or other interpolation methods may be used;

– illumination may change. Two choices may be made: we could either perform a chromatic adaptation (a process to adapt the model to another light source, often by a linear transform) or, if this approximation is not sufficient, perform a complete characterization of the camera (by estimating S).

The characterization data are normally stored within a standard *International Color Consortium* (ICC) file in order to be used by a color management system.

5.6. Conclusion

When capturing a scene using two cameras, the geometric correspondence between the two images is not evident for reasons such as the distance between the object and two or multiple cameras. Equally, the correspondence between the color of one object in pictures taken by several cameras is far from obvious for various reasons. Reasons for this include:

– the geometric and material properties of the scene, such as a change in illumination, shadows or specific bidirectional reflectance distribution function (BRDF) (where the material does not reflect light in the same way depending on the angle of view). We can also cite the presence of different light sources, each lighting a different part of the scene;

– a difference between the cameras due to the absence of calibration/characterization, or material differences that induce differences in metamerism (the two cameras cannot tell the difference between the same colors).

These considerations and adequate solutions can also be applied to capture a scene using multiple cameras as well as taking several images of the same scene using the same camera for a 3D effect or reconstruction, and even to recreate a panoramic view based on several images. It is therefore necessary to obtain correspondence between various parts in the scene and to define a transformation, which can standardize the sensation of color; a "color correction", also known as "color mapping".

5.7. Bibliography

[BAR 02] BARNARD K., FUNT B., "Camera characterization for color research", *Color Research & Application*, vol. 27, no. 3, pp. 152–163, 2002.

[CHE 02] CHEUNG V., WESTLAND S., "Color camera characterisation using artificial neural networks", *10th Color and Imaging Conference (CIC'10), 2002*, Scottsdale, Arizona, pp. 117 –120, 12–20 12–15 November 2002.

[CHE 04] CHEUNG V., WESTLAND S., CONNAH D., *et al.*, "A comparative study of the characterisation of colour cameras by means of neural networks and polynomial transforms", *Coloration Technology*, vol. 120, no. 1, pp. 19–25, 2004.

[CHE 05] CHEUNG V., WESTLAND S., LI C., *et al.*, "Characterization of trichromatic color cameras by using a new multispectral imaging technique", *Journal of the Optical Society of America A*, vol. 22, no. 7, pp. 1231–1240, July 2005.

[CIE 01] CIE, *142-2001, Improvement to Industrial Colour-Difference Evaluation*, Commission Internationale de l'Eclairage, 2001.

[CIE 04] CIE, *015:2004, Colorimetry, 3rd ed.*, Commission Internationale de l'Eclairage, 2004.

[DAY 03] DAY D.C., Spectral sensitivities of the sinarback 54 camera, Technical report, Spectral Color Imaging Laboratory Group Munsell Color Science Laboratory Chester F. Carlson Center for Imaging Science Rochester Institute of Technology, February 2003.

[DEB 97] DEBEVEC P.E., MALIK J., "Recovering high dynamic range radiance maps from photographs", *Proceedings of the 24th Annual Conference on Computer Graphics and Interactive Techniques*, SIGGRAPH '97, ACM Press/Addison-Wesley Publishing Co., New York, NY, pp. 369–378, 1997.

[DEV 01] DEVERNAY F., FAUGERAS O., "Straight lines have to be straight: automatic calibration and removal of distortion from scenes of structured environments", *Machine Vision and Application*, vol. 13, no. 1, pp. 14–24, August 2001.

[FAI 98] FAIRCHILD M.D., *Color Appearance Models*, Addison-Wesley, Reading, MA, 1998.

[FIN 98] FINLAYSON G.D., HORDLEY S., HUBELL P.M., "Recovering device sensitivities using quadratic programming", *IS&T and SID's Color Imaging Conference (CIC'6)*, Scottsdale, Arizona, 17–20 November 1998.

[GLU 01] GLUCKMAN J., NAYAR S.K., "Rectifying transformations that minimize resampling effects", *IEEE Computer Society Conference on Computer Vision and Pattern Recognition*, Scottsdale, Arizona, vol. 1, IEEE Computer Society, pp. 111–117, 7–10 November 2001.

[HAR 99] HARTLEY R.I., "Theory and practice of projective rectification", *International Journal of Computer Vision*, vol. 35, no. 2, pp. 115–127, 1999.

[HAR 04] HARTLEY R.I., ZISSERMAN A., *Multiple View Geometry in Computer Vision*, 2nd ed., Cambridge University Press, 2004.

[HON 01] HONG G., LUO M.R., RHODES P.A., "A study of digital camera colorimetric characterization based on polynomial modeling", *Color Research & Application*, vol. 26, no. 1, pp. 76–84, 2001.

[HUN 00] HUNG P.-C., "Comparison of camera quality indexes", *8th Color and Imaging Conference (CIC'8), 2000*, Scottsdale, Arizona, pp. 167–171, 7–10 November 2000.

[JOH 96] JOHNSON T., "Methods for characterizing colour scanners and digital cameras", *Displays*, vol. 16, no. 4, pp. 183–191, 1996.

[LOO 99] LOOP C., ZHANG Z., "Computing rectifying homographies for stereo vision", *Computer Vision and Pattern Recognition*, vol. 1, pp. 1125–1131, 1999.

[MAR 03] MARTÍNEZ-VERDU F., PUJOL J., VILASECA M., *et al.*, "Characterization of a digital camera as an absolute tristimulus colorimeter", in MARTÍNEZ-VERD U.F., PUJOL J., CAPILLA P. (eds), *Imaging Science and Technology*, vol. 47, no. 4, pp. 279–295, 2003.

[NIS 04] NISTÉR D., "An efficient solution to the five-point relative pose problem", *IEEE Transactions on Pattern Analysis and Machine Intelligence*, vol. 26, no. 6, pp. 756–777, June 2004.

[NOZ 11] NOZICK V., "Multiple view image rectification", *Proceeding of IEEE-ISAS 2011, International Symposium on Access Spaces*, Yokohama, Japan, pp. 277–282, 17–19 June 2011.

[ROB 95] ROBERT L., ZELLER C., FAUGERAS O., *et al.*, Applications of non-metric vision to some visually guided robotics tasks, Report no. RR-2584, INRIA, June 1995.

[SHA 96] SHARMA G., TRUSSELL H., "Set theoretic estimation in color scanner characterization", *Journal of Electronic Imaging*, vol. 5, no. 4, pp. 479–489, 1996.

[SHA 97] SHARMA G., TRUSSELL H., "Digital color imaging", *IEEE Transactions on Image Processing*, vol. 6, no. 7, pp. 901–932, 1997.

[SHA 00] SHARMA G., "Targetless scanner color calibration", *Journal of Imaging Science and Technology*, vol. 44, no. 4, pp. 301–307, 2000.

[STR 05] STRAND R., HAYMAN E., "Correcting radial distortion by circle fitting", *Proceedings of the British Machine Vision Conference*, BMVA Press, pp. 9.1–9.10, 2005.

[THO 11] THOMAS J.-B., BOUST C., "Colorimetric characterization of a positive film scanner using an extremely reduced training data set", *19th Color and Imaging Conference (CIC'19), 2011*, San Jose, California, 7–11 November 2011.

[TRI 00] TRIGGS B., MCLAUCHLAN P., HARTLEY R.I., *et al.*, "Bundle adjustment – a modern synthesis", *Vision Algorithms: Theory and Practice*, vol. 1883, pp. 153–177, 2000.

[VOR 93] VORA P.L., TRUSSELL H.J., "Measure of goodness of a set of color-scanning filters", *Journal of the Optical Society of America A*, vol. 10, no. 7, pp. 1499–1508, July 1993.

[WYS 00] WYSZECKI G., STILES W.S., *Color Science: Concepts and Methods, Quantitative Data and Formulae* (Wiley Series in Pure and Applied Optics), 2nd ed. Wiley-Interscience, 2000.

[ZHA 00] ZHANG Z., "A flexible new technique for camera calibration", *IEEE Transactions on Pattern Analysis and Machine Intelligence*, vol. 22, no. 11, pp. 1330–1334, November 2000.

PART 2

Description/Reconstruction of 3D Scenes

Chapter 6

Feature Points Detection and Image Matching

6.1. Introduction

Image matching is a part of many computer vision or image processing applications, such as object recognition, registration, panoramic images and image mosaics, three-dimensional (3D) reconstruction and modeling, stereovision or even indexing and searching for images via content. The problem is finding the geometric transformation (rigid, affine, homographic or projective) that best matches two images using visual information shared by the two images. The main hypothesis is that the visible part of the 3D object from a given angle is almost identical to that obtained from another angle.

The different matching techniques vary according to the number and the nature of visual information used. When all the pixels in the image are used, matching is dense. Within the context of medical image registration, this may relate to *landmarks*, precisely defined anatomical points [ROM 02], or "semi-landmarks" [BER 04]: the amount of visual information used is minimal but matching is preceded by a landmark recognition stage. This chapter will examine a general scenario where the amount of visual information is limited and where no prior knowledge is available.

Chapter written by Michel DESVIGNES, Lara YOUNES and Barbara ROMANIUK.

This visual information corresponds to regions detected using specific photometric or geometric properties such as edges, corners or "blobs". They are represented by local descriptors often based on local photometry. Matching visual elements shared by two images is carried out by comparing their descriptors using adapted metrics or by classification. The transformation between the two images can therefore be calculated using these elements. The current process consists of three main stages:

– detecting the feature points/regions;

– calculating the descriptors in this region, normalized if necessary;

– matching the feature points of the two images using their descriptors and estimating the geometric transformation by removing false matches.

6.2. Feature points

There is a wide body of literature   [AWR 12, GAU 11, MIK 05a, MOK 06, ROM 12, TUY 08] on feature point extraction. Local information searches have been largely motivated by the unreliability of global object representations in terms of occlusion, articulated objects, intensity or viewpoint variations, robustness to noise acquisition (sampling, shaking, blur, etc.) and detection accuracy. The detection and representation of obtained points/regions must be invariant, particularly in terms of:

– viewpoint: translation, rotation, scale, affine transformations and projection;

– lighting: changes in lighting conditions and shadows;

– occlusion;

– local deformations and articulations.

The local nature of detectors lends a certain robustness to occlusions and local deformations. It is the spatial organization of these points that characterizes an object. They must therefore be fairly numerous and discriminative in order to uniquely describe an object. Lighting and viewpoint invariant properties are ensured by detectors and descriptors. One feature point can be assimilated with a pattern in the image that differs from its neighbor due to its photometric properties (intensity, color, texture, etc.), its specific local form or even its segmentation stability. These include points, lines or regions.

6.2.1. *Point detection*

6.2.1.1. *Differential operators: autocorrelation, Harris and Hessian*

Historically, objectives of repeatability/robustness, precise location, and informative and understandable content have resulted in the study of variations in intensity function $S(x, y)$. In this context, the first corner detector was developed by Moravec [MOR 81].

It calculates the windowed neighborhood autocorrelation for each point as follows:

$$E(x, y) = \sum_{k=-n}^{n} \sum_{l=-m}^{m} w(x, y) \times (S(x+k, y+l) - S(x, y))^2$$

where $w(x, y)$ is a weighting function. It detects the local maxima in all directions of the autocorrelation function. However, this simple detector is very sensitive to noise in the image, particularly near its edges, thereby favoring directions in the image frame.

To overcome these disadvantages, Harris [HAR 88] used a second-order development of $S(x, y)$ and a Gaussian weighting function G to decrease its sensitivity to noise and suppress anisotropy. Therefore, the function E becomes:

$$E(x, y) = \sum_{k=-n}^{n} \sum_{l=-m}^{m} \frac{1}{2\pi\sigma^2} e^{-\frac{(k^2+l^2)}{2\sigma^2}} \times \left[x \frac{\partial S}{\partial x} + y \frac{\partial S}{\partial y} \right]^2 = (x\ y).A.(x\ y)^t$$

with:

$$A = G(\sigma) * \begin{pmatrix} \left(\frac{\partial S}{\partial x}\right)^2 & \frac{\partial S}{\partial x} \cdot \frac{\partial S}{\partial y} \\ \frac{\partial S}{\partial x} \cdot \frac{\partial S}{\partial y} & \left(\frac{\partial S}{\partial y}\right)^2 \end{pmatrix}$$

The matrix A is a positive-definite matrix whose eigenvectors v_{max} and v_{min} associated with the eigenvalues λ_{max} and λ_{min} provide information about the directions of local intensity changes: the vector v_{max} (respectively, v_{min}) is associated with rapid changes in intensity (respectively, slow) and is perpendicular (respectively, parallel) to the edge. When these two eigenvalues are small, the pixel is in a homogeneous zone; when only one eigenvalue is high, it is an edge; and when the two eigenvalues are large, it is a corner. To avoid the costly calculation of eigenvalues, Harris has proposed using the following measure:

$$E(x, y) = \det(A) - \alpha.(\text{trace}(A))^2 = \lambda_{min}.\lambda_{max} - \alpha.(\lambda_{min} + \lambda_{max})^2$$

The function $E(x, y)$ is high and positive for a corner and negative or nul otherwise. α is approximately 0.04 and is defined empirically. The feature points retained by the Harris detector correspond to the local maxima of the function $E(x, y)$. This corner detector is invariant to translations, rotations and affine intensity variations. However, it is affected by scale and affine transformations. An example using this detector is shown in Figure 6.1(a).

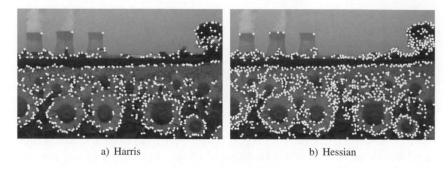

a) Harris   b) Hessian

Figure 6.1. *Feature points based on derivatives*

An alternative solution based on derivatives has been proposed by Beaudet [BEA 87]. It involves determining the maximal values of the Hessian matrix determinant.

Considering the image function S as a curved surface in $\mathbb{R}^3$, the curvatures of the surface [REN 00] are studied. This operator is invariant to rotation but is sensitive to noise as it is based on second-order differential operators. An improvement can be made by a Gaussian smoothing and a selection of points where the trace of the Hessian is a local extremum. This operator shows the same invariances as the Harrris operator and detects "blobs". An example of detection using the Hessian detector is illustrated in Figure 6.1(b).

6.2.1.2. *Scale invariance using multi-scale analysis*

Detectors such as Harris and Hessian use a measure based on a smoothing dependent parameter: scale. They detect the points or regions at a single scale. Since this scale is *a priori* unknown, multi-scale approaches automatically determine the correct scale. Lindeberg [LIN 08] has shown that the Gaussian kernel is the only linear smoothing that provides a correct multi-scale analysis. Current detectors construct a Gaussian pyramid, calculate a spatial measure of interest normalized for each pixel and level in the pyramid, and detect the extrema localized in x, y, σ in this 3D "spatial-scale space".

The Scale-Invariant Feature Transform (SIFT) detector [LOW 04] uses the Gaussian Laplacian as a saliency measure and looks for the extrema of this measure in the 3D space x, y, σ. Feature points with weak contrasts are eliminated along with feature points near low curvature edges (see Figure 6.2(a)). To reduce the algorithmic cost of this approach, the Gaussian Laplacian is approximated using Gaussian differences. This approximation only slightly affects the detector's performance and remains a reference tool. The Speeded-Up Robust Features (SURF) detector [BAY 08] uses a Hessian approximation as a spatial measure and is much faster. It replaces Gaussian smoothing with "box filters". These filters are composed of three or four blocks whose coefficients are identical inside a block. The integral image is then used to calculate convolution making this detector highly efficient (see Figure 6.2(b)).

a) SIFT: Multi-scale   b) SURF

Figure 6.2. *Scale invariant detection*

The main disadvantage of these detectors is that they respond fairly strongly to edges, particularly for the Laplacian. They are, however, a reference tool in feature point detection. To ensure scale invariance, they are included in multi-scale analysis. This analysis can be used to define the scale where saliency is maximal and therefore has the most stable value. They have good "blob" and corner detectability, with the exception of junctions between several areas.

6.2.1.3. *A corner intensity model*

The Smallest Univalue Segment Assimilating Nucleus (SUSAN) detector [SMI 97] uses the following property. We separate the circular neighborhood pixels from a sample pixel into two parts: those whose intensity is similar and those whose intensity is far from that in the sample pixel. Intuitively, a homogeneous region partitions the neighborhood into an empty zone and a zone of identical size to that of the neighborhood used, an edge partitions it into two zones of equal sizes and a corner partitions it into two zones of different sizes (25% and 75%). SUSAN calculates the number of pixels (known as USAN) similar to the pixel intensity by a given threshold. For a corner, this number must be less than half the window size. False positive detections are eliminated by checking the distance between the USAN center of gravity and the sample pixel that must be sufficient to validate the corner. SUSAN's primary disadvantage is its sensitivity to blurry edges whose corners are poorly detected. This detector is also less robust to noise near edges, depending on the size of the mask used. This is evident in Figure 6.3 where, despite adjusting the parameters, the number of corners detected is fewer in the right image that is blurred. The SUSAN detector is almost invariant to lighting conditions and is, in particular, fast.

Figure 6.3. *The SUSAN detector*

Designed for real-time applications, the Feature from Accelerated Segment Test (FAST) detector [ROS 05] uses the same principle with a lower computational cost. A point is considered to be a corner when a large number n of connected points situated on a circle (and not a disk, as with SUSAN) are clearer or darker than the central point. These connected sets are denoted as E_{clear} and E_{dark}. The initial parameters are fixed at a radius of 3 pixels for a discrete circle of 16 pixels and $n = 12$. To optimize the algorithmic cost, tests are prioritized following the order of the neighboring positions: north, south, east and west. At least three of them must be clear in order for the studied point to be a corner. The other pixels are only examined if this test is verified.

To eliminate multiple responses to a single corner, the non-maximal values of the following score V are removed:

$$V(p) = max(\sum_{p1\in E_{clear}} |S(p) - S(p1)| - \varepsilon, \sum_{p1\in E_{dark}} |S(p) - S(p1)| - \varepsilon)$$

An improved version [ROS 10] replaces the static tests with a decision tree that is automatically constructed by learning from a set of training images. Learning is based on an entropy criterion. This fast and high performing detector has become a standard for real-time applications. It is, however, sensitive to blurring, as is the SUSAN detector.

Scale invariance is achieved by the Binary Robust Invariant Scalable Keypoints (BRISK) detector, which combines the scale space and the FAST approaches [LEU 11]. In this algorithm, only a few scale levels are considered, thereby limiting computational cost. However, a quadratic interpolation between scales is applied. BRISK has detection performance comparable to SURF and SIFT, with the exception of blurred images, with six times lower computation cost than SURF and 10 times lower computation cost than SIFT.

6.2.2. *Edge-based feature points*

Another class of detectors operates according to image edges. Edges are fairly stable objects in relation to changes in perspective and lighting conditions. These detectors also use local geometry, straight lines and parallelograms for the "Edge-Based Regions" (EBR) operator or curvature for the "Curvature Scale Space" (CSS) operator.

6.2.2.1. *Shape detectors*

The EBR operator [TUY 04] initially locates corners using the Harris detector and then detects the edges using a Deriche operator. If we consider a corner p, then two points $p_1(l_1)$ and $p_2(l_2)$ move along the edges sharing this corner. The curvilinear distance $p_{1,2}$ of each point $p_{1,2}$ at the corner p is defined as the surface between the edge and straight line joining the corner and the point $p_{1,2}$. This distance is invariant to affine transformations. These two points and the corner define a family of parallelograms. The retained parallelogram Ω is the one that shows an extremum value of a photometric function based on the first-order points from the intensity function. This operator locates parallelepiped forms that are invariant to affine geometric and photometric transformations (see Figure 6.4(a)).

For these detectors, the primitives (parallelepiped) mean that they are designed primarily for structured scenes with artificial or well-contrasted elements.

6.2.2.2. *Curvature and scale space*

The CSS [MOK 98] searches for the extrema of the curvature function along the edges using a multi-scale approach. It detects edges using the Canny operator first, fuses the close edges and then calculates the curvature $K(t,\sigma)$ smoothed by a Gaussian defining the curvature space.

a) EBR   b) CSS

Figure 6.4. *Edge-based operators*

The corners are the extrema in this space. Those with a too low value or which are too close to other corners are not retained (see Figure 6.4(b)). A variant of this detector [MOK 01] differentiates the analysis of short curvatures and long curvatures by automatically determining the decision threshold. This operator is invariant to affine geometric and photometric transformations.

The primary disadvantage of these edge-based detectors is the definition of parameters used by the edge detector. As shown in Figure 6.4 some corners are not detected when edges are slightly contrasted.

6.2.3. *Stable regions: IBR and MSER*

The "Intensity Based Regions" (IBR) detector [TUY 04] detects the local extrema of the image intensity S. It then explores the area around each of the extrema by studying an intensity function along the rays traced from the local extremum. For each ray, the following function is evaluated:

$$f_S(t) = \frac{|S(t) - S(0)|}{max\left(d, \frac{1}{t}\int_0^t |S(x) - S(0)|\ \mathrm{d}x\right)}$$

where t is the curvilinear abscissa along the ray and d is a small number that prevents divisions by 0. The extrema of this function for each ray are chained and form a closed area. The denominator ensures good localization of extrema. This region is then approximated by an ellipse. This operator detects "blobs" and is robust, particularly for printed documents. It is invariant to affine geometric and photometric transformations.

a) IBR   b) MSER

Figure 6.5. *Stable region based operators*

The "Maximally Stable Extremal Regions" (MSER) detector [MAT 02] relies on the idea that as intensity varies strongly along the edges of objects, their subgraph varies slightly according to this intensity. The regions considered by this detector are therefore the connected components in the thresholded image whose surface varies slightly according to this threshold. They are detected by a watershed algorithm in which the watershed surfaces are analyzed. These regions are invariant to monotone intensity transformations (and not only photometric affine transformations), to homographic geometric transformations and to continuous geometric rather than linear transformations. This detector is robust, efficient and fast. Its only disadvantage is its sensitivity to blurring because in this case the intensity at the edges is poorly defined and the extremum threshold is very poorly localized.

Region-based detectors do not detect corners or angles in the image. They consider the center of the detected regions as the point of interest, as shown in Figure 6.5 (particularly in the sunflowers' heads).

6.3. Feature point descriptors

A description of the local content of these points is necessary to differentiate them within the context of image matching. In the literature, there are a number of description methods that have been proposed and have resulted in the computation of descriptor vectors for each point. A point's descriptor indicates the distribution of information in a region centered at the keypoint. These descriptors are designed to be invariant to small deformations, localization errors and rigid or affine transformations as well as illumination changes. The choice of a descriptor size is important and has a direct impact on the speed and cost of matching. Low dimensional descriptors are often less discriminatory but faster and are a source of complexity in the matching stage. They are therefore best suited to real-time contexts.

6.3.1. *Scale-invariant feature transform*

The SIFT descriptor was introduced by Lowe [LOW 99] and has become the most commonly used descriptor due to its robustness and computation time. It describes the distribution of gradients within a region around the feature point. This approach ensures scale and rotation invariance. The feature point detection scale is used to define the size of the window in which the descriptor is calculated. Each gradient magnitude $m(x, y)$ is weighted by a Gaussian distance centered on the feature point with a standard deviation $1, 5$ times the scale σ of the point:

$$wm(x, y) = m(x, y) * g_{1,5\sigma}(x, y) = m(x, y)\frac{1}{2\pi(1, 5\sigma)^2}e^{-\frac{dx^2+dy^2}{2(1,5\sigma)^2}}$$

where d_x and d_y are the distances in x and y directions at the feature point descriptor window.

The descriptor window is rotated relative to the dominant orientation to achieve a rotation invariance of the SIFT descriptor (see Figure 6.6). This orientation is obtained by analyzing the orientation histogram. The descriptor window is subdivided into 4×4 subregions. An orientation histogram (eight orientations at 45 degrees intervals) is then calculated for each subregion. The different histograms are stored in a $4 \times 4 \times 8 = 128$ dimensional vector. The descriptor vector is then normalized to achieve invariance to changes in illumination. To reduce the influence of large gradient magnitudes on the descriptor, the values in the descriptor vector are thresholded and those above 0.2 are restricted to this value. The descriptor vector is then normalized to unit length.

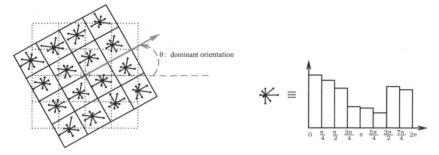

Figure 6.6. *The structure of a SIFT descriptor*

6.3.2. *Gradient Location and Orientation Histogram*

Similar to the SIFT descriptor, the Gradient Location and Orientation Histogram (GLOH) descriptor [MIK 05b] indicates the distribution of intensity gradients in the local region around the feature point. This is represented by a polar grid of 17 subregions (see Figure 6.7) (a central circular region and 8×2 subregions with intervals of $\frac{\pi}{4}$ on two circular regions centered on the keypoint).

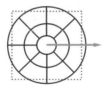

Figure 6.7. *The structure of the GLOH descriptor*

For each of these subregions, a gradient histogram of 16 bins is calculated to produce a descriptor vector with $17 \times 16 = 272$ dimensions.

6.3.3. *The DAISY descriptor*

The DAISY descriptor [TOL 10] is similar to the SIFT and GLOH descriptors. Their main difference lies in the shape of the region in which the descriptor vector is calculated. This region is composed of several overlapping circles centered on the feature point (see Figure 6.8): the radii of the circular grids are proportional to the distance of the Gaussian kernels from the feature point and the Gaussian smoothing is proportional to the circles' radii. For each circular region, an orientation histogram is calculated and normalized. The descriptor vector is the result of the concatenation of all the orientation histograms and is of high dimensionality (544). Due to the shape of its computation area, the DAISY descriptor is invariant to rotation. Therefore, it has the advantage of being able to be redefined in different orientations without the need to compute convolutions. This descriptor outperforms previously presented descriptors in dense matching applications and has a low computation time.

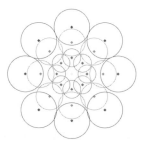

Figure 6.8. *The structure of the DAISY descriptor*

6.3.4. *Speeded-Up Robust Features (SURF)*

As was the case for the SIFT descriptor, the SURF descriptor [BAY 08] describes the distribution of intensities in the neighborhood of the feature point. It has a low computation time due to the use of the integral image [VIO 01], which is used to estimate derivatives using first-order Haar wavelet filters.

The SURF descriptor is scale and rotation invariant. The detection scale σ defines the size of the descriptor window (6σ) as well as the size of the filters. Within the descriptor window, the wavelet responses in the x and y directions are calculated and weighted by a Gaussian of 3.3 times the detection scale (see Figure 6.9). For each pixel in the descriptor window, an orientation vector is calculated and the maximal orientation is identified as the dominant orientation, sum of all responses within a sliding orientation window covering an angle of $\frac{\pi}{3}$, around the feature point. The descriptor window is orientated relative to the dominant orientation and subdivided into 4×4 subregions. For each of these subregions, the four following values are calculated: $\sum d_x, \sum d_y, \sum |d_x|$ and $\sum |d_y|$, where d_x and d_y are the responses to the first-order Haar filters in the x and y directions, respectively (see Figure 6.9).

The descriptor vector is obtained by concatenation of these four values for each of the subregions. The final vector with a size of 64 is normalized to ensure invariance to contrast.

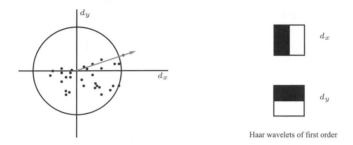

Figure 6.9. *Structure of the SURF descriptor*

6.3.5. *Multi-scale Oriented Patches (MOPS)*

As with the SIFT descriptor, the MOPS descriptor [BRO 05] uses gradients to extract a dominant orientation in a region centered at the feature point. A patch of 8×8 pixels using a spacing of 5 pixels is sampled. The 64 dimensional vector is then normalized. A Haar wavelet transform is applied to this vector to obtain a descriptor vector containing the wavelet coefficients. This descriptor is invariant to small changes in intensity.

6.3.6. *Shape context*

The "shape context" descriptor [BEL 00] is based on the idea that a shape is associated with an object. A shape is described by a discrete subset of its edge points. The "shape context" descriptor is a histogram of edges' coordinates. It is calculated in the log-polar grid centered on an edge point, sampled as 5 (rings) $\times 12$ (orientations) regions, similar to that of the GLOH descriptor. The shape context h_i in a selected edge point corresponds to the histogram of edge points that belong to each of these regions.

To provide a better description, the orientation is studied. In each region or "bin" k, the vectors t_j tangent to each of the points q_j in the edge Q belonging to the grid are calculated. The orientation is given by the sum of tangent vectors on each bin as follow:

$$h_i(k) = \sum_{q_j \in Q} t_j, \text{where } Q = \{q_j \neq p_i, (q_j - p_i) \in bin(k)\}$$

This descriptor is known as "Generalized Shape Context" and contains the dominant histogram orientations of each region k of the grid.

6.4. Image matching

Matching between image feature descriptors is the classic approach used to compare or match images.

6.4.1. *Descriptor matching*

The matching process involves comparing the descriptors taken from an unknown image with the descriptors taken from a target image. This comparison relies on calculating a dissimilarity measure (for example, Euclidean or Mahalanobis distances) and a selection criterion. Maintaining correspondence between a new descriptor and one or several potential descriptors requires the implementation of a threshold on similarity distances. In practice, three approaches are used to determine the matching between two descriptors [MIK 05b]. All three approaches rely on an empirical threshold definition. The simplest approach uses a direct threshold on similarity distances. This threshold depends directly on the deformation in a scene, which means that this approach is not particularly robust. The following approaches use the spatial nearest neighbor concept to establish valid correspondences. In the second approach, a match is validated if the nearest neighbor has a similarity distance lower than the threshold. The last approach, introduced by Lowe [LOW 04], involves calculating a ratio between the similarity distances to the two nearest neighbors. This last approach penalizes descriptors that have several similar correspondents. For data sets of significant size, Lowe [LOW 04] proposes accelerating the search using "k-d tree" [FRI 77] or even "Best-Bin-First" [BEI 97] algorithms.

The weakness of these three matching approaches is the use of an empirically determined threshold. Each descriptor can be associated with exactly one known candidate descriptor. An alternative approach proposed by [RAB 08] avoids setting the detection threshold and the usual restriction to the nearest neighbor, allowing multiple matches between images. This means a match is rejected according to a dissimilarity criterion using the "Earth Mover Distance" [RUB 00], since this is less sensitive to the descriptor quantification problem. This matching criterion, called *a contrario*, involves detecting groups of descriptors whose existence is barely probable according to the hypothesis that these descriptors are independent of each other.

6.4.2. *Correspondence group detection*

Matching descriptors between two images, based on calculating their similarity, produces strong matches. However, a number of false correspondences are also retained. To remove them, a second stage is necessary. It involves estimating a geometric transformation by detecting groups of consistent correspondences. Therefore, the objective is to eliminate false matches while keeping a maximum of good correspondences. This task is, in particular, difficult when images are subject to strong deformations, occlusions or ambiguities (repetitive structures and absences of texture). Several groups of global methods provide a solution to this problem.

The complexity of matching is closely related to the geometric transformation that is considered during the matching stage. It is therefore necessary to study groups with a minimum of two correspondences to estimate similarity, three correspondences for affine transformation, four correspondences for homography and seven correspondences in the case of epipolar geometry.

6.4.2.1. *Generalized Hough transform*

The Hough transform [BAL 81, HOU 59] is an approach used to recognize small objects or strongly occluded objects within a scene. In these circumstances, Lowe [LOW 04] recommends using three descriptors to optimize the recognition process. The use of the generalized Hough transform identifies consistent groups of descriptors using the vote principle over samples of n correspondences. These votes are made within the scale space assigned to the parameters of the geometric transformation using quantified accumulators.

Given the high algorithmic complexity of this approach, it cannot be used for geometric transformations requiring more than three descriptor matches to be estimated, particularly if there is a particularly large amount of prior data. Estimating complex geometric transformations such as epipolar or even projective geometry can therefore be excluded from these approaches.

6.4.2.2. *Graph matching*

Graph matching is another tool that can be used to determine correspondences between descriptors using a global consistency criterion. The main idea here is to construct a graph whose descriptors correspond to the nodes and relations connecting them to the edges. High-level constraints, involving more than two descriptors, can be represented by hyperedges.

Graph matching involves determining correspondences between the nodes of two graphs using matching constraints or by optimizing a global score.

Leordeanu and Hebert [LEO 05] have used the distance between two points as a constraint to ensure invariance to rotations. Berg *et al.* [BER 05] have used a criterion combining the notion of distance and angle, thereby ensuring scale and rotation invariance. Zheng and Doermann [ZHE 06] have proposed using the concept of neighborhood to define the matching constraint.

The more the order of graphs increases, the more precise the matching process and the more resistant it is to noise. In contrast to the previous approaches based on matching couples of points, point triplets are compared for third-order hypergraphs. [DUC 11, LEO 07, ZAS 08] have ensured invariance to various projections that include transformations in the similarity group, affine transformations or even projective transformations.

These approaches identify objects that have been subject to rigid deformation but can also be easily adapted to recognizing deformable objects. These methods, however, also prove fairly costly in terms of algorithmic complexity that increases with the order of the graph. They give good results provided the "inliers" represent at least 50% and the number of considered descriptors is relatively low (fewer than 100 in [DUC 11]).

6.4.2.3. *RANSAC and its variants*

The most commonly used method within the context of object recognition is the *random sample consensus* (RANSAC) algorithm [FIS 81] and its variants. This method effectively separates inliers from outliers. This approach consists of randomly sampling subsets of matches in order to predict the geometric transformation. The correspondences that are coherent with this model are therefore recorded. The set of correspondences yielding the maximal consensus is retained and represents the inliers while the others are considered to be outliers.

The strategy adopted in the RANSAC algorithm is fast and robust but requires regulating sensitive parameters. It is inconclusive when the rate ρ of inliers is low. This is due to the fact that the number of iterations required to randomly sample subsets is in the order of $1/\rho^n$, where n is the number of matches required to define the model. The strategies proposed in the MLESAC variants [TOR 00] or even PROgressive SAmple Consensus (PROSAC) [CHU 05] considerably reduce the number of models estimated while simplifying parameter regulation. However, they are not more effective than the original algorithm when $\rho < 50\%$. The Optimized RANSAC

(ORSA) algorithm [MOI 04] provides an effective solution to this problem when ρ is close to 10%. This last variant relies on the "a contrario" detection principle and overcomes the parameter setting in the case of epipolar geometry.

The previous RANSAC-derived methods all have a disadvantage when estimating the fundamental matrix. The outliers, whose corresponding features are near their epipolar line but far from their correct location, cannot be eliminated.

Moreover, the major disadvantage of these methods is the fact that detection is limited to a single object. To detect several objects, two strategies have been adopted in the literature: a simultaneous search of all object [TOR 00] or a sequential search that successively detects each object separately [STE 95]. In this context, MAC-RANSAC algorithm [RAB 08] proposes an automatic variant of the algorithm that sequentially detects multiple objects based on the "a contrario" detection principle.

6.5. Conclusion

Image matching is the key aspect in a number of applications such as image mosaics, panoramic images, object indexing or recognition, stereoscopy and 3D multiview reconstruction. In this context, the matching process is based on a non-dense approach that first requires the detection of feature points in the image, and then describes them using invariant descriptors. These descriptors are matched between images to estimate the transformation.

To detect corresponding points in images with different geometric and photometric characteristics, the detection stage uses a multi-scale approach of invariant spatial measurements. These measurements detect corners, blobs or even areas of random forms that are assimilated at their center. They are generally invariant to similarity group transformations and sometimes even to affine transformations. All detectors use one or several thresholds or parameters that affect the number and quality of detected points. Some have adopted a compromise that reduces algorithmic cost to respect the constraints of real-time applications while minimizing loss in terms of robustness.

Descriptors synthesize information contained in the neighborhood of a feature point. In most cases, the information coded is the histogram of gradients' orientations calculated in this neighborhood. Descriptors vary in terms of neighborhood shape and their computation time. However, the SIFT algorithm remains the most commonly used descriptor.

The last stage involves matching the descriptors of the feature points detected in two different images. It is based on the similarity between two descriptors, on the one hand, and coherence in the transformation calculated using all couples of matched points, on the other hand. Different strategies, including RANSAC and its variants, which are today the most popular, can be used to effectively estimate homography or projective transformation.

6.6. Bibliography

[AWR 12] AWRANGJEB M., LU G., FRASER C.S., "Performance comparisons of contour-based corner detectors", *IEEE Transactions on Image Processing*, vol. 21, no. 9, pp. 4167–4179, 2012.

[BAL 81] BALLARD D.H., "Generalizing the Hough transform to detect arbitrary shapes", *Pattern Recognition*, vol. 13, no. 2, pp. 111–122, January 1981.

[BAY 08] BAY H., ESS A., TUYTELAARS T., *et al.*, "Speeded-up robust features (SURF)", *Computer Vision and Image Understanding*, vol. 110, no. 3, pp. 346–359, 2008.

[BEA 87] BEAUDET P.R., "Rotationally invariant image operators", *International Joint Conference on Artificial Intelligence*, pp. 579–583, 1987.

[BEI 97] BEIS J.S., LOWE D.G., "Shape indexing using approximate nearest-neighbour search in high-dimensional spaces", *Proceedings of the 1997 Conference on Computer Vision and Pattern Recognition (CVPR '97)*, IEEE Computer Society, Washington, DC, pp. 1000–1006, 1997.

[BEL 00] BELONGIE S., MALIK J., PUZICHA J., "Shape context: a new descriptor for shape matching and object recognition", *NIPS*, IEEE Computer Society, pp. 831–837, 2000.

[BER 04] BERAR M., DESVIGNES M., BAILLY G., *et al.*, "3D meshes registration: application to statistical skull model", in CAMPILHO A., KAMEL M. (eds), *Image Analysis and Recognition*, Lecture Notes in Computer Science, vol. 3212, Springer, Berlin, Heidelberg, pp. 100–107, 2004.

[BER 05] BERG A.C., BERG T.L., MALIK J., "Shape matching and object recognition using low distortion correspondences", *Proceedings of the 2005 IEEE Computer Society Conference on Computer Vision and Pattern Recognition (CVPR' 05)*, vol. 1, pp. 26–33, 2005.

[BRO 05] BROWN M., SZELISKI R., WINDER S., "Multi-image matching using multi-scale oriented patches", *IEEE Computer Society Conference on Computer Vision and Pattern Recognition 2005* (CVPR 2005), vol. 1, IEEE, pp. 510–517, 2005.

[CHU 05] CHUM O., MATAS J., "Matching with PROSAC – progressive sample consensus", *IEEE Computer Society Conference on Computer Vision and Pattern Recognition, 2005* (CVPR 2005), IEEE Computer Society, vol. 1, pp. 220–226, 2005.

[DUC 11] DUCHENNE O., BACH F., KWEON I.-S., *et al.*, "A tensor-based algorithm for high-order graph matching", *IEEE Transactions on Pattern Analysis and Machine Intelligence*, IEEE Computer Society, Los Alamitos, CA, USA, vol. 33, no. 12, pp. 2383–2395, 2011.

[FIS 81] FISCHLER M.A., BOLLES R.C., "Random sample consensus: a paradigm for model fitting with applications to image analysis and automated cartography", *Communications of the ACM*, vol. 24, no. 6, pp. 381–395, June 1981.

[FRI 77] FRIEDMAN J.H., BENTLEY J.L., FINKEL R.A., "An algorithm for finding best matches in logarithmic expected time", *ACM Transactions on Mathematical Software*, vol. 3, no. 3, pp. 209–226, September 1977.

[GAU 11] GAUGLITZ S., HOLLERER T., TURK M., "Evaluation of interest point detectors and feature descriptors for visual tracking", *International Journal of Computer Vision*, vol. 94, no. 3, pp. 335–360, 2011.

[HAR 88] HARRIS C., STEPHENS M., "A combined corner and edge detector", *4th Alvey Vision Conference*, Manchester, United Kingdom, pp. 147–152, 1988.

[HOU 59] HOUGH P.V.C., "Machine analysis of bubble chamber pictures", *International Conference on High Energy Accelerators and Instrumentation*, CERN, 1959.

[LEO 05] LEORDEANU M., HEBERT M., "A spectral technique for correspondence problems using pairwise constraints", *International Conference of Computer Vision (ICCV)*, vol. 2, pp. 1482–1489, October 2005.

[LEO 07] LEORDEANU M., HEBERT M., SUKTHANKAR R., "Beyond local appearance: category recognition from pairwise interactions of simple features", *Proceedings of the CVPR*, Minneapolis, Minnesota, USA, June 2007.

[LEU 11] LEUTENEGGER S., CHLI M., SIEGWART R.Y., "BRISK: Binary Robust invariant scalable keypoints", *IEEE International Conference on Computer Vision*, IEEE Computer Society, Los Alamitos, CA, USA, pp. 2548–2555, 2011.

[LIN 08] LINDEBERG T., "Scale-space", in BENJAMIN W. (ed.), *Encyclopedia of Computer Science and Engineering*, vol. IV, John Wiley & Sons, Hoboken, NJ, pp. 2495–2504, 2008.

[LOW 99] LOWE D., "Object recognition from local scale-invariant features", *Proceedings of the 7th IEEE International Conference on Computer Vision, 1999*, vol. 2, IEEE, pp. 1150–1157, 1999.

[LOW 04] LOWE D., "Distinctive image features from scale-invariant keypoints", *International Journal of Computer Vision*, vol. 60, no. 2, pp. 91–110, 2004.

[MAT 02] MATAS J., CHUM O., MARTIN U., *et al.*, "Robust wide baseline stereo from maximally stable extremal regions", *British Machine Vision Conference*, vol. 1, London, pp. 384–393, 2002.

[MIK 05a] MIKOLAJCZYK K., TUYTELAARS T., SCHMID C., *et al.*, "A comparison of affine region detectors", *International Journal of Computer Vision*, vol. 65, no. 1, pp. 43–72, 2005.

[MIK 05b] MIKOLAJCZYK K., SCHMID C., "A performance evaluation of local descriptors", *IEEE Transactions on Pattern Analysis & Machine Intelligence*, vol. 27, no. 10, pp. 1615–1630, 2005.

[MOI 04] MOISAN L., STIVAL B., "A probabilistic criterion to detect rigid point matches between two images and estimate the fundamental matrix", *International Journal of Computer Vision*, vol. 57, no. 3, pp. 201–218, May 2004.

[MOK 98] MOKHTARIAN F., SUOMELA R., "Robust image corner detection through curvature scale space", *IEEE Transactions on Pattern Analysis and Machine Intelligence (T-PAMI)*, vol. 20, pp. 1376–1381, 1998.

[MOK 01] MOKHTARIAN F., MOHANNA F., "Enhancing the curvature scale space corner detector", *Scandinavian Conference on Image Analysis*, Bergen, Norway, pp. 145–152, 2001.

[MOK 06] MOKHTARIAN F., MOHANNA F., "Performance evaluation of corner detectors using consistency andaccuracy measures", *Computer Vision and Image Understanding*, vol. 102, no. 1, pp. 81–94, April 2006.

[MOR 81] MORAVEC H.P., *Robot Rover Visual navigation*, UMI Research Press, 1981.

[RAB 08] RABIN J., DELON J., GOUSSEAU Y., "A contrario matching of SIFT-like descriptors", *19th International Conference on Pattern Recognition (ICPR 2008)*, IEEE, Tampa, FL, pp. 1–4, December 8–11, 2008.

[REN 00] RENAULT C., DESVIGNES M., REVENU M., "3D curves tracking and its application to cortical sulci detection", *IEEE International Conference on Image Processing (ICIP2000)*, vol. 2, pp. 491–494, September 2000.

[ROM 02] ROMANIUK B., DESVIGNES M., REVENU M., *et al.*, "Linear and non-linear model for statistical localization of landmarks", *International Conference of Pattern Recognition (ICPR)*, vol. 4, pp. 393–396, 2002.

[ROM 12] ROMANIUK B., YOUNES L., BITTAR E., "First steps toward spatio-temporal Rheims reconstruction using old postcards", *International Conference on Signal Image Technology & Internet Based Systems (SITIS)*, pp. 374–380, 2012.

[ROS 05] ROSTEN E., DRUMMOND T., "Fusing points and lines for high performance tracking", *IEEE International Conference on Computer Vision*, vol. 2, pp. 1508–1511, October 2005.

[ROS 10] ROSTEN E., PORTER R., DRUMMOND T., "Faster and better: a machine learning approach to corner detection", *IEEE Pattern Analysis Machine Intelligence*, vol. 32, no. 1, pp. 105–119, January 2010.

[RUB 00] RUBNER Y., TOMASI C., GUIBAS L.J., "The earth mover's distance as a metric for image retrieval", *International Journal of Computer Vision*, Hingham, MA, USA, vol. 40, no. 2, pp. 99–121, November 2000.

[SMI 97] SMITH S.M., BRADY J.M., "SUSAN – a new approach to low level image processing", *International Journal of Computer Vision*, vol. 23, no. 1, pp. 45–78, 1997.

[STE 95] STEWART C.V., "MINPRAN: a new robust estimator for computer vision", *IEEE Transactions on Pattern Analysis and Machine Intelligence*, vol. 17, no. 10, pp. 925–938, 1995.

[TOL 10] TOLA E., LEPETIT V., FUA P., "DAISY: an efficient dense descriptor applied to wide-baseline stereo", *IEEE Transactions on Pattern Analysis and Machine Intelligence*, vol. 32, no. 5, pp. 815–830, 2010.

[TOR 00] TORR P.H.S., ZISSERMAN A., "MLESAC: a new robust estimator with application to estimating image geometry", *Computer Vision and Image Understanding*, vol. 78, no. 1, pp. 138–156, 2000.

[TUY 04] TUYTELAARS T., GOOL L.V., "Matching widely separated views based on affine invariant regions", *International Journal of Computer Vision*, vol. 59, no. 1, pp. 61–85, August 2004.

[TUY 08] TUYTELAARS T., MIKOLAJCZYK K., *Local Invariant Feature Detectors: A Survey*, Now Publishers, Inc. cop., Hanover, MA, 2008.

[VIO 01] VIOLA P., JONES M., "Rapid object detection using a boosted cascade of simple features", *Proceedings of the 2001 IEEE Computer Society Conference on Computer Vision and Pattern Recognition, 2001 (CVPR 2001)*, vol. 1, IEEE, pp. 1–511, 2001.

[ZAS 08] ZASS R., SHASHUA A., "Probabilistic graph and hypergraph matching", *Proceedings of the 2008 Conference on Computer Vision and Pattern Recognition (CVPR '08)*, Anchorage, Alaska, USA, pp. 1–8, 2008.

[ZHE 06] ZHENG Y., DOERMANN D., "Robust point matching for nonrigid shapes by preserving local neighborhood structures", *IEEE Transactions on Pattern Analysis and Machine Intelligence*, vol. 28, no. 4, pp. 643–649, April 2006.

Chapter 7

Multi- and Stereoscopic Matching, Depth and Disparity

7.1. Introduction

Three-dimensional (3D) reconstruction using stereo-correlation relates to the automatic extraction of data about the scene's 3D structure from 2 to N images acquired simultaneously. In this context, in order to estimate depth within a scene, the 3D points are triangulated using their projections in the images taken from different viewpoints and the characteristics of the capture system. This problem therefore relates to matching[1] homologous pixels (i.e. projections of the same 3D point in images). This research can be based on specific geometric constraints, including the epipolar constraint that creates a first-order indeterminacy, reducing the search space to a segment. The photometry compared between pixels from different images is therefore used to match homologues although anomalies (similar photometries or variations in brightness) may occur, requiring the use of more complex heuristics or information redundancy. Matching pixels, in this context, are known as stereoscopic matching.

Chapter written by Stéphanie PRÉVOST, Cédric NIQUIN, Sylvie CHAMBON and Guillaume GALES.
1 In stereoscopy, dedicated terms are "pair/pairing", i.e. to match two things together. In multiscopy, the number of elements is not limited to 2, so we will use, in this case, the term match/matching namely to bind similar and multiple elements.

We will first introduce the difficulties related to homologue searches as well as primitives and capture geometry. We will then examine the generic algorithms of two existing approaches with the most commonly used constraints and costs. Second, we will concentrate on the occlusion problem by describing two approaches, the first being stereoscopic and the other being multiscopic.

7.2. Difficulties, primitives and stereoscopic matching

7.2.1. *Difficulties*

The quality of the 3D reconstruction is highly dependent on the quality of matching. Regardless of the number of points used to estimate depth and the matching method used, the same difficulties and limitations can affect the obtained results. Gales [GAL 11] highlights two categories of problems, as illustrated in Figure 7.1:

– Missing information: absent from certain viewpoints, they generate matching uncertainty. They are of three types:

- Occlusions: areas visible in some images and hidden in others (see Figure 7.1(d), black edges).

- Depth discontinuities: found at the edges of objects and in front of distant object, they create color variations in the neighborhood of the pixel studied in different images (see Figure 7.1(b)).

- Shortening effect: different points in the scene are projected onto several pixels in an image but onto a single pixel in another image (for example it occurs when a surface's tangent plane is close to the optical center).

– Ambiguous information: information that does not select the best correspondent without ambiguity, which are of three types:

- Homogeneous areas: all pixels have similar photometric attributes and it is therefore difficult to distinguish them from one another in order to match them (see Figure 7.1(c)).

- Repetitive textures: photometric attributes of the neighborhoods of several correspondent candidates do not allow them to be distinguished (see Figure 7.1(a)).

- Changes in lighting: such changes can induce photometric differences in corresponding pixels, which therefore make them difficult to match.

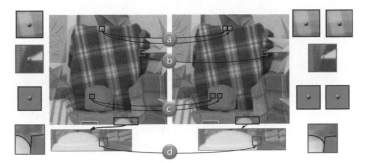

Figure 7.1. *Difficulties in matching. a) Repetitive textures,*
b) discontinuity, c) homogeneous areas and d) occlusions

7.2.2. *Primitives and density*

As shown in   [JON 92], two kinds of primitives can be used in stereoscopic matching: pixels for the first kind of primitives and feature types, also called points of interest (see Chapter 6), for the second. The pixel-based approach uses the whole set and provides dense results because it is possible to estimate as many matches (3D points) as there are pixels in the N images and use attributes such as lightness, color, gradient, etc. to identify them. These characteristics are subject to noise and only necessarily provide a small amount of information about the scene, thereby generating a number of false matches (decoys). In contrast, methods within the second category are feature based [JAW 02], uses a partial set of pixels and focus on structured and more discriminative primitives to remove ambiguity and limit combinatorics during matching. However, the detection of these primitives does not provide any consistency between images or a particularly dense reconstruction for some images. There are two approaches to this type of conflict (decoy/inconsistency and dense/sparse): hybrid and multiscopic methods. The first method combines the advantages of these two categories, for example, by matching and segmenting the image at the same time (see section 7.5.1) while the second method is based on the redundancy of information created by adding images into the series (see section 7.5.2).

7.3. Simplified geometry and disparity

Chapters 3 and 4 introduced the geometry of a sensor and specifically the epipolar constraint. We will concentrate specifically on multiscopic matching ($N \geq 2$) where the capture system respects the shooting conditions with a (non) off-axis parallel geometry (see Figure 4.4). Since this includes both cases, we will focus on off-axis geometry, which we will term "parallel geometry" in the rest of this chapter. This configuration places matching within a simplified epipolar geometric framework, which ensures, among other things, that the co-epipolar straight lines are horizontals of the same rank as N images. As such, two homologous pixels have the same y-coordinate and the gap between one pixel in an image i and its homologue in the following image $i + 1$ will be a simple horizontal translation, which limits the search area in the image $i + 1$ to a horizontal segment. In this chapter, we will systematically use this configuration. If the geometry of capture used does not match this configuration, it is possible to make a prior image rectification by reprojecting the initial images using parallel geometry [HAR 03] (see Chapter 5).

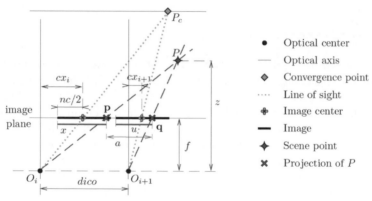

Figure 7.2. *Relation between disparity and depth*

In parallel geometry, a point P in the 3D scene is projected in the numbered images i and $i + 1$, if it is not occluded on either of these images, at the positions connected by simplified epipolar geometry, respectively, $\mathbf{p} = (x, y)$ and $\mathbf{q} = (u, y' = y)$ (see Figure 7.2). The homologue of the pixel $(\mathbf{p}, i)$ in the following image $(\mathbf{q}, i + 1)$ can therefore be identified by the single difference of the abscissa $\bar{\delta} = x - u$, defined as a "horizontal disparity", which will be called disparity in the rest of this chapter. For a pixel $(\mathbf{p}, i)$, the disparity $\bar{\delta}$ (see Figure 7.2) calculates the depth z of the point P, which is projected in it, according to the following:

$$\left.\begin{array}{l} (z - f)/a = z/dico \Leftrightarrow z = (f.dico)/(dico - a) \\ a = u - \frac{nc}{2} + cx_{i+1} + dico - cx_i + \frac{nc}{2} - x \\ a = dico - \bar{\delta} - (cx_i - cx_{i+1}) \end{array}\right\} \Rightarrow z = \frac{f.dico}{\bar{\delta} + \hat{\delta}} \qquad [7.1]$$

where $dico$ is the distance between the optical centers, f is the darkroom depth of the virtual sensors and $\hat{\delta}$ is the off-axis difference $cx_i - cx_{i+1}$ (in pixels) in the two images (gaps between the center of the image and the optical image–axes intersection).

In multiscopic stereovision, if the parallel geometry is regular (regular spacing ($dico$) between the optical centers and convergences between all lines of sight), the disparity $\bar{\delta}$ of the pixel $(\mathbf{p}, i)$ connects the abscissa of the N potential projections in P, which are the possible homologues in $(\mathbf{p}, i)$. Indeed, the application between pairs of successive images in Figure 7.2 and the equation [7.1] shows that the disparity related to a depth z is equal for all pairs of images $(i, i + 1)$. The position in an image $j \in \mathbb{N}_N$ of the homologue of the pixel $(\mathbf{p} = (x, y), i)$, assigned the disparity $\bar{\delta}$, can therefore only be:

$$\mathbf{h}^j_{(x,y),i,\bar{\delta}} = (x - (j - i).\bar{\delta}, y) = \mathbf{p} + (i - j)\bar{\delta}.\mathbf{x} \qquad [7.2]$$

As such, in this simplified geometry, the depth estimation problem can be seen as a disparity estimation problem, formulated as the search of a multimap $\delta \in \mathbb{X}[\Upsilon^2]$ attributing to each pixel $(\mathbf{p}, i)$ its disparity $\delta[\mathbf{p}, i] \equiv \delta_{\mathbf{p},i}$ in $\mathbb{X} = \{\bar{\delta}_m, \ldots, \bar{\delta}_M\} = \mathbb{Z}$ or $\mathbb{R}$, depending on whether real or integer disparities are needed. The next section will introduce the generic algorithm for matching methods more precisely, including a description of the constraints and the cost functions that can be used.

7.4. A description of stereoscopic and multiscopic methods

7.4.1. *Local and global matching algorithms*

As we discussed in section 7.2, application of the epipolar constraint alone is not sufficient to generate a multimap of disparities in qualities. To remove ambiguity due to decoys, heuristics must be used. Stereovision methods are characterized by the following criteria (see [SCH 02]):

– "primitives" to be matched (see section 7.2.2), and their "attributes";

– the expression of the "cost of matching or energy" on a support made up of all primitives to be matched: N images, one image or a subimage;

– the "optimization method" used to find the solution with the minimum cost.

These latter elements depend on the method's classification. Brown *et al.* have [BRO 03] distinguished local methods from global methods. Both seek to minimize the cost function: the first pixel by pixel and the second in relation to all pixels.

Algorithm 7.1. Local pixel matching with $\mathrm{Cor}^{D,\mathcal{F}}(\mathcal{M}, \mathbf{p}, i, \mathbf{q}, j)$ defined in section 7.4.3.

Data: $\mathcal{M} \in \mathcal{E}[\Upsilon^2]$, all N images to be matched
Result: $\delta \in \mathbb{X}[\Upsilon^2]$, all N disparity maps
foreach *number of image* $i \in \mathbb{N}_N$ **do**

> **foreach** *position* $\mathbf{p} = (x, y) \in \Omega^2$ **do**
>
> > **foreach** *disparity* $\bar{\delta} \in \{\delta_m, \ldots, \delta_M\}$ **do**
> > Score = 0
> > **foreach** *number in the image* $j \in \mathbb{N}_N$ **do**
> > > Score += $\mathrm{Cor}^{D,\mathcal{F}}(\mathcal{M}, \mathbf{p}, i, \mathbf{h}^j_{\mathbf{p},i,\bar{\delta}}, j)$
> >
> > **end**
> >
> > **end**
>
> $\delta_{\mathbf{p},i} = \arg\min_{\bar{\delta}}$ Score (estimation according to the WTA approach)
>
> **end**

end

In local approaches, the energy used is a "similarity measure" also known as "correlation" or, in contrast, a "dissimilarity measure" that evaluates the degree of photometric similarity/dissimilarity between two homologous pixels and their respective neighborhoods (see section 7.4.3), and the maximization/minimization strategy is the "winner takes all" (WTA). Algorithm 7.1, a generic version of this approach, requires the choice of a Cor energy function, a neighborhood form $\mathcal{F}$ and a distance D.

In global approaches, the difficult aspects are the initialization of disparities (most of the time with a correlation-based approach), the choice of the stop conditions, the update of the disparity function and cost, which are dependent on the optimization method used. In algorithm 7.2, the conditions chosen are simple in order to facilitate its understanding. For further information on optimization methods, see [FEL 11].

Algorithm 7.2. Global pixel matching.

Data: $\mathcal{M} \in \mathcal{E}[\Upsilon^2]$, all N images to be matched
Result: $\delta \in \mathbb{X}[\Upsilon^2]$, all N disparity maps
Initialization of the disparity multimap δ using a local method
cost' $= E_{\text{tot}}^{global}(\mathcal{M}, \delta)$ cost $= \infty$
repeat
$\quad$ cost $=$ cost'
$\quad$ $\delta' =$ evolution of δ according to a heuristic designed to minimize
$\quad$ E_{tot}^{global}
$\quad$ cost' $= E_{\text{tot}}^{global}(\mathcal{M}, \delta')$
$\quad$ **if** cost' $<$ cost **then** $\delta = \delta'$;
until (cost' $\geq$ cost);

The reliability of a match is evaluated using an energy function. This function, to be minimized for a set of matches, integrates terms measuring dissimilarities in neighborhoods of homologous pixel E_{dis} and violations of constraints on the estimated disparities E_{cont}. Its general form is:

$$E_{\text{tot}}^{global}(\mathcal{M}, \delta) = (1 - \lambda)E_{\text{dis}}^{global}(\mathcal{M}, \delta) + \lambda E_{\text{cont}}^{global}(\mathcal{M}, \delta) \qquad [7.3]$$

where $\lambda \in [0, 1]$ influences the weight of each of the two terms E_{dis} and E_{cont}. The dissimilarity cost E_{dis} or the term related to the data is often a sum of "local costs" of all matches where each local cost measures the dissimilarity between two homologous primitives. The constraint costs, E_{cont}, is used to choose between several potential homologues and to limit the combinatorial. This cost models the interactions between the pixels considered. It quantifies the respect of the constraints used for all matches and corresponds to the sum of "neighborhood costs".

In 2002, the Middlebury evaluation protocol[2] was introduced [SCH 02]. Widely used, it compares matching methods and proposes different data sets. At this moment in time, however, only stereoscopic or multiscopic methods in convergent capture are considered in this protocol.

---

2 http://vision.middlebury.edu/stereo.

7.4.2. *Principal constraints*

A constraint is related to a match taken from hypotheses based on the geometry of capture and of the scene as well as a reflection on objects' surfaces. The geometry of the scene is described by different constraints and costs, which we will detail here and in the following section. For each of these constraints, two aspects need to be considered: the definition (the rule) itself and their objectives of use.

– Uniqueness constraint. Widely used in stereovision, it is defined by:

$$\forall\, x_1, x_2, y, i, j, \quad x_1 \neq x_2 \;\Rightarrow\; \mathbf{h}^{j}_{(x_1,y),i,\delta[(x_1,y),i]} \neq \mathbf{h}^{j}_{(x_2,y),i,\delta[(x_2,y),i]} \quad [7.4]$$

where two pixels in the image i cannot have the same homologue in the image j.

– Ordering constraint. Occasionally used in stereovision, it is defined by:

$$\forall\, x_1, x_2, y, i, j \quad \left.\begin{array}{l} \mathbf{h}^{j}_{(x_1,y),i,\delta[(x_1,y),i]} = (u_1, y) \\ \mathbf{h}^{j}_{(x_2,y),i,\delta[(x_2,y),i]} = (u_2, y) \end{array}\right\} \Rightarrow (x_1 - x_2)(u_1 - u_2) \geq 0$$

$$[7.5]$$

It indicates that the order of the pixels in the image i along the epipolar line y must be the same as their correspondents in the image j. However, the presence of a shortening effect (see section 7.2.1) transgresses these two constraints. Kostková, and Šára have therefore proposed a variant in [KOS 03] called weak consistency.

– Symmetry constraint, or bidirectional verification, is written as:

$$\forall\, \mathbf{p}, i, j \quad \mathbf{h}^{j}_{\mathbf{p},i,\delta[\mathbf{p},i]} = \mathbf{q} \;\Rightarrow\; \mathbf{h}^{i}_{\mathbf{q},j,\delta[\mathbf{q},j]} = \mathbf{p} \quad [7.6]$$

It is respect for a pixel $(\mathbf{p}, i)$ when it is a homologue of its homologous pixel[3] $(\mathbf{q}, j)$. In addition, this constraint ensures the uniqueness constraint.

These three constraints reduce the ambiguities induced by homogeneous areas, repetitive textures or changes in lighting between the different views (see section 7.2.1). However, none of the constraints examined here limit the

---

3 The term "homologue" normally refers to pairwise symmetry but in a formal mathematical formulation when creating disparity maps, this property is not systematic. Depending on the constraints applied, it must be explicitly mentioned when this property is required.

effect of a lack of information because accounting for this difficulty is strongly dependent on the type of method used, as discussed in section 7.5.

7.4.3. *Energy costs*

Energy costs also rely as heavily on photometric aspects as on geometric aspects based on disparities in neighboring pixels. In the literature, there are a large number of energy functions but only the most significant ones are presented: dissimilarity costs, smoothing costs and costs that explicitly take into account occlusion problems.

Local cost functions (denoted by the indices xxx) in a pixel $(\mathbf{p}, i)$ and for a predicted disparity value $\bar{\delta}$ are as follows:

$$E_{xxx} \in \mathbb{R}^{\mathcal{E}[\Upsilon^2] \times \Omega^2 \times \mathbb{N}_N \times \mathbb{X}} \quad : \quad E_{xxx}(\mathcal{M}, \mathbf{p}, i, \bar{\delta}) \tag{7.7}$$

The global costs, for a given disparity multimap $\delta \in \mathbb{X}^{\Upsilon^2}$, are obtained by creating the sum on all pixels of their disparity in the multimap according to the following general formula:

$$E_{xxx}^{global}(\mathcal{M}, \delta) = \sum_{(\mathbf{p}, i) \in \Upsilon^2} E_{xxx}(\mathcal{M}, \mathbf{p}, i, \delta_{\mathbf{p}, i}) \tag{7.8}$$

7.4.3.1. *Photometric dissimilarity cost*

According to the hypothesis that all objects are matte and without any specular effect, a match must be penalized if the photometric or colorimetric components of homologous pixels involved are dissimilar within a given neighborhood, regardless of the color space chosen (generally RGB; Bleyer *et al.* [BLE 08] have studied the influence of this choice on the estimated disparities). The function of this cost E_{dis} can generally be defined by:

$$E_{dis}(\mathcal{M}, \mathbf{p}, i, \bar{\delta}) = \sum_{j \in \mathbb{N}_N, j \neq i} \mathrm{Cor}^{D, \mathcal{F}}(\mathcal{M}, \mathbf{p}, i, \mathbf{h}_{\mathbf{p}, i, \bar{\delta}}^{j}, j) \tag{7.9}$$

In binocular stereovision, the sum in equation [7.9] on j is not carried out. As such, Cor measures the photometric differences in M between two pixels $(\mathbf{p}, i)$ and $(\mathbf{q}, j)$ by cumulating the distance $D \in \mathbb{R}^{\mathcal{E}^2}$ (generally L_1 or L_2

[CHA 11]) used in $\mathcal{E}$ between their respective neighbors in correlation form $\mathcal{F}$, such that:

$$\text{Cor}^{D,\mathcal{F}}(\mathcal{M},\mathbf{p},i,\mathbf{q},j) = \sum_{\mathbf{v}\in\mathcal{F}} D(\mathcal{M}_{\mathbf{p}+\mathbf{v},i},\mathcal{M}_{\mathbf{q}+\mathbf{v},j}) \qquad [7.10]$$

Within the context of multi-ocular stereo ($N > 2$) where problems of changes in illumination can be accentuated by the cumulative distances between the capture systems, Niquin [NIQ 11] has restricted the calculation of dissimilarity costs to homologues of two successive images in the scene. The energy function, therefore, becomes:

$$E_{dis}(\mathcal{M},\mathbf{p},i,\bar{\delta}) = \sum_{j\in\mathbb{N}_{N-1}} \text{Cor}^{D,\mathcal{F}}(\mathcal{M},\mathbf{h}^j_{\mathbf{p},i,\bar{\delta}},j,\mathbf{h}^{j+1}_{\mathbf{p},i,\bar{\delta}},j+1) \qquad [7.11]$$

7.4.3.2. Geometric and/or photometric smoothing costs

To limit noise sensitivity to dissimilarity costs, a photometric and/or geometric smoothing is often added to the term $E_{\text{cont}}^{global}(\mathcal{M},\delta)$ in equation [7.3]. Except for areas near depth discontinuities, smoothing is applied to the attributes (disparity, intensity and color) of the pixels in the smoothing form $\mathcal{F}$. Centered or not, composed of several configurations or adapted to the scenes contents, $\mathcal{F}$ has a significant influence on the quality of this smoothing [FUS 97]: when $\mathcal{F}$ is too small, noise sensitivity persists and, when too large, depth discontinuities tend to be smoothed. A photometric smoothing $E_{lissPhoto}$ and a geometric $E_{lissGeom}$ smoothing or a combination of both $E_{lissGeomPhoto}$ can be written as follows:

$$E_{lissPhoto}(\mathcal{M},\mathbf{p},i,\bar{\delta}) = \sum_{\mathbf{v}\in\mathcal{F}_{\mathbf{p}}} E_{dis}(\mathcal{M},\mathbf{v},i,\bar{\delta}) \qquad [7.12]$$

$$E_{lissGeom}(\mathcal{M},\mathbf{p},i,\delta) = \sum_{\mathbf{v}\in\mathcal{F}_{\mathbf{p}}} |\delta_{\mathbf{p},i} - \delta_{\mathbf{v},i}| \qquad [7.13]$$

$$E_{lissGeomPhoto}(\mathcal{M},\mathbf{p},i,\delta) = \sum_{\mathbf{v}\in\mathcal{F}_{\mathbf{p}}} |\delta_{\mathbf{p},i} - \delta_{\mathbf{v},i}|\, D(\mathcal{M}_{\mathbf{p},i},\mathcal{M}_{\mathbf{v},i}) \qquad [7.14]$$

In contrast to the previous costs, those with geometric constraints are based on disparities of the pixels in the neighborhood, previously estimated, and therefore require, at a local level, having the multimap δ. As a result, their global formulation becomes:

$$E_{xxx}^{global}(\mathcal{M},\delta) = \sum_{(\mathbf{p},i)\in\Upsilon^2} E_{xxx}(\mathcal{M},\mathbf{p},i,\delta) \qquad [7.15]$$

These smoothing constraints therefore can take into account the depth discontinuity problems but remain ineffective for reducing errors caused by occlusion.

7.4.3.3. *Geometric and/or photometric occlusion costs*

Occlusions can be detected by using the constraints discussed in section 7.4.2, in post-processing with local approaches or directly in the constraint cost formulation. In this case, the dissimilarity cost is replaced by an expensive cost when the detection is positive. Some fast methods, such as [MIN 08], exploit colorimetric dissimilarities to indicate a potential occlusion. However, they are sensitive to changes in illumination that can be improved by considering the disparities. One of the most common methods in binocular stereovision relies on "Left Right Checking", denoted by LRC where, using an arbitrary threshold, the difference in disparities between the pixel and its homologue indicates an occlusion. However, within the context of multiscopic stereovision, occlusions are rarely present in $N - 1$ images. A 3D point is therefore generally visible in several images even if it is occluded in others. As a result, this method and its extensions [INC 05, JOD 06] are not applicable in this state. A method based on this observation is introduced in section 7.5.2.

7.5. Methods for explicitly accounting for occlusions

7.5.1. *A local stereoscopic method – seeds propagation*

Using a parallel configuration, a homologous pixel is found in a horizontal segment with a maximal width of δ_M (maximal disparity). The seed propagation, a binocular technique, reduces this search area and therefore the risk of selecting a wrong correspondent. In accordance with the observation that generally two neighboring points on the same surface are projected as two neighboring pixels in each of the two images, implying that the two neighboring pixels have similar disparities, the search area can be reduced as follows: a pixel's correspondent, the neighbor of a previously matched neighbor, is searched within the neighborhood of the homologue of the latter. Propagation is iterative and requires the selection of an initial set of reliable matches, known as "seeds". At each iteration, the newly calculated matches are added to the set of seeds and the process continues as new matches are found.

However, this assertion is not verified in relation to depth discontinuities. A threshold on the correlation scores is therefore necessary to prevent propagation near depth discontinuities. The more tolerant the threshold, the denser the result but the higher the risk of propagating errors. An alternative may be to carry out a prior segmentation of the reference image in homogeneous color regions, most commonly using the mean-shift technique [COM 02]. By way of hypothesis, each region corresponds to the same surface and does not present depth discontinuities. The propagation of seeds occurs within each region, thereby preventing propagation above the depth discontinuities. Using this segmentation also has an advantage in that it accounts for occlusions. Indeed, in a single region there may be occluded pixels and non-occluded pixels that correspond to the same surface. As a result, after propagation, it is possible to carry out a regularization, stage that will be described later, in order to estimate disparities for occluded pixels from non-occluded pixels.

7.5.1.1. *Calculating the initial seeds*

There are two families of automatic seed selection methods:

– Matching feature points: these pixels are different from others (see Chapter 6), and matching them can be done fairly reliably. As such, the homologue of a feature point in the image 0 can be found within the set of feature points in the image 1. This technique first involves detecting the points of interest in the two images and then matching them. To do so, the dissimilarity measures and the geometric constraints presented in sections 7.4.3 and 7.4.2 can be used.

– Matching interests: the correspondents of the pixels in the image 0 are found in image 1 according to algorithm 7.1. Only the matches satisfying a series of strong constraints are then retained, as discussed in section 7.4.2.

It is not necessary to have a large number of seeds. However, in cases where scenes present depth discontinuities, there should be a "good" distribution with at least one correct initial seed per region of homogeneous depths to carry out propagation in these regions. In [GAL 11], a hybrid "completion-validation" approach is proposed to improve this distribution.

7.5.1.2. *Propagation approaches*

Depending on the means of using seeds within each iteration, there are two propagation approaches:

– A simultaneous approach: all seeds are considered simultaneously to find correspondents, meaning that each iteration can be carried out in parallel but at the risk of errors if wrong seeds exist.

– A sequential approach: a single seed is considered. It is selected according to a predefined criterion in order to propagate the "best" seed first. This approach, summarized in algorithm 7.3 limits the propagation of errors. The cost of dissimilarity (see section 7.4.3) between the neighborhoods of the seed's correspondents is generally used as a selection criterion. However, this cost does not provide information about the seed's reliability. In [GAL 12], the criterion used is the correspondence probability, calculated during the matching of each pixel. It is given by the dissimilarity cost between the two correspondents, divided by the sum of costs of all other candidates. As such, the more different a candidate is from others, the greater is its probability of correspondence (and vice versa).

Algorithm 7.3. Matching pixels using sequential propagation with prior segmentation, see section 7.5.1 where $\mathcal{V}_8$ is an 8-connected form

---

Data: $\mathcal{M} \in \mathcal{E}[\Upsilon^2]$, set of 2 to be matched
Data: $\mathcal{G}$, initial set of seeds (see section 7.5.1.1)
Result: $\delta \in \mathbb{X}[\Upsilon^2]$, the disparity map $(0 \to 1)$
$\mathcal{R} \leftarrow$ set of homogeneous color regions in the image 0 by mean-shift
foreach *region r in $\mathcal{R}$* **do**
    repeat
        select the best seed $g = \left((\mathbf{p}, 0), h^1_{\mathbf{p}, 0, \delta_{\mathbf{p}, 0}} \right) \in r$
        $\mathcal{G} \leftarrow \mathcal{G} \setminus g$
        foreach *neighbor $(\mathbf{p} + v, 0)$ unmatched with $v \in \mathcal{V}_8$* **do**
            $Score \leftarrow \{\}$
            foreach *disparity $\bar{\delta}$ induced by each candidate in the search area neighboring $h^1_{\mathbf{p}+v, 0, \bar{\delta}}$* **do**
                $Score \leftarrow Score \cup Cor^{\mathcal{D}, \mathcal{F}} \left(\mathcal{M}, \mathbf{p} + v, 0, h^1_{\mathbf{p}+v, 0, \bar{\delta}}, 1 \right)$
            end
            if $\min(Score) < threshold$ **then**
                $\delta_{\mathbf{p}+v, 0} \leftarrow \arg \min_{\bar{\delta}} Score$
                $\mathcal{G} \leftarrow \mathcal{G} \cup \left((\mathbf{p} + v, 0), h^1_{\mathbf{p}+v, 0, \bar{\delta}} \right)$
            end
        end
    until $\mathcal{G} \neq \varnothing$;
end

7.5.1.3. *Adjustment by region-based voting scheme*

The result of a local stereoscopic matching by propagation is only slightly dense, notably in the occluded areas (see section 7.2.1). Adjustment by the region-based voting scheme is therefore required to densify it (see Figure 7.3). This stage is used to estimate disparities among occluded pixels as well as to correct some errors. It is based on homogeneous color regions in the reference image (mean-shift). In a region, there are occluded and non-occluded pixels, presumed to correspond to the same surface. An approach using estimated disparities for each region has been proposed by Gales [GAL 11].

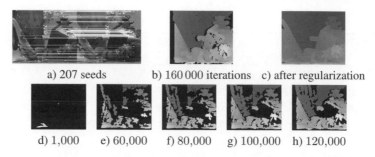

a) 207 seeds   b) 160 000 iterations   c) after regularization

d) 1,000   e) 60,000   f) 80,000   g) 100,000   h) 120,000

Figure 7.3. *Initial set of seeds a) and disparity maps obtained from different iterations (b, d, e, f, g and h) during sequential propagation. The image c) shows the result obtained after regularization*

7.5.2. *A global multiscopic method*

The approach explored in this section uses an original pixel matching formulation, specifically constructed around a multiscopic context and simplified epipolar geometry. It simultaneously calculates the N disparity maps while ensuring geometric consistency, managing occlusions and precisely identifying information redundancy. An in-depth study of this solution and different global and local methods can be found in [NIQ 11].

7.5.2.1. *Multiscopic matching formulation*

When searching for 3D points in a finite number of constant depth planes, this formulation restricts its disparity search, for pixels with N views, to a set of integer values $\mathbb{D} = \mathbb{Z}_{\bar{\delta}, \bar{\delta}_{M+1}} = \left\{ \bar{\delta}_m, \ldots, \bar{\delta}_M \right\} \subset \mathbb{Z}$. Indeed, the density of this reconstruction is reduced but a refinement of disparities can solve this at a later stage. This choice is justified, at least for initializing maps, by the assurance that co-homologues[4] are all pixels (and not subpixels) in the images, which clarifies redundancies and the consistency search. As such, potential 3D points are discrete in number and correspond to intersections of optical-pixel center rays with the planes $\Pi_{\bar{\delta}}$, for $\bar{\delta} \in \mathbb{D}$.

The idea proposed in [NIQ 11] relates to all the co-homologous pixels with the disparity $\bar{\delta}$, representing the same 3D point $\mathcal{P}$ in an entity known as a *match*, which intrinsically codes the matching redundancies and partially ensures consistency. Denoted as $m_{\mathbf{p}_0,\alpha}^{\bar{\delta}}$, a match is identified by the position $\mathbf{p}_0 = (x, y)$ of the pixel in the reference image $(i = 0)$, a disparity $\bar{\delta}$ and a boolean vector α where $\alpha[j] = 1$ if $\mathbf{h}_{\mathbf{p}_0,0,\bar{\delta}}^{j}$ is a projection of $\mathcal{P}$. As such, it contains at most 1 pixel per image and 0 in some images when there is occlusion. Multiscopic matching therefore involves finding a set $\mathcal{L}_m$ of matches that form a consistent partition of all the pixels Υ^2.

7.5.2.2. *Energy function and geometric consistency constraint*

By defining a match, the uniqueness and symmetry constraints are ensured if the number of pixels composing it is N. However, an inferior value indicates that the pixels in certain views are not co-homologous $(\alpha[i] = 0)$ with the others $(\alpha[i] = 1)$, the symmetry constraint is therefore no longer verified and this indicates the presence of either an occlusion or a geometric inconsistency in the partition as a distant object that will mask a near object. To prevent the creation of such partitions, a new constraint is integrated into the cost function: the "geometric consistency constraint" (see [NIQ 10]).

The aim of this multiscopic formulation is to explicitly use the notions of "match" and partition without questioning all existing stereocorrelation methods. It is for this reason that the energy function $E_{tot}^{global}(\mathcal{L}_m)$ evaluating a partition can be reduced to a classic energy function in the form of equation [7.3], in order to use the existing constraints (see section 7.4.2). The first part integrates dissimilarity, occlusion and geometric costs at once using the energy E_{doc} :

$$E_{doc}\left(\mathcal{M}, \delta, \mathbf{p}, \bar{\delta}, \alpha\right) = \sum_{k \in (-1;1)} \sum_{(\mathbf{p}',i) \in m_{\mathbf{p},\alpha}^{\bar{\delta}}/\alpha_i=1} C_{doc}\left(\mathcal{M}, \mathbf{p}', i, \bar{\delta}, \mathbf{h}_{\mathbf{p}',i,\bar{\delta}}^{i+k}, i+k, \delta\right)$$

with $\quad E_{doc}(\mathcal{L}_m) = \sum_{((\mathbf{p},\bar{\delta}),\alpha) \in \mathcal{L}_m} E_{doc}\left(\mathcal{M}, \delta, ((\mathbf{p}, \bar{\delta}), \alpha)\right)$ [7.16]

---

4 Co-homologue is the term used to refer to homologous pixels between them in the N images taken from projections in different images from the same 3D point.

The cost C_{doc} only evaluates the colorimetric dissimilarity Cor between the pixel $(\mathbf{p}, i)$ and its unproven homologue $(\mathbf{q}, j)$ if they have the same disparity (i.e. both belonging to the same match). Its form is:

$$C_{doc}(\mathcal{M}, \mathbf{p}, i, \bar{\delta}_1, \mathbf{q}, j, \delta) = \begin{cases} \mathrm{Cor}^{D,\mathcal{F}}(\mathcal{M}, \mathbf{p}, i, q, j) & \text{if } \delta_{\mathbf{q},j} = \bar{\delta}_1 \text{ (dissimilarity)} \\ K^{occ} & \text{if } \delta_{\mathbf{q},j} < \bar{\delta}_1 \text{ (occlusion)} \\ K^{coh} & \text{if } \delta_{\mathbf{q},j} > \bar{\delta}_1 \text{ (inconsistency)} \end{cases}$$

$$[7.17]$$

where K^{coh} is a constant large enough representing the cost of a geometric inconsistency and K^{occ} is the cost of an occlusion.

7.5.2.3. *Global selection and partition construction*

In addition to the energy function, Niquin [NIQ 11] has also proposed the "Near-Far" selection method in order to find a visibility function at minimal cost. The principle is, after an initialization of the partition g with the matches from $m^{\bar{\delta}_M}$, to consider the disparity planes from the largest $(\bar{\delta}_M - 1)$ to the smallest $(\bar{\delta}_m)$, in order to find the 3D target points from the nearest to the farthest. Figures 7.4(a) and (b) illustrate this progression. At each studied disparity $\bar{\delta}$, the previous matches are reexamined so that they are retained or removed after evaluation. As a result, the suppression of a match involve redistributing its constituent pixels in other disparity matches $\bar{\delta}$. To solve this binary choice problem, a graph cut method is used with a "min-cuts/max-flows" algorithm where each arc is valuated using the energy function. Therefore, for each disparity studied $\bar{\delta}$, a graph is constructed and minimized, creating a new partition g' that replaces g if its cost is inferior to that of g. In contrast to other existing work, this formulation reduces the graph to one node per match instead of one node per pixel. In practice, their number only rarely exceeds twice the number of pixels in a single image, regardless of the number of images N, thereby reducing the computation time. For a complete examination of the graph cut technique and for a detailed study of costs for each arc, please refer to [BOY 99] and [NIQ 11].

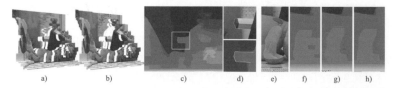

Figure 7.4. *Different stages in reconstruction (a and b), the disparity map and occlusion zoom in with $K^{occ} = 100$ and $K^{liss} = 20$ (c and d), increasing from (e) with $K^{occ} = 200$ (f), $K^{liss} = 0$ (g) and $K^{liss} = 40$ (h)*

7.5.2.4. *Results*

The cost C_{doc} (see equation [7.17]) detects the edges of objects with precision (see Figure 7.4(d)) but penalizes, with K^{occ}, not only occlusions but also each change in disparity along an epipolar line. The value K^{occ} must therefore be small enough to avoid the type of problems illustrated in Figure 7.4(f), but large enough to maintain an effective colorimetric comparison. E_{cont}^{global} (see equation [7.3]) contains a smoothing constraint of which the influence of the coefficient K^{liss} reinforces (or not) the robustness of matches (see Figures 7.4(g) and (h). However, the reduction in the size of the graph to one node by matching means that the number of nodes in the graph (i.e., the complexity of the cutting) remains almost unchanged. As a result, the computation times obtained (see Table 7.1) increase linearly according to the number of images where the algorithms using one node per pixel have an exponential evolution time.

Nb images	2	3	4	5	6	7	8
Times	1,226	1,333	1,527	1,812	2,085	2,360	2,634

Table 7.1. *Computation time (in ms) according to the number of images for the "Teddy" Middlebury scene (Intel Core i5-3470 @ 3.2 GHz, 8 GB RAM)*

7.6. Conclusion

In the context of binocular and multiocular stereovision, the configuration of (un)centered parallel geometry capture allows the use of the simplified epipolar geometry constraint in order to reduce the homologue search area. However, it cannot be used to solve the pixel matching problem. Therefore, there are two approaches: hybrid methods combining the advantages of pixel and feature point matching, and multiscopic methods, which exploit information redundancy. Both are based on cost constraints and functions that include photometric as well as geometric or smoothing characteristics, either locally or globally. Among the known difficulties, this chapter has focused on occlusions by describing two approaches that can be used to account for them. One is hybrid, local and stereoscopic, based on seed propagation (using previously established and reliable matches), whereas the other is global and multiscopic, ensuring geometric consistency while highlighting and exploiting information redundancy.

7.7. Bibliography

[BLE 08] BLEYER M., CHAMBON S., POPPE U. *et al.*, "Evaluation of different methods for using colour information in global stereo matching approaches", *The Congress of the International Society for Photogrammetry and Remote Sensing*, vol. XXXVII, Part B3a, Beijing, China, pp. 415–420, July 2008.

[BOY 99] BOYKOV Y., VEKSLER O., ZABIH R., "Fast approximate energy minimization via graph cuts", *IEEE Transactions on Pattern Analysis and Machine Intelligence*, vol. 23, pp. 1222–1239, 1999.

[BRO 03] BROWN M., BURSCHKA D., HAGER G., "Advances in computational stereo", *IEEE Transactions on Pattern Analysis and Machine Intelligence*, vol. 25, no. 8, pp. 993–1008, August 2003.

[CHA 11] CHAMBON S., CROUZIL A., "Similarity measures for image matching despite occlusions in stereo vision", *Pattern Recognition*, vol. 44, no. 9, pp. 2063–2075, September 2011.

[COM 02] COMANICIU D., MEER P., "Mean shift: a robust approach toward feature space analysis", *IEEE Transactions on Pattern Analysis and Machine Intelligence*, vol. 24, no. 5, pp. 603–619, May 2002.

[FEL 11] FELZENSZWALB P.F., ZABIH R., "Dynamic programming and graph algorithms in computer vision", *IEEE Transactions on Pattern Analysis and Machine Intelligence*, vol. 33, no. 4, pp. 721–740, April 2011.

[FUS 97] FUSIELLO A., ROBERTO V., TRUCCO E., "Efficient stereo with multiple windowing", *IEEE Conference on Computer Vision and Pattern Recognition*, San Juan, Porto Rico, p. 721–740, June 1997.

[GAL 11] GALES G., Mise en correspondance de pixels pour la stéréovision binoculaire par propagation d'appariements de points d'intérêt et sondage de régions, PhD Thesis, University of Toulouse, July 2011.

[GAL 12] GALES G., CHAMBON S., CROUZIL A. *et al.*, "Reliability measure for propagation-based stereo matching", *International Workshop on Image Analysis for Multimedia Interactive Services*, Dublin, Ireland, May 2012.

[HAR 03] HARTLEY R., ZISSERMAN A., *Multiple View Geometry in Computer Vision, 2nd ed.*, Cambridge University Press, 2003.

[INC 05] INCE S., KONRAD J., "Geometry-based estimation of occlusions from video frame pairs", *IEEE International Conference on Acoustics, Speech, and Signal Processing*, Philadelphia, PA, USA, vol. 2, pp. 933–936, March 2005.

[JAW 02] JAWAHAR C., NARAYANAN P., "Generalised correlation for multi-feature correspondence", *Pattern Recognition*, vol. 35, no. 6, pp. 1303–1313, June 2002.

[JOD 06] JODOIN P.-M., ROSENBERGER C., MIGNOTTE M., "Detecting half-occlusion with a fast region-based fusion procedure", *British Machine Vision Conference*, Edinburgh, United Kingdom, pp. 417–426, September 2006.

[JON 92] JONES D., MALIK J., "A Computational framework for determining stereo correspondence from a set of linear spatial filters", *International Journal of Image and Vision Computing*, vol. 10, no. 10, pp. 699–708, December 1992.

[KOS 03] KOSTKOVÁ J., ŠÁRA R., "Stratified dense matching for stereopsis in complex scenes", *British Machine Vision Conference*, vol. 1, Norwich, United Kingdom, pp. 339–348, September 2003.

[MIN 08] MIN D., KIM D., SOHN K., "Virtual view rendering system for 3DTV", *3DTV-Conference 2008: The True Vision - Capture, Transmission and Display of 3D Video (3DTV-CON)*, Istanbul, Turkey, pp. 249–252, May 2008.

[NIQ 10] NIQUIN C., PRÉVOST S., REMION Y., "An occlusion approach with consistency constraint for multiscopic depth extraction", *International Journal of Digital Multimedia Broadcasting*, vol. 2010, 8 pages, 2010.

[NIQ 11] NIQUIN C., Reconstruction du relief et mixage réel virtuel par caméras relief multi-points de vues, PhD Thesis, University of Reims Champagne-Ardenne, March 2011.

[SCH 02] SCHARSTEIN D., SZELISKI R., "A taxomomy and evaluation of dense two-frame stereo correspondence algorithms", *International Journal of Computer Vision*, vol. 47, no. 1, pp. 7–42, 2002.

Chapter 8

3D Scene Reconstruction and Structuring

8.1. Problems and challenges

The cinema and video games industries increasingly combine real images with computer-generated images. Today, there is a tendency to mix these techniques at the point of recording so that producers can use a three-dimensional (3D) result to judge whether a scene will appear in the final production and to guide actors, as well allowing the results of filming to be directly inserted into a traditional computer graphics production chain.

To satisfy this growing demand in the image industry, a large body of research has focused on multiview reconstruction. The approaches proposed until now can be split into two families of methods:

– "Model-free methods" that can be distinguished by the fact that no prior knowledge (relating to the nature and number of objects, characters' morphologies) is given to the system. The silhouette-based reconstruction technique belongs to this family. Due to their general nature, these techniques generate results without any temporal coherence. Indeed, a reconstruction is calculated for each "frame" independently of others.

– "Model-based methods" use a reference geometric description (for example triangular mesh) of the object to be reconstructed. This prior knowledge can be manually constructed or obtained using an acquisition

Chapter written by Ludovic BLACHE, Muhannad ISMAEL and Philippe SOUCHET.

system (for example a 3D scanner). Reconstruction involves evolving the reference form in relation to data taken from multiview capture (silhouettes, optical waves, etc.). These methods are more reliable than model-free methods and present the strong advantage of generating data with strong temporal coherence. However, due to the use of a reference form, they are, in the majority of cases, restricted to reconstructing a single individual with human morphology.

This chapter will focus on silhouette-based reconstruction and its improvement. Following a detailed description of the different stages in this method, its implementation in an industrial context will also be examined. Finally, we will conclude the chapter with an overview of multiview reconstruction analysis techniques to extract temporally stable semantic information to facilitate their integration into the computer graphics production chain.

8.2. Silhouette-based reconstruction

A silhouette is a binary mask associated with a given perspective that includes all pixels corresponding to the projection of a point of the 3D object to be reconstructed. In Figure 8.1, the colored pixels in the images taken by cameras C_1, C_2 and C_3 correspond to silhouettes of the 3D object in each view. Silhouette-based reconstruction [SNO 00] therefore involves estimating the visual hull of the 3D object, represented by a polygon in Figure 8.1.

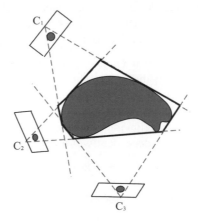

Figure 8.1. *Silhouette-based reconstruction*

8.2.1. *Silhouette extraction methods*

The extraction of a silhouette involves isolating the region of projection for the object to be reconstructed from the scene's background. There are several methods for arriving at this result, grouped according to the following categories:

– "Color difference-based methods" that use an image from the scene's background. To extract an object from the background, the technique uses image differences. To overcome the problem of variations in lighting in the background, the "chroma-keying" technique is often favored. "Chroma-keying" is one of the most common and most frequently used semantic segmentation techniques within audiovisual contexts. Video acquisition takes place against a "key color" background, generally blue or green. The problem of shadowing in the background is solved using learning techniques such as Gaussian mixture model or "k-means" [STA 99, ZIV 04].

– "Region based methods" aggregate, step-by-step, pixels with shared colorimetric properties. They establish region filling heuristics within an image by propagating local criteria, often based on the image's gradient (higher at the edges and lower in the middle of the area). The most commonly used methods in this category include histogram segmentation, region growing and region merging. For a more detailed presentation of region-based segmentation methods, Caillet's [CAI 06] doctoral thesis is an interesting resource.

– "Contour-based methods" involve extracting the connected components using a threshold of the image's gradient. Using these methods, the silhouette is characterized by its edge with the background of the scene.

8.2.2. *Reconstruction methods*

Surface methods deduce the object's visual hull using the intersection of silhouette cones from each camera. The silhouette cone associated with a camera is defined by the set of infinite triangles delimited by half-lines connecting the optical center with two neighboring pixels in the contour of the silhouette. The reconstructed object is therefore described by its surface, represented in the form of a triangular mesh [LAZ 07].

Volumic methods subdivide the capture space according to a regular grid of basic cells, known as voxels (volume elements). In this approach, the visual hull corresponds to the set of voxels projected into the silhouettes of each camera. The reconstructed object is described by its volume within the discrete grid [SZE 93].

8.2.3. *Improving volume reconstruction*

The main disadvantage of silhouette-based reconstruction lies in its inability to reconstruct certain details on the object's surface. The techniques examined in this chapter use color information from each view to select voxels within the bounding volume.

8.2.3.1. *Voxel coloring*

This technique, proposed by Seitz and Dyer [SEI 99], involves subdividing the regular grid of voxels into successive layers, from the nearest to the farthest in relation to the cameras (the cameras being set out in a semicircle around the object to be reconstructed). Voxel coloring is based on the hypothesis that a voxel on the surface of an object must have the same color in each view, known as a photo-consistent voxel. For example, the object in Figure 8.2 shows a concavity ignored by the silhouette-based reconstruction technique. The voxel v_1 found on the visual hull, is projected on different color pixels in views taken from cameras C_1 and C_2. On the basis of this statement, the voxel coloring algorithm is composed of the following stages:

Algorithm 8.1. The voxel coloring algorithm

Input: the sequence of calibrated images
Output: the volume of voxels representing the object being modeled
Initialize a bounding box and divide it into layers;
for *each layer* c *from the nearest to the farthest from the cameras* **do**
    for *each voxel* v *in* c **do**
        Projection if **v** on all the image planes where it is not occluded
        by a previously validated voxel;
        if v *is not visible or photo-consistent in certain images* **then**
            v is eliminated from the volume;
        end
    end
end

However, this method requires a slight modification to effectively handle the occlusion problem. Two voxels, taken from different layers, can be projected onto the same pixel in a given view. The voxel from the nearest layer occludes the other. To solve this problem, the method takes into account the fact that a voxel from a layer i cannot block a voxel from a layer j when $j > i$, as illustrated in Figure 8.3.

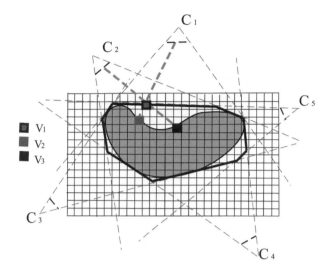

Figure 8.2. *Improvement of the visual hull by identifying concave zones*

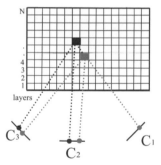

Figure 8.3. *The occlusion manipulation proposed by Seitz and Dyer [SEI 99]*

8.2.3.2. *Space carving*

The disadvantage of voxel coloring lies in its inability to completely reconstruct the object due to the arrangement of cameras in a semicircle. The space carving algorithm, introduced by Kutulakos *et al.* [KUT 00], can be seen as an extension of the previous method adapted to an arbitrary camera arrangement. This relies on sweep planes aligned with the three principal axes **x**, **y** and **z**. Only the cameras behind the sweep plane are used to manage the occlusion. For example, in Figure 8.4(a), the voxels in the plane highlighted in bold are visible via cameras C_1 and C_2. According to Kutulakos, a voxel is not visible by a camera if it is out of the view frustum or if it is occluded. We will consider a sweep plane in the positive direction from the axis x, the voxel v occludes the voxel w if $v_x < w_x$. As a result, v is evaluated before w in order to visit the blocking voxel.

Algorithm 8.2. The space carving algorithm

Input: the sequence of calibrated images
Output: a volume of voxels representing the object we want to model
Initialize the volume with the bounding box;
repeat
　for *each sweep plane in the 6 main directions* **do**
　　for *each voxel* v *in the current plane* **do**
　　　Project v onto the cameras in the sweep plane background;
　　　(v not out of view frustum, not occluded);
　　　if v *is neither visible nor photo-consistent* **then**
　　　　v is eliminated from the volume;
　　　end
　　end
　end
until *until there are no more voxels to eliminate*;

8.3. Industrial application

The industrial use of motion capture requires a real-time visualization of animations on shooting location. This allows directors to guide their actors accurately. In the case of crowded virtual scenery where an avatar can, for example, walk around a virtual table and chairs that are not on the real set, actors can have a video equivalent allowing them to better understand the environment they are supposed to be in, and adapt their performance accordingly. Today, industry is using multiview reconstruction filming sets, seeking to reiterate what has been done for motion capture, to provide the same facilities to teams using the next generation of images.

8.3.1. *Hardware acceleration*

Real-time visualization now seems possible, despite the significantly more complex algorithms required for animated volume reconstruction than for motion capture. This involves the multiplication of processing units and the use of new parallel calculation possibilities provided by heterogeneous processing systems, composed of central processing units (CPUs) and graphic processing units (GPU) for non-graphic purposes. These computation techniques form part of the "general-purpose processing on graphics processing units" (GPGPU) approach. In this field, Open Calculating Language (OpenCL) [KHR 11] is an example of an emerging technology, combining API[1] with a C-derived programming language. OpenCL, suggested as an open standard by the Khronos™ Group, is designed to program heterogeneous parallel systems and proposes a programming model including features from both CPUs and GPUs, the former being increasingly parallel with the latter being more and more programmable.

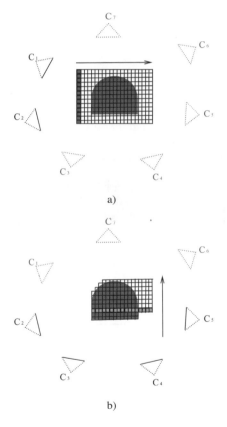

Figure 8.4. *Configuration of cameras for space carving*

---

1 Application programming interface.

8.3.2. *Results*

A complete pipeline of silhouette-based multiview reconstruction must be composed of several processes in order to improve the quality of the obtained result. The process chain may include the following steps that can all be implemented in the form of OpenCL "kernels" such as reconstruction, "marching cubes", refinement, relaxation and texturing.

The intensive use of hardware acceleration along the whole processing chain satisfies the real-time constraints for reasonable sizes of grids of voxels and a reasonable number of viewpoints. XD Productions has achieved real time (25 images per second) for voxel grids of 128^3, covered by 24 red, green and blue (RGB) high-definition cameras (1920×1080 pixels).

As explained in section 8.2, reconstruction involves subdividing the acquired 3D space into a multitude of voxels split into two categories: voxels inside the object to be reconstructed and voxels outside. The resolution of the subdivision, and therefore the number of voxels, directly influences the quality of the result. Figure 8.5 illustrates a model reconstructed according to three different resolutions.

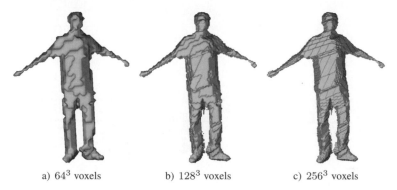

a) 64^3 voxels   b) 128^3 voxels   c) 256^3 voxels

Figure 8.5. *Models of voxel grids with three different solutions*
(© 2013 XD Productions)

The "Marching Cubes" algorithm deduces a set of triangles describing the surface of the reconstructed model from volumetric data, i.e. voxels. The advantage of a polygonal representation lies in its compatibility with the large majority of 3D modeling software. Figure 8.6(a) shows the triangles obtained on a plane similar to the model in Figure 8.5(c). As explained previously, voxels can be compared to cubes that are clearly visible in the generated models. This lack of quality is not suitable for the broadcast industry, it is thus necessary to refine the model.

"Refining" the model means adapting the position of triangles generated by the "marching cubes" algorithm in order to place them as closely as possible to the character's silhouette. This removes the step effect created by voxels as illustrated in Figure 8.6(b), compared with Figure 8.5(c). However, a slight aliasing is visible due to the used pixel-mask. To overcome this problem, the 3D model needs to be relaxed.

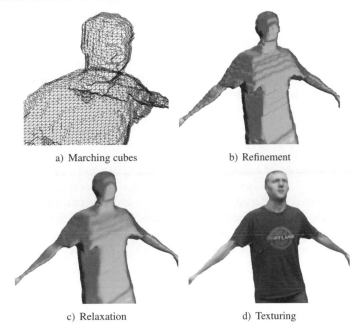

a) Marching cubes   b) Refinement

c) Relaxation   d) Texturing

Figure 8.6. *Stages in the graphic pipeline after reconstruction*
(© 2013 XD Productions)

The "relaxation" of the model corrects small defects by smoothing its surface. Figure 8.6(c) illustrates a model after relaxation. Once this stage has been completed, the model is ready to receive a texture.

"Texturing" involves calculating the texture attached to the reconstructed model. To achieve this, source images are projected onto the polygonal surface that connects the nearest pixels with the vertices of the model's triangles. By interpolation, the triangle is then filled with pixels present in the identified region of interest. Figure 8.6(d) illustrates the result of this last stage.

The reconstruction of a moving person is stored in the form of a sequence of triangular meshes (with the same frequency as the video source) and their associated textures. The use of such sequences in a traditional production "pipeline" is made difficult by the volume of data and their lack of temporal coherence. Triangular meshes have a temporally variable topology that may cause flickering in lighting and shading. Finally, this lack of temporal coherence makes impossible physics simulations involving body parts (for example speed, acceleration, points of inertia, collision volumes to simulate clothes). Solutions can therefore be proposed to structure what is commonly called a "polygon soup". This has a double objective. First, to provide logically organized mesh data to commercial production tools, in order to relight or redress them and to insert them into a controlled virtual universe. Second, it is also designed to produce coherent sequences qualifying the 3D scene, no longer statically but dynamically. Different stages in the solutions to these problems are examined in the following section.

8.4. Temporally structuring reconstructions

Since a sequence of meshes can take a variety of forms, Arcila [ARC 11] has proposed a formalism for describing the different types of mesh sequences, identifying the following categories: dynamic meshes, stable mesh sequences and unconstrained mesh sequences (see Figure 8.7). These distinctions are based on the existence of temporal coherence in meshes, at both a topological and structural level, as shown in Table 8.1.

Number of vertices	Connectivity	Topology	Name
Constant	Constant	Constant	Dynamic mesh
Variable	Variable	Constant	Stable mesh sequence
Variable	Variable	Variable	Unconstrained mesh sequence

Table 8.1. *Classification of mesh sequences*

Model-free reconstruction methods generally produce stable or unconstrained mesh sequences. The content of the scene is reconstructed in each frame individually. In these conditions, a geometric primitive (a vertex or a triangle) in a given frame does not have any correspondence in the following frame. Indeed, only dynamic meshes with temporal coherence can treat the sequence of meshes as a single animated object. As such, the challenge of temporally structuring multiview reconstructions lies in converting a sequence of meshes into a normalized representation when animating characters. Among the different character animation techniques, skeletal animation and vertex animation seem to be particularly well adapted to mesh sequences.

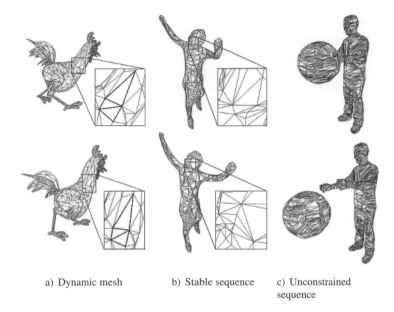

a) Dynamic mesh   b) Stable sequence   c) Unconstrained
                                                       sequence

Figure 8.7. *Different types of mesh sequences. a) © 1996 Microsoft Corporation; b) and c) source: GRImage INRIA Rhône-Alpes & 4DView Solutions, http://4drepository.inrialpes.fr*

"Skeletal animation" relies on identifying a hierarchized set (generally a tree) of articulations whose configuration is characterized by a rigid transformation (rotation and translation) in relation to their parent in the hierarchy. This skeleton guides the deformation of all the vertices in the mesh. To do so, a "skinning" method is necessary. This involves allocating to each vertex the influence weight of each bone in the skeleton. A bone is a segment linking two adjacent articulations in the tree. As a result, the movement of a vertex is obtained by averaging the movements of all bones in the skeleton, weighted by influence weights. These influence weights can be automatically calculated according to the distance from the vertex to each bone. As such, the "linear blend skinning" (LBS) system calculates the movement of each vertex using a linear interpolation of bone movement. A number of other "skinning" techniques have also been developed such as "skeletal subspace deformation" (SSD) and "multi-weight enveloping" (MWE). A comparison of these methods has been carried out by Jacka *et al.* [JAC 07]. Unfortunately, the movement obtained by this method is quasi-rigid (rotation and translation component), which means that significant movements with loose clothing, for example, cannot be reproduced.

However, "morph target animation" involves defining the character's movement using the trajectories of each of the vertices. This technique, equivalent to "morphing", is much more subtle than skeletal animation and can reproduce far more complex movements. Nevertheless, the animation produced can become unstable due to distortions created by interpolating key positions from the two consecutive frames. Above all, however, this method generates vast quantities of data (3D positions of mesh vertices at each frame), compared with skeletal animation.

8.4.1. *Generalized skeletal extraction*

The use of skeletons for animation is a modern procedure. It is therefore natural to represent mesh sequences taken from reconstructions in this form in order to make them easier to use within a traditional computer graphics chain. A skeletal model can be provided at the start. However, if we prefer a more generic approach, not restricted to the character's morphology, several automatic skeleton extraction techniques have been proposed, notably by Baran and Popovic [BAR 07]. However, it should also be noted that the reliability of this kind of approach is less than that of methods using *a priori* knowledge. Among the methods used to convert a sequence of animated skeleton meshes, the method proposed by De Aguiar *et al.* [DE 08] has provided convincing results. However, this technique cannot be applied to dynamic meshes.

Skeletal extraction generally involves identifying the different rigid components in a mesh. A rigid component is defined as a set of vertices whose movements are governed by the same rigid transformation. These rigid components define the bones in the skeleton. Searching for rigid components involves a segmentation operation that partitions the object into several subsets of vertices. Segmentation may be based on a convexity criterion. In the case of human morphology, bones are generally described by convex sets and articulations by concave sets.

While the automatic segmentation of static meshes is a problem that has been widely examined, the segmentation of sequences of meshes has been far less so, particularly for unconstrained sequences. In terms of dynamic meshes, some techniques analyze individual vertex trajectories during the sequence in order to regroup them into homogeneous motion "clusters". Due to their lack of temporal coherence, this approach is difficult to apply to unconstrained sequences. In the latter case, Arcila [ARC 11] has proposed segmenting each mesh independently and then moving the "clusters" from one frame to another.

8.4.2. *Calculating displacement fields*

Recently, a number of "morphing"-based approaches have led to the development of animation techniques without using skeletons. This change in paradigm has been accompanied by the appearance of new multiview reconstruction approaches using deformable meshes rather than models made up of rigid components. However, while these mesh-based approaches provide a greater degree of flexibility at the point of animation than skeletal-based algorithms, they produce more voluminous representations of dynamic scenes composed of sets of moving positions for each vertex.

To generate an animation of vertices using multiview reconstruction, equivalent to a dynamic mesh, it is necessary to calculate the individual trajectories of the reference vertices using the series of poses in the sequence. A number of articles examine motion extraction methods within multiview videos using traditional motion capture constraints. These include image-based methods that directly analyze videos and model-based approaches that apply poses taken from videos to an articulated human body in order to generate its movement. This latter approach, restricted to human morphology, is evidently less generic. Weinland *et al.* provide an overview of current research in this field in [WEI 11]. The extraction of a character's movement within a 3D scene is generalized by the notion of "scene flow", introduced by Vedula *et al.* [VED 05], of which an example is shown in Figure 8.8. It is a 3D version of the optical flow that describes movements identified in a series of images. Two types of methods are most commonly used: image-based techniques and object-based techniques.

Image-based methods calculate the scene flow using all the optical flows that can be extracted from the videos recorded by the cameras. The optical flow is the result of the projection of the scene flow toward the cameras, the method calculates the optical flow of sequences of images from each camera and then retroprojects the obtained vectors in the 3D scene space. Finally, a readjustment obtains the field of 3D vectors that constitute the scene flow.

Object-based methods (also known as "mesh tracking") establish a correspondence between the vertices in a sequence of meshes in order to follow the evolution of the object. This correspondence, based on criteria such as curvature or texture color, estimate each vertex's place within the following frame, and as such calculates its trajectory. A good example of this can be found in [PET 11] in which Petit *et al.* provide an examination of current research in this subject. Finally, the method proposed by Matsuyama *et al.* [MAT 04] uses sequences of discrete volumes (sets of binary voxels) taken from a silhouette-based reconstruction. Based on a pixel correspondence method for "morphing" images, the method is designed to match voxels from two consecutive frames.

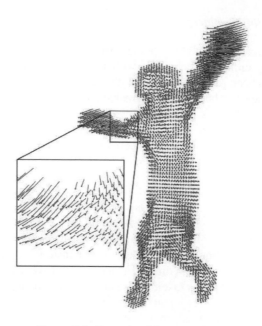

Figure 8.8. *Example of the scene flow*

8.5. Conclusion

Multiview reconstruction using several synchronized video sequences appears to be the future of the audiovisual industry, providing new perspectives in terms of creating hybrid content, combining live action and image synthesis. It is for this reason that a number of companies have attempted to create industrial solutions to this problem. In addition, within the context of a fragmented TV audience due to the increase in the number of channels and competition between new modes of reception (video on demand (VOD), Internet, etc.), providers and producers are, more than ever, focused on creating quality content, produced in optimal economic conditions. Multiview reconstruction can therefore be used to satisfy these strategic demands.

According to the research carried out by various companies such as 4DViews (Grenoble, France) and XD Productions (Paris, France), it is easy to imagine production studios soon being able to film from all axes using virtual cameras, with actors completely cloned within 3D sets, based on real or completely synthesized objects. We could also create images for a film or television program using virtual video. Technologically, multiview reconstruction will create new dedicated processing tools for the image industry through adjusting and hybridizing multiscopic video streams for reconstructive, relighting and composition purposes. These tools will provide:

– freedom and precision for scaling and movement of recording devices;

– an unlimited number of cameras;

– infinite possibilities for slow motion analysis, production and composition.

8.6. Bibliography

[ARC 11] ARCILA R., Séquences de maillages: classification et méthodes de segmentation, PhD Thesis, Université Claude Bernard - Lyon I, November 2011.

[BAR 07] BARAN I., POPOVIĆ J., "Automatic rigging and animation of 3D characters", *ACM SIGGRAPH 2007 Papers, SIGGRAPH '07*, ACM, New York, NY, 2007.

[CAI 06] CAILLETTE F., Real-time markerless 3D human body tracking, PhD Thesis, University of Manchester, 2006.

[DE 08] DE AGUIAR E., THEOBALT C., THRUN S., *et al.*, "Automatic conversion of mesh animations into skeleton-based animations", *Computer Graphics Forum*, vol. 27, pp. 389–397, 2008.

[JAC 07] JACKA D., REID A., MERRY B., *et al.*, "A comparison of linear skinning techniques for character animation", *Proceedings of the 5th international conference on Computer graphics, virtual reality, visualisation and interaction in Africa, AFRIGRAPH '07*, ACM, New York, NY, pp. 177–186, 2007.

[KHR 11] KHRONOS™ GROUP, "OpenCL – the open standard for parallel programming of heterogeneous systems", available at http://www.khronos.org/opencl/ 2011.

[KUT 00] KUTULAKOS K.N., SEITZ S.M., "A theory of shape by space carving", *International Journal of Computer Vision*, vol. 38, no. 3, pp. 199–218, July 2000.

[LAZ 07] LAZEBNIK S., FURUKAWA Y., PONCE J., "Projective visual hulls", *International Journal of Computer Vision*, vol. 74, no. 2, pp. 137–165, August 2007.

[MAT 04] MATSUYAMA T., WU X., TAKAIT T., *et al.*, "Real-time 3D shape reconstruction, dynamic 3D mesh deformation, and high fidelity visualization for 3D video", *Computer Vision and Image Understanding*, vol. 96, no. 3, pp. 393–434, 2004.

[PET 11] PETIT B., LETOUZEY A., BOYER E., "Flot de surface à partir d'indices visuels", *ORASIS–Congrès des jeunes chercheurs en vision par ordinateur*, INRIA Grenoble Rhône-Alpes, Praz-sur-Arly, France, 2011.

[SEI 99] SEITZ S.M., DYER C.R., "Photorealistic scene reconstruction by voxel coloring", *International Journal of Computer Vision*, vol. 35, no. 2, pp. 151–173, 1999.

[SNO 00] SNOW D., VIOLA P., ZABIH R., "Exact voxel occupancy with graph cuts", *Proceedings of IEEE Conference on Computer Vision and Pattern Recognition, 2000*, vol. 1, pp. 345–352, 2000.

[STA 99] STAUFFER C., GRIMSON W., "Adaptive background mixture models for real-time tracking", *IEEE Computer Society Conference on Computer Vision and Pattern Recognition, 1999*, vol. 2, pp. 637–663, 1999.

[SZE 93] SZELISKI R., "Rapid octree construction from image sequences", *CVGIP: Image Understanding*, vol. 58, no. 1, pp. 23–32, July 1993.

[VED 05] VEDULA S., BAKER S., RANDER P., *et al.*, "Three-dimensional scene flow", *IEEE Transactions on Pattern Analysis and Machine Intelligence*, vol. 27, no. 3, pp. 475–480, March 2005.

[WEI 11] WEINLAND D., RONFARD R., BOYER E., "A survey of vision-based methods for action representation, segmentation and recognition", *Computer Vision and Image Understanding*, vol. 115, no. 2, pp. 224–241, 2011.

[ZIV 04] ZIVKOVIC Z., "Improved adaptive Gaussian mixture model for background subtraction", *Proceedings of the 17th International Conference on Pattern Recognition, 2004, ICPR 2004*, vol. 2, pp. 28–31, August 2004.

Chapter 9

Synthesizing Intermediary Viewpoints

9.1. Introduction

Virtual or intermediary viewpoint synthesis relates to creating a different perspective from that in the acquisition bank. A number of applications such as robotized navigation, object recognition or even free navigation, more commonly known as "free-viewpoint navigation", require the use of a virtual view point. In this chapter, we focus on the synthesis of this viewpoint using depth information, a method known as "depth-image-based rendering", (DIBR).

In this chapter, we begin with a general overview of the basic requirements for virtual viewpoint synthesis. In particular, we examine viewpoint synthesis with interpolation and extrapolation. Section 9.3 focuses on a specific problem in viewpoint synthesis, completing discovered zones. Finally, section 9.4 concludes the chapter.

9.2. Viewpoint synthesis by interpolation and extrapolation

In this section, we examine the general principal underlying DIBR viewpoint synthesis, specifically by means of direct projection and inverse projection. The typical distortions and the methods used in viewpoint

---

Chapter written by Luce MORIN, Olivier LE MEUR, Christine GUILLEMOT, Vincent JANTET and Josselin GAUTIER.

synthesis which allow us to correct or avoid these distortions will be examined in section 9.2.2. We examine viewpoint interpolation and extrapolation, particularly texture fusion methods in section 9.2.3. View synthesis by DIBR in the context of 3DTV has been well studied in the European project ATTEST [FEH 04]. For a more general overview, the book by Schreer *et al.* provides a classification of viewpoint synthesis methods, with or without depth information [SCH 05].

9.2.1. *Direct and inverse projections*

9.2.1.1. *Direct projection equations*

Let us consider an image I with an associated depth map Z. We want to synthesize a new image I_v which corresponds to a new viewpoint. This process, known as viewpoint synthesis or viewpoint transfer, allows us to create an image known as a virtual view because it does not correspond to a real acquisition. The simplest method is that of direct projection. For each pixel in the image I, the coordinates of the 3D point associated with this pixel are estimated and then projected in the virtual viewpoint. $\mathbf{K}, \mathbf{R}, \mathbf{T}$ are the intrinsic and extrinsic parameters of the camera associated with the original viewpoint and $\mathbf{K}_v, \mathbf{R}_v, \mathbf{T}_v$ are the parameters which define the virtual viewpoint. The depth map $Z(p)$ provides the coordinate $Z - c$ of each pixel p in the camera's coordinate system.

With these definitions, the projection equations are written as follows:

$$\mathbf{p}^h \sim \mathbf{K}\,[\mathbf{R}|\mathbf{T}]\,\mathbf{M}^h \qquad\qquad [9.1]$$

where $\mathbf{M}^h = (x_o, y_o, z_o, 1)^T$ indicates the homogeneous coordinates of a 3D point M in the absolute coordinate system and $\mathbf{p}^h = (u, v, 1)^T$ is the homogeneous coordinates of its projection p in the image's coordinate system. By expressing the Cartesian coordinates $\mathbf{M}_c = (x_c, y_c, z_c)$ of the 3D point in the camera coordinate system in relation to $\mathbf{p}^h$ and $\mathbf{M}^h$, we obtain the equality:

$$z_c\mathbf{K}^{-1}\mathbf{p}^h = \mathbf{R}(\mathbf{M} - \mathbf{C}) \qquad\qquad [9.2]$$

where $\mathbf{M} = (x_o, y_o, z_o)^T$ indicates the Cartesian coordinates of the 3D point M in the absolute coordinate system and where $\mathbf{C} = -\mathbf{R}^T\mathbf{T}$ is the position of the camera in this absolute coordinate system. We then deduce the reconstruction equations which give the Cartesian coordinates of the 3D point:

$$\mathbf{M} = z_c \mathbf{R}^T \mathbf{K}^{-1} \mathbf{p}^h + \mathbf{C} = z_c \mathbf{R}^T \mathbf{K}^{-1} \mathbf{p}^h - \mathbf{R}^T \mathbf{T} \qquad [9.3]$$

The 3D point M is then projected onto the virtual viewpoint I_v according to equation [9.1] which is applied to the virtual viewpoint and expressed in Cartesian coordinates:

$$\mathbf{p}_v^h \sim \mathbf{K}_v \mathbf{R}_v \mathbf{M} + \mathbf{K}_v \mathbf{T}_v$$

The final transfer equation is written as:

$$\mathbf{p}_v^h \sim \mathbf{K}_v \mathbf{R}_v (z_c \mathbf{R}^T \mathbf{K}^{-1} \mathbf{p}^h + \mathbf{C} - \mathbf{C}_v) = z_c \mathbf{K}_v \mathbf{R}_v \mathbf{R}^T \mathbf{K}^{-1} \mathbf{p}^h$$
$$-\mathbf{K}_v \mathbf{R}_v \mathbf{R}^T \mathbf{T} + \mathbf{K}_v \mathbf{T}_v \qquad [9.4]$$

By applying this equation to each of the points in an acquired image, a new viewpoint can be generated. To do so, the information required includes the original image and its associated depth map as well as the camera parameters associated with the original and virtual viewpoints. This general principal of direct projection is illustrated in Figure 9.1.

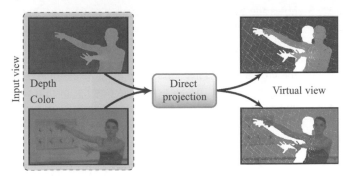

Figure 9.1. *Direct projection*

9.2.1.2. *Distortions associated with direct projection*

Direct projection generates typical distortions found in viewpoint synthesis, i.e. uncovered zones, cracks and a ghost effect, as illustrated in Figure 9.2. The uncovered zones correspond to areas which are not visible from the original viewpoint and which become visible in the virtual image due to parallax. Cracks are tiny versions of these uncovered zones, principally due to resampling. Ghost contours are caused by projecting pixels whose color is a mixture of the background and foreground colors. Other typical distortions in viewpoint synthesis are caused by errors in the depth map produced within the estimation or compression stages. An error in the depth value creates a displacement in the point projected in the synthesized viewpoint. Resulting geometric deformations are typically texture stretches, deformed edges or isolated points (a crumbling effect).

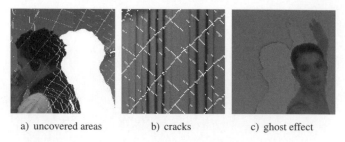

a) uncovered areas   b) cracks   c) ghost effect

Figure 9.2. *Distortions commonly associated with direct projection*

9.2.1.3. *Inverse projection*

Figure 9.3 shows the general outline of inverse projection. Each pixel in the virtual viewpoint is retro-projected in the original viewpoint where its color is determined by interpolation. Retro-projection applies the transfer equation [9.4] from the virtual view to the original view using a depth map associated with the virtual view. This depth map must be estimated in advance using direct projection. Inverse projection is generally preferred to direct projection because it prevents distortions caused by resampling [MOR 09, NDJ 11, TSU 09]. Resampling distortions (such as cracks) and uncovered zones are present in the depth map estimated by direct projection but, given the typical appearance of a depth map (lack of texture, large uniform zones and soft variations), it is better to avoid and correct these distortions in the depth maps than to correct them in the synthesized viewpoints themselves.

9.2.2. *Reducing distortions in viewpoint synthesis*

Several methods have been proposed to prevent or correct synthesis distortions.

9.2.2.1. *Cracks*

Post-production can be used to suppress cracks while preserving contours. Nguyen *et al.* have proposed a bilateral color-depth filter to improve the estimated depth map by direct projection [NGU 09]. "Splatting" methods [RUS 00, PFI 00, ZWI 02] eliminate cracks. To reduce the algorithmic complexity in viewpoint synthesis, Zinger *et al.* [ZIN 10] have proposed applying inverse projection only to pixels labeled as cracks, i.e. those significantly modified by filtering. These cracks can also be treated via an evenly spaced ordering of projected pixels which will be described in section 9.2.2.3.

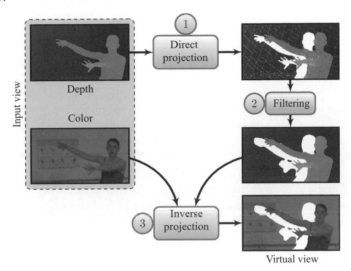

Figure 9.3. *Inverse projection*

9.2.2.2. *Ghost contours*

Ghost contours are prevented by detecting discontinuities in depth (or overlap contours) in the depth map. The pixels near to these edges are projected during a secondary phase but only if they are visible, i.e. if their depth is inferior to that in the pixels projected previously [ZIT 04]. Müller *et al.* [MÜL 08] have proposed dividing the edge pixels into two classes: foreground edges and background edges. The pixels in this latter category are only projected if they fill empty zones. Ghost effects can also be avoided by evaluating the contribution of the background and foreground at the discontinuity edges using advanced background and foreground matting separation methods [CHU 02, HAS 06, WAN 07].

9.2.2.3. *Uncovered areas*

Uncovered areas are caused by unavailable information in the original image. Part of the uncovered area can be completed by information from other input viewpoints, if available, as is the case in viewpoint interpolation, examined in section 9.2.3. When information about the uncovered area is not present in any acquired image, a filling method is required. A low-pass depth image filter can be used to preprocess the image around the overlapping areas in order to limit gaps. Chen *et al.* [CHE 05] have also proposed a correction of vertical contours next to the edges in the depth map. These pretreatments are particularly well suited to small-scale uncovered areas. Large uncovered areas must be completed using texture "inpainting" methods which will be described in section 9.3.

However, in direct projection it is possible to propagate background values in the uncovered and cracked zones using a specific pixel order. The order proposed by McMillan [MCM 95] is based on epipolar geometry. This order applies the painting algorithm, i.e. it synthesizes the image first for the furthest points and then nearest points, with each new value potentially replacing a previously estimated value. As such, the uncovered areas are correctly treated with the visible point being the last to be examined. Uncovered zones and cracks can also be detected and treated using this ordering: if two successive points p and q in the path are projected onto the non-adjacent points p' and q', it is therefore a crack or an uncovered zone. The path order ensures that the first point p' is in the background in relation to q' and its value can be immediately propagated in the interval $[p', q']$. Jantet's [JAN 12] Joint Projection Filling (JPF) algorithm distinguishes between cracks, where interpolation is carried out, and large uncovered zones where only the background value is propagated:

$$m = \begin{cases} (1-\alpha)p' + \alpha q' & if \ d \le K \\ q' & if \ d > K \end{cases} \quad with \quad \begin{cases} d = q'_x - p'_x \\ \alpha = \frac{1}{d}(m_x - p'_x) \end{cases} \quad [9.5]$$

This method, which propagates or interpolates the background values at the point of projection, is well adapted for generating depth maps in direct projection, as shown in Figure 9.4.

9.2.3. *Viewpoint interpolation*

Viewpoint interpolation, introduced by Manning and Dyer [MAN 96], relates to synthesizing an intermediary viewpoint using several pieces of input data. Here, we consider input as view + depth (i.e. MVD, multi-view plus depth) data. The standard approach involves projecting two or three input

views nearest to the virtual point and then fusing the obtained virtual
viewpoints.

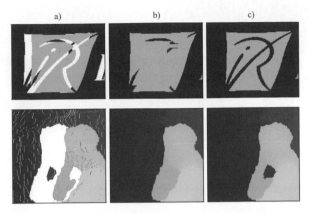

Figure 9.4. *Depth map obtained by direct projection using different methods:*
a) point-based basic direct projection; b) median filtering and directional
"inpainting" [NGU 09]; and c) JPF with a path order[JAN 12]

9.2.3.1. *Fusing virtual viewpoints*

The simplest fusion method is a linear combination, associated with a
binary visibility map [CAR 03] or soft, i.e. non-binary visibility
map [EIS 08]. The weighting factor is obtained using either the distance of
input viewpoints from the virtual viewpoints or the relative angle of view
between input viewpoints and the virtual viewpoints [BUE 01, DEB 96], or
on the disparity using a different weighting for each point [DEV 11, TAN 08].

9.2.3.2. *Detecting and smoothing distortions during interpolation*

To limit the impact of geometric deformations due to errors in the depth
map, several approaches have been proposed. Devernay *et al.* [DEV 11] have
detected geometric distortions by comparing texture indices (intensity,
gradient and Laplacian) at the point considered between the synthesized
viewpoint and one of the original images. The idea is that the synthesized
viewpoint must have the same kind of texture as the original viewpoints. A
confidence map is then created and an anisotropic smoothing is applied to the
low confidence zones, corresponding to strong differences in textures indices.
The results obtained by this method are shown in Figure 9.5.

Figure 9.5. *Detecting and suppressing synthesis distortions using anisotropic smoothing [DEV 11]*

9.2.3.3. *Floating textures*

Depth errors and imprecision in the camera models and parameters cause geometric shifts during viewpoint synthesis. The linear combination of slightly offset textures produces a blurred effect. To avoid this, Eisemann *et al.* [EIS 08] have proposed the principle of floating textures. Synthesized viewpoints are readjusted, prior to linear combination, by estimation and dense motion compensation. A version which provides a quicker result applies an estimation/motion compensation to the original viewpoints and interpolates the motion fields obtained for the virtual viewpoint at the point of viewpoint synthesis.

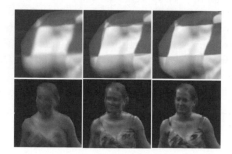

Figure 9.6. *Removing blurring when fusing textures using floating textures [EIS 08]*

9.2.3.4. *Viewpoint extrapolation*

Viewpoint extrapolation is the synthesis of a viewpoint outside the acquired viewpoint interval. When we only have a single input viewpoint, viewpoint synthesis is necessarily an extrapolation. This is also the case when we want to increase the interocular distance between images in a stereoscopic pair.

The main difficulty is therefore filling the uncovered zones because no original information is available. The methods cited previously are sufficient for small uncovered zones. When uncovered zones are larger, more complicated inpainting methods must be used to obtain a realistic effect.

9.3. Inpainting uncovered zones

As examined previously, the projection of a real viewpoint, also known as the reference viewpoint in another virtual camera referential, uncovers certain areas in the background. When the distance between the reference viewpoint and the virtual viewpoint becomes large, the uncovered zone increases in size and therefore becomes more difficult to fill. This problem is associated with both types of viewpoint synthesis methods, both interpolation and extrapolation methods. In order fill the background area, inpainting techniques are used. Before examining these methods, a few notations are necessary. Let us take an image I, defined by:

$$I : \begin{vmatrix} \Omega^2 \subset \mathcal{R}^2 \to \mathcal{R}^m \\ \mathbf{p} \to I(\mathbf{p}) \end{vmatrix} \qquad [9.6]$$

For color images, each pixel $\mathbf{p}$ is defined on three color planes ($m = 3$). The problem of inpainting in a viewpoint synthesis context can be formalized as follows: the image definition domain after projection is made up of two parts $\Omega^2 = S \cup U$, with S being the known part (or source) of I and the unknown part U representing the uncovered zones. It is this part which we want to fill. In this section, we first provide a brief overview of 2D inpainting techniques. We then examine several extensions of 2D inpainting to 3D contexts.

9.3.1. *Overview of 2D inpainting techniques*

This section is not designed to be an exhaustive examination of current work in this field but to provide a general overview of the two main families of inpainting methods: diffusion-based methods and those based on searching for similarities using block (or patch) correspondences.

9.3.1.1. *Diffusion-based methods*

To fill the unknown zones in an image, one solution entails diffusing the information from the outside to the inside of the missing area. This type of approach, proposed by Bertalmio *et al.* [BER 00], uses the physical phenomenon of heat diffusion. The heat spreads until it reaches an equilibrium, i.e. a uniform distribution of temperature. By applying this phenomenon to reconstructing a missing zone in an image, it is possible to gradually spread the color information throughout the image. The diffusion procedure involves solving a partial derivative equation from the heat equation using an initial condition and Neumann limit conditions:

$$\frac{\partial I}{\partial t}(\mathbf{p}, t) = div(f(\nabla I(\mathbf{p}))) \tag{9.7}$$

with $div()$ being the divergence operators and f being a decreasing function of the gradient at the point $\mathbf{p}$ in the image I, noted as $\nabla I(\mathbf{p})$. This function affects the spread which can, for example, be linear isotropic, linear anisotropic or even nonlinear anisotropic.

These diffusion-based inpainting methods obtain very good results for filling uniform zones. However, they are not well adapted to textured images.

9.3.1.2. *Similarity search based methods*

More recent inpainting methods based on similarity searches can be summarized as five stages applied iteratively as follows:

– calculate a filling priority;

– identify a block with the strongest priority;

– search for similar blocks in the image;

– complete the local copy from the exterior towards the interior;

– update the filling priority and repeat stage two.

Before discussing the mathematical purpose and details of these stages, a number of notations are required. A texture block centered on the point $\mathbf{p}$ is noted as $\psi(\mathbf{p})$. It is potentially composed of one part belonging to S and an unknown part belonging to U. Formally, the texture block can be written as: $\psi = \psi^S \cup \psi^U$. The "filling property" p proposed in a precursor article [CRI 04] is designed to hierarchize filling to begin with the structure in the scene. Criminisi *et al.* have defined the priority $P(\mathbf{p})$ at the point $\mathbf{p}$ as a product of two terms: a confidence term $C(\mathbf{p})$ and a data term $D(\mathbf{p})$. These terms are given by:

$$C(\mathbf{p}) = \frac{\sum_{q\in\psi_p^S} C(\mathbf{q})}{|\psi_p|} \qquad\qquad [9.8]$$

$$D(\mathbf{p}) = \frac{|\nabla I(\mathbf{p})^\perp \cdot \mathbf{n}_p|}{\alpha} \qquad\qquad [9.9]$$

with $|\psi(\mathbf{p})|$ in the patch area $\psi(\mathbf{p})$, with α being a normalization factor guaranteeing a dynamic between 0 and 1, $\mathbf{n}_p$ is the orthogonal unitary vector along the edge $\delta\Omega$ at the point $\mathbf{p}$ and $\nabla I(\mathbf{p})$ is the gradient vector at the point $\mathbf{p}$. Figure 9.7 shows the notations used in the previous formulas.

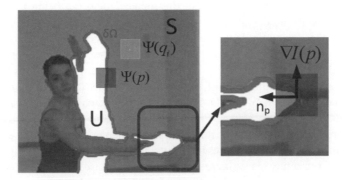

Figure 9.7. *Notations used for 2D inpainting 2D*

The term $C(\mathbf{p})$, defined recursively (the term $C(\mathbf{p})$ being initialized at 1 in the source region S), gives greater importance to the pixels $\mathbf{p}$ which are surrounded by known pixels. This term is the ratio between the numbers of known pixels compared to the total number of pixels in the block. The term $D(\mathbf{p})$ attributes a strong importance to structured zones with an orthogonally oriented isophote along the edge $\partial\Omega$ (i.e. the boundary between S and U). The "similar block search" determines the $\mathcal{W}$ most similar block in a local neighborhood, noted as $\psi^*(\mathbf{p})$, corresponding to the known part in the block $\psi^S(\mathbf{p})$:

$$\psi^*(\mathbf{p}) = \arg\min_{\mathbf{q}\in\mathcal{W}} d(\psi^S(\mathbf{p}), \psi^S(\mathbf{q})) \qquad\qquad [9.10]$$

with $d(\cdot)$ being the similarity metric. Often, we use the mean square error (MSE) calculated between the known pixels of the block considered and the candidate block. "Filling the unknown zone" of the block $\psi^U(\mathbf{p})$ is achieved by recopying the co-localized pixels of the most similar block in the block $\psi^U(\mathbf{p})$.

There are a number of variations for these stages. For the filling priority, Le Meur *et al.* [LEM 11] have used a structure tensor to analyze the local geometry along the boundary line. For each point $\mathbf{p}$, the structure tensor $T_\sigma(\mathbf{p})$, spatially regularized by a Gaussian filter, is calculated as follows:

$$T_\sigma(\mathbf{p}) = \left(\sum_{i=1}^{m} \nabla I_i(\mathbf{p}) \times \nabla I_i(\mathbf{p})^T \right) * G_\sigma \qquad [9.11]$$

with I_i being the component i^{nth} of the image I, and G_σ being a Gaussian filter with a standard deviation of σ. The $\times$ operator is a matrix product and $*$ represents the convolution operator. Analysis of the eigenvalues in the structure tensor $\lambda_\pm$ determines the type of geometric structure. The orthogonal basis composed of the eigenvectors $\theta_\pm$ indicates the dominant orientations with θ_- being oriented in the direction of the isophote (parallel to the contour). When λ_+ is much greater than λ_-, the local geometric structure appears similar to a contour. This is illustrated in Figure 9.8 in the form of an ellipsis oriented in the direction of the contour (θ_-). A homogeneous zone is represented by highly similar values. It is represented by a circle, as illustrated in Figure 9.8. The use of a structure tensor increases resilience to noise and locates important structures in the scene better. In reference to work by [WEI 99], the term $D(\mathbf{p})$ is written as:

$$D(\mathbf{p}) = \alpha + (1 - \alpha) \exp\left(-\eta/(\lambda_+ - \lambda_-)^2 \right) \qquad [9.12]$$

with η being a positive constant value and α being a value between 0 and 1. When there is a strong structure ($\lambda_+ \gg \lambda_-$), the term $D(\mathbf{p})$ tends toward $1 - \alpha$. If not, the term $D(\mathbf{p})$ tends toward α. By choosing a low *alpha* value, a high value of the term $D(\mathbf{p})$ indicates the presence of strongly structured zones. Another recent variation has been proposed by Xu *et al.* [XU 10]. In contrast to previous studies, Xu *et al.* [XU 10] have not used gradients to determine priority but the singular nature of the zone to be filled. The data term $D(\mathbf{p})$ is calculated by measuring the redundancy of the known part in the block to be filled with its local neighborhood $\mathcal{V}$. The term is defined by:

$$D(\mathbf{p}) = \|\mathbf{w}\|_2 \times \sqrt{|N_s(\mathbf{p})|/|N(\mathbf{p})|} \qquad [9.13]$$

with $\| \cdot \|_2$ being the norm L_2 and the vector $\mathbf{w}$ indicating the degrees of similarity between the known part of the block to be filled $\psi^S(\mathbf{p})$ and the potential candidates $\psi^S(\mathbf{i})$, with $\mathbf{i} \in \mathcal{V}$. The similarity coefficient between the known part of the current block and a candidate block located at the point $\mathbf{i}$ is denoted by w_i and is given by: $w_i = \exp\left(-d\left(\psi^S(\mathbf{p}), \psi^S(\mathbf{i})\right) / (2\sigma^2)\right)$, with $\sum_i w_i = 1$ and the term σ modulating the exponential decrease. To illustrate this method, we suppose that the vector $\mathbf{w}$ has a size of 4. For a block situated in a uniform zone, the similarity coefficients w_i have a quasi-uniform distribution: $\mathbf{w} = \begin{bmatrix} \frac{1}{4} & \frac{1}{4} & \frac{1}{4} & \frac{1}{4} \end{bmatrix}$, resulting in $\|\mathbf{w}\|_2 = \frac{1}{2}$. However, if the block being filled shows a structure, the distribution of the coefficients w_i is no longer uniform. For example, $\mathbf{w} = \begin{bmatrix} \frac{1}{10} & \frac{1}{50} & \frac{4}{5} & \frac{2}{25} \end{bmatrix}$ results in $\|\mathbf{w}\|_2 = 0,6568$. The priority of this site is therefore superior to the previous site.

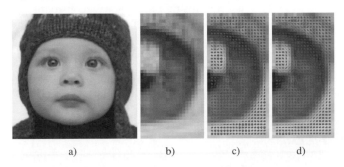

a)   b)   c)   d)

Figure 9.8. *Illustration of a structure tensor field: a) the original image; b) close-up of part of the eye; c) a non-regularized tensor field; and d) a tensor field regularized by Gaussian smoothing*

The second term of equation [9.13] weights the result according to the number of valid blocks N_s (all the pixels in the block are known) and the total number of blocks N in the neighborhood $\mathcal{V}$.

In relation to the nearest neighbor search (stage 3 of inpainting, a variation of the algorithms based on a similarity search using block correspondences involves determining not only the best candidate but also the N best first candidates. These candidates are therefore combined linearly to obtain the value of the pixels in the unknown zone from the block to be filled $\psi^U(\mathbf{p})$:

$$\psi^U(\mathbf{p}) = \sum_{i=1}^{K} \alpha_i \times \psi^U(\mathbf{q}_i) \tag{9.14}$$

with α_i being the linear combination coefficients (respecting the constraint $\sum_{i=1}^{K} \alpha_i = 1$) and $\psi(\mathbf{q}_i)$ the i^{nth} block centered on the point $\mathbf{q}_i$ being the most similar to the known zone of the current block centered on $\mathbf{p}$. Here, the difficulty lies in determining the coefficients of the linear combination. There are several methods for doing this:

– calculate the coefficients using a "non-local means" approach [BUA 05, WON 06];

– calculate the coefficients by least squares optimization under constraint or not. When the sum of coefficients in the linear combination is restricted to being equal to 1, the problem is written as the "locally linear embedded" type approach [ROW 00]. However, the previous constraint does not impose coefficient positivity which can be surprising, given that the input data are either positive or null. To account for this point, a new constraint can be used by forcing the coefficients to be positive. It is therefore a "non-negative matrix factorization" problem [LIN 07].

When the coefficients of the linear combination are determined, the unknown part of the block is synthesized via equation [9.14].

9.3.2. *3D Inpainting*

9.3.2.1. *An extension of Criminisi et al.'s approach to 3D contexts*

The use of 2D inpainting algorithms in a 3D context is not necessarily easy. This is because the uncovered areas are most of the time close or connected to foreground areas. The aim of inpainting is to reconstruct these zones using information the background. This is a fundamental constraint since only background zones should be used when inpainting. Additionally, the foreground should not be propagated in the uncovered zone. Figure 9.9 illustrates this problem. The zones uncovered by projection (white areas in Figure 9.9(a)) are filled using Criminisi *et al.*'s method [CRI 04]. We can see that the background zones are recopied into the background, providing poor visual quality.

To solve such problems, Daribo and Pesquet-Popescu [DAR 10] suppose that the depth map is known and apply two modifications to the Criminisi algorithm. The first relates to calculating the priority $P(\mathbf{p})$. In addition to the terms $C(\mathbf{p})$ and $D(\mathbf{p})$, Daribo and Pesquet-Popescu have integrated a term $L(\mathbf{p})$ which is depth dependent:

$$L(\mathbf{p}) = \frac{|Z(\mathbf{p})|}{|Z(\mathbf{p})| + \sum_{r \in \psi^s(\mathbf{p})} (Z(\mathbf{r}) - Z(\mathbf{p}))^2} \qquad [9.15]$$

with $Z(\mathbf{p})$ being a block in the depth map z centered on the point $\mathbf{p}$. $L(\mathbf{p})$ is equal to 1 when the candidate is selected at the same depth as the block to be filled.

The second modification concerns the best candidate search which includes a depth term (with the notations of algorithm [9.10]) and a factor β in order to adjust the contribution of the second term as given below:

$$\psi^*(\mathbf{p}) = \arg \min_{\psi(\mathbf{q}) \in \mathcal{W}} \left[d(\psi^S(\mathbf{p}), \psi^S(\mathbf{q})) + \beta \times d(Z^S(\mathbf{p}), Z^S(\mathbf{q})) \right] [9.16]$$

These two modifications restrict the filling procedure but do not necessarily remove the possibility of taking background blocks. The results of this approach are illustrated in Figure 9.9.

a)   b)   c)   d)

Figure 9.9. *Examples of "inpainting": a) original viewpoint projected in two camera referentials, uncovered zones are shown in white; b) examples of inpainting by Criminisi et al. [CRI 04] c), Daribo and Pesquet-Popescu [DAR 10], and Gautier d) et al. [GAU 11]*

In 2011, Gautier *et al.* [GAU 11] extended the 2D approach by [LEM 11] to a 3D context. First, the term priority, which uses the structure tensor from the [LEM 11] method described in the previous section, is modified by adding the notion of depth. It is completed by a directional term which annuls the priority on part of the boundary line. The idea here is to consider the movement of the camera during viewpoint synthesis. A virtual viewpoint positioned to the right of the reference viewpoint requires a movement of the camera to the right. In this context, uncovered zones appear on the right of the foreground object. As a result, the order of filling must move from right to left (an illustration is shown in Figure 9.9). For a left camera movement, the order of filling is from left to right. This simple directional term effectively restricts filling in the uncovered zone. Thanks to this "trick" the foreground object can no longer be propagated in the background. These two modifications of the term priority and the candidate block search using depth, similarly carried out by [DAR 10], result in an improvement in quality, as shown in Figure 9.9. Note, however, that these two algorithms rely on the assumption that we know the depth map for the viewpoint in which the viewpoint references are projected.

9.3.2.2. *Global optimization inpainting*

Sun *et al.* [SUN 12] have filled uncovered zones with texture and depth while minimizing the overall energy defined in the unknown region U. The idea here is to determine the block belonging to the known region S which minimizes quadratic error in terms of texture E_p^{Text} and depth E_p^{Prof}. It is defined by:

$$E_p(I, Z, C|I_o, Z_o) = E_p^{Text}(I, C|I_o) + \lambda E_p^{Prof}(Z, C|Z_o) \qquad [9.17]$$

with I_o and Z_o being the reference viewpoint and its depth map, respectively. I and Z represent the synthesized viewpoint and its depth map. C represents the coordinates of the point where the energy is lowest. The first term E_p^{Text} represents the energy when the block centered on the point **p** is paired with a block in the known region S. The second represents the pairing of the block in terms of depth.

To infer the texture and depth of blocks in uncovered zones, Sun *et al.* used iterative Gauss–Seidel optimization by presupposing that either the depth or the texture of the block to be inpainted is successively known. Figure 9.10 shows the results of inpainting using this method, showing that the small uncovered areas are reconstructed well.

a)   b)   a)   b)

Figure 9.10. *Examples of inpainting by [SUN 12]: a) and c) virtual viewpoints with inpainting; b) and d) result (taken from [SUN 12])*

9.4. Conclusion

In this chapter, we have examined the problem of DIBR viewpoint synthesis using image plus depth information. A number of approaches have been proposed to avoid or limit problem distortions generated in the obtained virtual viewpoint, with the most difficult being the treatment of uncovered zones. The strategy which appears the most effective in this case is the combined used of information related to texture and depth. The evaluation of the quality of synthesized viewpoints remains an ongoing problem; on the one hand, we do not generally have a reference viewpoint and, on the other, existing quality measures (PSNR, SSIM, etc.) are not well suited to measuring the visual impact of geometric distortions and evaluating the visual quality of filling in an uncovered zone. However, the use of viewpoint synthesis to control view synthesis optimization quantification parameters has been recently proposed, normalizing the MVD data code, with a research group currently working on studying quality objective metrics in synthesizes viewpoints. If these studies are successful, they will open up new horizons for the use of viewpoint synthesis techniques to predict coding schemas in images and video.

9.5. Bibliography

[BER 00] BERTALMIO M., SAPIRO G., CASELLES V., *et al.*, "Image inpainting", *Proceedings of SIGGRAPH 2000*, 2000.

[BUA 05] BUADES A., COLL B., MOREL J., "A non local algorithm for image denoising", *IEEE Computer Vision and Pattern Recognition (CVPR)*, vol. 2, pp. 60–65, 2005.

[BUE 01] BUEHLER C., BOSSE M., MCMILLAN L., *et al.*, "Unstructured lumigraph rendering", in FIUME E., (ed.), *SIGGRAPH 2001, Computer Graphics Proceedings*, ACM Press/ACM SIGGRAPH, pp. 425–432, 2001.

[CAR 03] CARRANZA J., THEOBALT C., MAGNOR M.A., *et al.*, "Free-viewpoint video of human actors", *ACM Transactions on Graphics*, vol. 22, no. 3, pp. 569–577, 2003.

[CHE 05] CHEN W.-Y., CHANG Y.-L., LIN S.-F., *et al.*, "Efficient depth image based rendering with edge dependent depth filter and interpolation", *ICME*, pp. 1314–1317, 2005.

[CHU 02] CHUANG Y.-Y., AGARWALA A., CURLESS B., *et al.*, "Video matting of complex scenes", *AACM Transactions on Graphics*, vol. 21, no. 3, pp. 243–248, 2002.

[CRI 04] CRIMINISI A., PÉREZ P., TOYAMA K., "Region filling and object removal by examplar-based image inpainting", *IEEE Transactions on Image Processing*, vol. 13, pp. 1200–1212, 2004.

[DAR 10] DARIBO I., PESQUET-POPESCU B., "Depth-aided image inpainting for novel view synthesis", *IEEE International Workshop on Multimedia Signal Processing*, 2010.

[DEB 96] DEBEVEC P.E., TAYLOR C.J., MALIK J., "Modeling and rendering architecture from photographs: a hybrid geometry- and image-based approach", *SIGGRAPH '96: Proceedings of the 23rd Annual Conference on Computer Graphics and Interactive Techniques* ACM, New York, pp. 11–20, 1996.

[DEV 11] DEVERNAY F., DUCHÊNE S., RAMOS-PEON A., "Adapting stereoscopic movies to the viewing conditions using depth-preserving and artifact-free novel view synthesis", in ANDREW J., WOODS N.S.H., DODGSON N.A., (eds), *Proceedings of SPIE-IS&T Electronic Imaging, SDA 2011 – Stereoscopic Displays and Applications XXII*, vol. 7863, SPIE and IST, San Francisco, CA, pp. 786302–786302, 12 March 2011.

[EIS 08] EISEMANN M., DECKER B.D., SELLENT A., *et al.*, "Floating textures", *Computer Graphics Forum (Proceedings of the Eurographics EG'08)*, vol. 27, no. 2, pp. 409–418, 2008.

[FEH 04] FEHN C., "Depth-image-based rendering (DIBR), compression, and transmission for a new approach on 3D-TV", *SPIE 5291, Stereoscopic Displays and Virtual Reality Systems XI*, pp. 93–104, 2004.

[GAU 11] GAUTIER J., LE MEUR O., GUILLEMOT C., "Depth-based image completion for view synthesis", *3DTV Conference*, 2011.

[HAS 06] HASINOFF S.W., KANG S.B., SZELISKI R., "Boundary matting for view synthesis", *Computer Vision and Image Understanding*, vol. 103, no. 1, pp. 22–32, 2006.

[JAN 12] JANTET V., Layered depth images for multi-view video coding, PhD Thesis, University of Rennes, 1 November 2012.

[LEM 11] LE MEUR O., GAUTIER J., GUILLEMOT C., "Examplar-based inpainting based on local geometry", *ICIP*, pp. 3401–3404, 2011.

[LIN 07] LIN C., "Projected gradient methods for non-negative matrix factorization", *Neural Computation*, vol. 19, pp. 2756–2779, 2007.

[MAN 96] MANNING R.A., DYER C.R., "Dynamic view morphing", *Proceedings of the SIGGRAPH 96*, pp. 21–30, 1996.

[MCM 95] MCMILLAN L., A list-priority rendering algorithm for redisplaying projected surfaces, Technical Report 95-005, University of North Carolina at Chapel Hill, NC, 1995.

[MOR 09] MORI Y., FUKUSHIMA N., YENDO T., *et al.*, "View generation with 3D warping using depth information for FTV", *Signal Processing: Image Communication*, vol. 24, nos. 1–2, pp. 65–72, 2009.

[MÜL 08] MÜLLER K., SMOLIC A., DIX K., *et al.*, "View synthesis for advanced 3D video systems", *Eurasip Journal on Image and Video Processing - EURASIP J Image Video Process*, vol. 2008, pp. 1–12, 2008.

[NDJ 11] NDJIKI-NYA P., KÖPPEL M., DOSHKOV D., *et al.*, "Depth image-based rendering with advanced texture synthesis for 3-D video", *IEEE Transactions on Multimedia*, vol. 13, no. 3, pp. 453–465, 2011.

[NGU 09] NGUYEN Q.H., DO M.N., PATEL S.J., "Depth image-based rendering with low resolution depth", *Image Processing (ICIP), 2009 16th IEEE International Conference on*, pp. 553–556, 2009.

[PFI 00] PFISTER H., ZWICKER M., VAN BAAR J., *et al.*, "Surfels: surface elements as rendering primitives", *SIGGRAPH '00: Proceedings of the 27th Annual Conference on Computer Graphics and Interactive Techniques*, ACM Press/Addison-Wesley Publishing Co., New York, pp. 335–342, 2000.

[ROW 00] ROWEIS S., SAUL L., "Nonlinear dimensionality reduction by locally linear embedding", *Science*, vol. 290, pp. 2323–2326, 2000.

[RUS 00] RUSINKIEWICZ S., LEVOY M., "QSplat: a multiresolution point rendering system for large meshes", *SIGGRAPH '00: Proceedings of the 27th Annual Conference on Computer Graphics and Interactive Techniques*, ACM Press/Addison-Wesley Publishing Co., New York, pp. 343–352, 2000.

[SCH 05] SCHREER O., KAUFF P., SIKORA T., *3D Videocommunication: Algorithms, Concepts and Real-time Systems in Human Centred communication*, John Wiley & Sons, 2005.

[SUN 12] SUN W., AU O.C., XU L., *et al.*, "Texture optimization for seamless view synthesis through energy minimization", *ACM Multimedia*, 2012.

[TAN 08] TANIMOTO M., FUJII T., SUZUKI K., *et al.*, *Reference Softwares for Depth Estimation and View Synthesis*, ISO/IEC JTC1/SC29/WG11MPEG2008/M15377, April 2008.

[TSU 09] TSUNG P.-K., LIN P.-C., DING L.-F., *et al.*, "Single iteration view interpolation for multiview video applications", *Proceedings of the 3DTV-Conference 2009, The True Vison, Capture, Transmission and Display of 3D Video*, Postdam, Germany pp. 1–4, May 2009.

[WAN 07] WANG J., COHEN M.F., "Image and video matting: a survey", *Foundations and Trends in Computer Graphics and Vision*, vol. 3, no. 2, pp. 97–175, 2007.

[WEI 99] WEICKERT J., "Coherence-enhancing diffusion filtering", *International Journal of Computer Vision*, vol. 32, pp. 111–127, 1999.

[WON 06] WONG A., ORCHARD J., "A nonlocal-means approach to examplar-based inpainting", *Proceedings of the IEEE International Conference Image Processing (ICIP)*, pp. 2600–2603, 2006.

[XU 10] XU Z., SUN J., "Image inpainting by patch propagation using patch sparsity", *IEEE Transactions on Image Processing*, vol. 19, no. 5, pp. 1153–1165, 2010.

[ZIN 10] ZINGER S., DO L., DE WITH P.H.N., "Free-viewpoint depth image based rendering", *Journal of Visual Communication and Image Representation*, vol. 21, nos. 5–6, pp. 533–541, 2010.

[ZIT 04] ZITNICK C., KANG S., UYTTENDAELE M., *et al.*, "High-quality video view interpolation using a layered representation", *ACM Transactions on Graphics*, vol. 23, no. 3, pp. 600–608, 2004.

[ZWI 02] ZWICKER M., PFISTER H., VAN BAAR J., *et al.*, "EWA splatting", *IEEE Transactions on Visualization and Computer Graphics*, vol. 8, no. 3, pp. 223–238, 2002.

PART 3

Standards and Compression of 3D Video

Chapter 10

Multiview Video Coding (MVC)

10.1. Introduction

Compression is today a fundamental part of digital communications. Different technological advances in screen devices, both in terms of their resolution (increasing use of ultra HD formats) and their refresh rate, have produced increasingly large volumes of data. This phenomenon is even more significant given the appearance of 3DTV which allows viewers to watch stereoscopic (two views) or multiview (N views) media.

To demonstrate the indispensable nature of compression within the context of multiview video (MVV), we will use a simple example. We will consider a video sequence of eight views in full HD resolution ($1,920 \times 1,080$), at 30 images per second with a duration of 5 min in which each pixel is coded on 24 bits. The memory required for this multiview sequence is therefore ($8 \times 1,920 \times 1,080 \times 30 \times 300 \times 24)/8 = 417.13$ Go with a rate of 11 Gbits/s. It is soon apparent that, without this crucial compression stage, disseminating and storing these kinds of sequences is almost completely impossible.

However, one of the fundamental characteristics of multiview media is the fact that there is a strong correlation between each view. This correlation is therefore used by compression schemes using 3D formats (for both stereoscopy and multiview) and specific coding techniques. This chapter is

Chapter written by Benjamin BATTIN, Philippe VAUTROT, Marco CAGNAZZO and Frédéric DUFAUX.

designed to present 3D formats and coding techniques for stereoscopic vision as well as for multiview examples.

10.2. Specific approaches to stereoscopy

10.2.1. *Formats*

10.2.1.1. *Frame-compatible formats*

Frame-compatible (FC) formats involve sub-sampling and multiplexing the images from left- and right-hand views into a single image or sequence of images [VET 10]. As a result, the new resulting stream has the same number of samples as a monoscopic video sequence. It can therefore be effectively encoded with a standard compression method such as H.264/MPEG-4 AVC [ITU 10].

With FC formats, multiplexing can be carried out spatially or temporally. With spatial multiplexing, left- and right-view images are first sub-sampled and then combined into a single image. The two views can, for example, be broken down vertically or horizontally and arranged in configurations side by side, as illustrated in Figure 10.1(a) and (b).

Equally, the data can be interleaved by columns, lines or diagonally, in line with the motifs in Figure 10.1(c), (d) and (e).

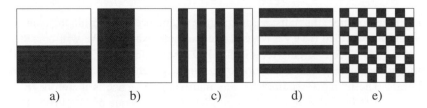

Figure 10.1. *FC formats side by side: a) top to bottom b) left-right; interleaved FC formats: c) by columns, d) by lines, and e) diagonally*

With temporal multiplexing, left- and right-hand images are sub-sampled and combined alternately into a single sequence, as shown in Figure 10.2.

In order to interpret and deinterleave samples, the FC formats require auxiliary information. As such, supplementary enhancement information (SEI) messages are normalized within the context of H.264/MPEG-4 AVC [ITU 10].

Figure 10.2. *Temporal FC formatting*

The inherent advantage of FC formats is their backward compatibility with distribution infrastructures as well as with current equipment. Deploying this kind of solution is therefore quick and easy. For this reason, these formats have been widely adopted for the first stereoscopic 3DTV services by reusing encoders, transmission channels, receivers and current decoders. Generally, side by side FC formats are the most frequently used due to their high visual quality after compression.

However, FC formats have two major disadvantages. First, spatial or temporal resolution is reduced which can result in a loss of quality even if the impact is partially limited by binocular fusion properties in human visual systems. Second, even if formats are backward compatible, current receivers are not yet able to correctly decode SEI messages and cannot therefore correctly interpret multiplexed data. This is a major obstacle for transmission given that equipment is difficult to upgrade.

10.2.1.2. *Mixed resolution stereo*

The mixed resolution stereo (MRS) format is based on the principle of binocular suppression in the human visual system. Symmetric representation, where one view has significantly reduced quality in relation to another, has a slight effect on overall perceived quality [STE 98, STE 00]. This property can be exploited using a low-pass filter on one of the views, a rough quantification or even sub-sampling.

The MRS format involves a spatial sub-sampling of one of the views from the stereoscopic pair. The horizontal and vertical resolutions can, for example, be divided into two in relation to the basic view (see Figure 10.3). As a result, the number of samples (and consequently the output rate) is strongly reduced.

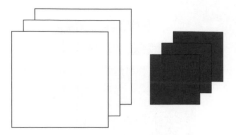

Figure 10.3. *Format MRS*

10.2.1.3. *2D-plus-depth*

The 2D-plus-depth format takes a single view (left or right) from the stereoscopic pair as well as a depth (or disparity, the two concepts being closely related), in line with Figure 10.4. A depth map is a gray scale image which shows the position of different objects relative to the scene in relation to the image plane. This can be obtained using specific devices (such as "time-of-flight" cameras or can be generated from the stereoscopic pair via depth estimation algorithms, such as those proposed by [NIQ 10]. During the reconstruction stage, the missing view can be reconstructed using depth-image-based rendering (DIBR) methods [FEH 02, FEH 04, SMO 08].

a) 2D view   b) Associated depth map

Figure 10.4. *The 2D-plus-depth format*

Use of the 2D-plus-depth format significantly reduces the volume of data. It is generally estimated that the rate associated with depth or disparity information generally represents 10–20% of the total budget [EKM 08, MAR 06]. This format also provides backward compatibility in relation to display for standard 2D screens. However, the quality of the synthesized view is strongly related to the precision of depth maps as well as the presence (or not) of overlap zones in the scene which are too large (in this case, the missing information cannot be deduced).

10.2.2. *Associated coding techniques*

10.2.2.1. *Simulcast*

Simulcast multiview video coding involves encoding N video streams ($N = 2$ in the case of a stereoscopic sequence) independently using standard tools (such as H.264/MPEG-4 AVC, for example). Figure 10.5 illustrates the simulcast coding process for a stereoscopic sequence.

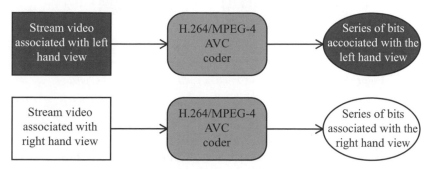

Figure 10.5. *Simulcast encoding of a stereoscopic sequence*

This technique does not require the deployment of infrastructures specific to stereoscopic and multiview sequence compression. However, the strong inter-view correlation is not used and the output rate is generally equal to N times that associated with single-view coding. It is for this reason that simulcast encoding is considered more as a basis for comparison for multiview compression algorithms than as a viable coding technique.

10.2.2.2. *MPEG-C and H.264/MPEG-4 AVC auxiliary picture syntax*

The MPEG-C and H.264/MPEG-4 AVC (auxiliary picture syntax) standards (presented in [BOU 06] and [MER 09]) are specifically designed for coding 2D-plus-depth multiscopic media. These standards allow an auxiliary video stream to be combined with a standard video stream. However, they still differ from one another, as detailed below.

MPEG-C part 3 provides a high-level syntax which allows the decoder to interpret data contained in the auxiliary stream. At this point in time, there are just two types of data: depth maps and disparity maps. The coding process, compatible with the majority of current codecs such as H.264/MPEG-4 AVC, is illustrated in Figure 10.6. The two streams, corresponding to 2D video and depth information, respectively, are coded independently in order to produce two distinct binary streams. These two streams are then recombined as a single stream by temporally interleaving their constituent frames. Another possibility with MPEG-C part 3 is the ability to sub-sample (spatially or temporally) the auxiliary stream, thereby adapting to small rates.

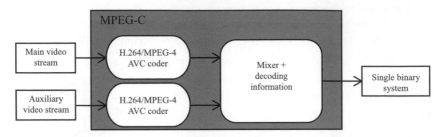

Figure 10.6. *2D-plus-depth stream encoding using MPEG-C part 3*

H.264/MPEG-4 AVC APS, in addition, associates an auxiliary component with a standard video stream and encodes these two sequences simultaneously but independently in order to produce a single binary stream (see Figure 10.7). No additional information is added and the interpretation of the data is left to the user (in contrast to MPEG-C part 3). In addition, each frame in the auxiliary stream must contain the same number of macro blocks as a frame belonging to the main stream (sub-sampling depth information is not allowed).

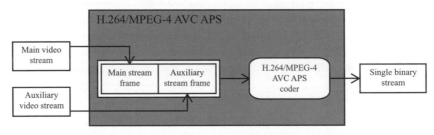

Figure 10.7. *Encoding a 2D-plus-depth stream using H.264/MPEG-4 AVC APS*

10.2.2.3. *H.264/MPEG-4 multiview video coding stereo profile*

The multiview video coding (MVC) extension of the H.264/MPEG-4 AVC standard [ITU 10, VET 11] was introduced to provide a standardized representation of stereoscopic and MVV while maintaining the same structure as the original standard as far as possible. The basic idea is to allow images taken from other cameras as a reference to predict the current image. Its coding and syntax are also very similar to H.264/MPEG-4 AVC in order to obtain the best possible level of quality while limiting complexity and demands on memory.

The MVC norm is used to code stereoscopic and MVV using two profiles known as "stereo high" and "multiview high", based on the H.264/MPEG-4 AVC "high" profile. More precisely, the MVC extension has been developed so that it is always possible to decode a specific view (known as the base view) from a stereo or multiview stream with an ordinary H.264/MPEG-4 AVC decoder. A binary MVC stream is composed of a part specifically for the base view (identical to an H.264/MPEG-4 AVC stream) and a part for other views. The two parts can be identified using two types of Network Abstraction Layer (NAL) unit; an H.264/MPEG-4 AVC decoder which correctly recognizes the base view while ignoring others, as well as an MVC decoder which can decode all views.

In terms of the compression algorithm, MVC introduces a major tool: inter-view prediction. Using this tool, it is possible to construct a prediction of a block of pixels from a current image, not only using past or future images from the same camera but also images from other views. This is made possible by modifying the list of H.264/MPEG-4 AVC reference images. For the base view, this list is not modified because backwards compatibility must be maintained. For the remaining views, images belonging to other views can be inserted into the reference list but must necessarily correspond to the same temporal point as the current image.

The stereo high profile encodes a video with two views which is either progressive or interleaved. The base view, generally the left-hand view, can therefore be decoded independently of the right-hand view. However, for each image in this right-hand view, the list of references will also contain the image corresponding to the left-hand view. An example of this prediction structure is shown in Figure 10.8, where the lines in bold represent temporal predictions and the dotted lines are inter-view predictions. The stereo high profile is currently used for 3D Blu-ray formats.

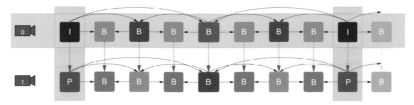

Figure 10.8. *Possible prediction structure for MVC, with stereo profile. The base view is shown above*

10.3. Multiview approaches

10.3.1. *Formats*

10.3.1.1. *Multivew video and Multiview plus depth*

Multiview video (MVV) and multiview-plus-depth (MVD) formats are two basic MVV formats. When a sequence with N views is stored using MVV format, the user has access to different viewpoints from N recording systems (either real or virtual), noted as $\mathcal{M}[i]$ (where $0 \leq i < N$) in Figure 10.9. The MVD format, on the other hand, takes the information in MVV format but adds the associated depth/ disparity maps (represented by $\mathcal{D}[i]$ in Figure 10.9).

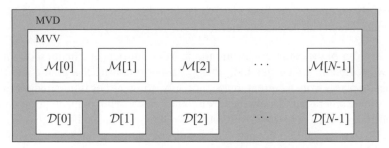

Figure 10.9. *MVV and MVD formats*

As with the 2D-plus-depth format, the MVD format only considers a restricted set of viewpoints in order to reduce the amount of data to be transmitted. The missing views are then reconstructed using DIBR algorithms.

10.3.1.2. *Layered-depth image and layered-depth video*

The concept of layered-depth image (LDI) was first introduced by Shade *et al.* [SHA 98], reused by Yoon *et al.* [YOO 05] and adapted to multiview compression by Yoon *et al.* [YOO 07].

An LDI is an image which contains several layers in which each pixel contains not only colorimetric information but also depth information. The LDI format is principally designed to only keep non-redundant information in the scene. Using depth/disparity information, available for each pixel from the N original views, it is possible to project the pixels from a view i toward a view j. This 3D format is sometimes mistaken for layered-depth video (LDV), which is a temporal extension of the LDI format.

The general construction process for LDI is as follows: we first select a reference view $\mathcal{M}_{ref}$ from the N original views (generally $\mathcal{M}[0]$) and take this as our layer 0 in the LDI (noted as $\mathcal{L}[0]$). For each of the pixels $\mathbf{r} = (\mathbf{p}, i)$ from the other views (where $0 < i < N$ indicates the view number and $\mathbf{p} = (x, y)$ indicates the pixel in position (x, y)), we project this onto $\mathcal{M}_{ref}$ using its depth/disparity information.

This pixel (situated in position (x', y') in the referential of view $\mathcal{M}_{ref}$ after projection) is compared to pixel $\mathbf{r'} = (\mathbf{p'}, ref)$, where $\mathbf{p'} = (x', y')$, using a decision function (based on a relative comparison of colorimetry, depth, mixed or other), thereby qualifying (or not) the redundant pixel. If the pixel is not considered redundant (generally due to an overlapping area in view $\mathcal{M}_{ref}$ but shown in view i) a new layer is added to the position (x', y') in the LDI and we add colorimetric and depth/disparity information associated with $\mathbf{r}$. If it is judged redundant, it is simply ignored. The LDI is completely constructed when all the pixels from the multiview set have been processed. Figure 10.10 illustrates three LDI layers obtained for the break dancers multiview sequence using the approach proposed by [YOO 07].

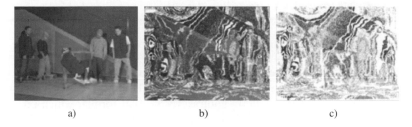

a)   b)   c)

Figure 10.10. *Three layers taken from the LDI generated for the break dancers sequence using the method proposed by [YOO 07]:*
a) $\mathcal{L}[0]$, b) $\mathcal{L}[2]$ and c) $\mathcal{L}[4]$

There are a number of variations on the LDI format [BAT 11, JAN 09, SMO 09] which differ in terms of their level of decision function as well as LDI formatting. Table 10.1 describes the different approaches for the LDI format.

Figure 10.11 shows three LDI layers obtained using the approach proposed by [BAT 11] on the image with seven views of dolls shown on the Middlebury Stereo Vision site[1].

---

1 http://vision.middlebury.edu/stereo/.

Approach	Decision function	Number of layers	Decorrelation
Yoon *et al.* (2007)	Depth/disparity comparison	$\leq N$	+
Jantet *et al.* (2009)	XOR operator	$\leq N$	+++
Battin *et al.* (2011)	Mixed and geometric constraints	N	+++
Smolic *et al.* (2009)	XOR operator	2	++

Table 10.1. *Different possible approaches for the LDI format*

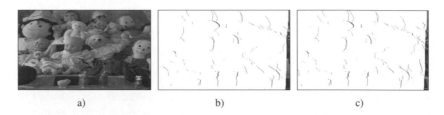

a)   b)   c)

Figure 10.11. *Three layers taken from the LDI generated for the doll sequence using the method proposed by [BAT 11]: a) $\mathcal{L}[0]$, b) $\mathcal{L}[2]$ and c) $\mathcal{L}[4]$*

10.3.1.3. Depth-enhanced stereo

The depth-enhanced stereo (DES) format was introduced by [SMO 09] in 2009. This format is a compromise between the LDI and MVD formats (see sections 10.3.1.2 and 10.3.1.1). In contrast to the LDI format, which takes a reference view from the N input views to generate the LDI, the DES format takes two adjacent reference views and for each generates an associated LDI. Figure 10.12 (taken from [SMO 09]) presents the information related to the DES format.

a)   b)

Figure 10.12. *DES format (figure taken from [SMO 09]): a) LDI associated with the first reference view, and b) LDI associated with the second reference view*

Thanks to these reference views, the DES format provides direct backward compatibility in relation to stereoscopic display (using the first layer from each LDI). In addition, it has a greater amount of information and therefore provides better quality intermediary viewpoints. However, the amount of information is greater than in the LDI approach while the memory required for the DES format is also greater.

10.3.2. *Associated coding techniques*

10.3.2.1. *H.264/MVC multiview profile*

The MVC extension of H.264/MPEG-4 AVC [ITU 10, VET 11] is designed for multiview videos with a number of views between 2 and 1,024. As for stereo, a single view is encoded as a base view and can also be decoded independently of other views. Every other view can only be decoded after the base view and potentially other views. However, the inter-view prediction structure is very flexible. With the "view-progressive" configuration, for example, only the first image from a group of pictures (GOP) is coded with the inter-view prediction, while the others use only temporal prediction. In addition, inter-view prediction uses only a single reference per image ("P" type prediction). This configuration provides access to each view with minimal computational cost (in relation to the number of images from other views to be decoded).

A more complex configuration is "fully hierarchical coding" where bi-directional inter-view predictions are possible. The aim is to optimize rate-distortion performance by exploiting the maximum amount of correlations between views. An example of this structure is shown in Figure 10.13, where the base view is shown in the center (view 2). Views 0 and 4 are coded using inter-view prediction only for the first image in the GOP. This is therefore the same prediction structure as that used for "progressive view" examples. Finally, for intermediary views 1 and 3, each image can use bi-directional inter-view prediction (dotted line).

The multiview profile can be used to code a stereoscopic video but, in contrast to the stereo profile, it does not allow interleaving. A bitstream may therefore be compatible with both profiles if it is coded with the stereo high profile but without interleaving.

10.3.2.2. *LDI format coding*

There is currently no specific coding standard for the 3D LDI format (nor for the DES format). However, there are several possible approaches used in the literature to solve this problem.

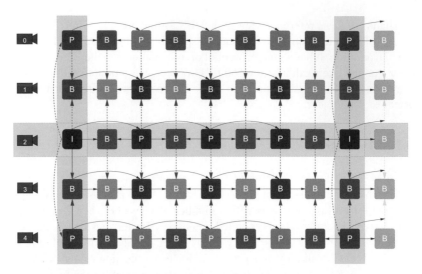

Figure 10.13. *Possible prediction structure for the multiview profile using the H.264/MVC format. The base view is shown in the center*

In [YOO 07], two methods are used to compress an LDI. The first involves bringing together the pixels from the LDI in a single texture by horizontal successive aggregation of its different layers. This texture as well as the additional information required to reconstruct the original views is then coded using H.264/MPEG-4 AVC. This first approach has a major disadvantage: horizontally aggregating the pixels taken from different layers of the LDI prevents spatial correlation which is critical for H.264/MPEG-4 AVC block coding. To overcome this problem, Yoon [YOO 07] has proposed a second method which entails combining the missing information from different locations in the $N - 1$ final layers with those present in layer 0 and then encoding them using H.264/MPEG-4 AVC. This second method obtains compression ratios two times greater than the first approach.

Another possible approach, proposed by [JAN 10], involves compressing the I-LDI (variation of the LDI proposed in [JAN 09] which uses redundancy between different views more effectively) using H.264/MVC. Jantet [JAN 10] proposes creating an MVD using only the two first I-LDI layers where information missing from the second layer is combined with that found in the first layer. This solution is possible because the I-LDI approach provides strong decorrelation and the first two layers contain the majority of information present in the original multiview set (around 90%). This MVD is then compressed using H.264/MVC with the same coding parameters. This approach has a significant reduction in output with equal quality in relation to the original MVD coding with H.264/MVC with output being less than 3 Mbits/s.

Finally, [BAT 11] proposes a real-time compression of his LDI approach based on DCT 3D. A horizontal aggregation of the pixels taken from each layer is carried out first and the different layers are then reassembled within a 3D volume. This 3D volume is then compressed using a DCT/quantification/entropic coding pipeline in order to obtain the compressed stream. While this last approach does not obtain the compression rates found in [JAN 10], it can carry out LDI generation and coding in real-time using the GPU.

10.4. Conclusion

With a view to improving performance and increasing functionality in current formats, MPEG has recently undergone a new normalization phase for 3D video coding (3DVC).

There are two main objectives of this. First, 3DVC combines the video format with display technology. It specifically includes advanced processing techniques to adjust the stereoscopic reference base and therefore control the perception of depth according to the visualization environment. This aspect is crucial in order to minimize visual fatigue and maximize the user's experience. Then, 3DVC must also take into account multiview autostereoscopic screens which are beginning to appear on the market. More specifically, 3DVC must allow the synthesis of several high-quality views with a strongly limited bit rate. As such, 3DVC uses depth map coding in order to separate the coding bit rate from the number of viewpoints.

Three approaches are currently used in normalization. The first is a backward compatible extension of MVC [VET 11]. More specifically, a second stream encodes the depth information independently of the stream representing textural information. The high-level syntax is adapted in order to signal this additional information, although there is no change regarding syntax or the decoding process in the macro block. A second approach involves a backward compatible extension of H.264/MPEG-4 AVC [ITU 10, WIE 03]. A basic video stream encodes texture information of a view with H.264/MPEG-4 AVC. For other views, as well as for depth maps, the syntax and decoding process in the macro block are modified in order to improve the efficacy of compression. A significant benefit must be evident in order to justify the normalization of this approach. Finally, a third approach uses a backward compatible extension of high efficiency video coding (HEVC) [BRO 12, OHM 13]. First, a simple multiview extension of HEVC is made using a schema identical to MVC. The depth map coding as well as the improvement in view resolution using scalability are then considered.

10.5. Bibliography

[BAT 11] BATTIN B., NIQUIN C., VAUTROT P., *et al.*, "Multiview image compression based on LDV scheme", *Proc. SPIE 7863, Stereoscopic Displays and Applications XXII*, 78630G, February 15, 2011.

[BOU 06] BOURGE A., GOBERT J., BRULS F., "MPEG-C part 3: enabling the introduction of video plus depth contents", *Proceedings of IEEE Workshop on Content Generation and Coding for 3D-Television*, 2006.

[BRO 12] BROSS B., HAN W.-J., OHM J.-R., *et al.*, "High efficiency video coding (HEVC) text specification draft 9", ITU-T SG16 WP3 & ISO/IEC JTC1/SC29/WG11 JCTVC-K1003, October 2012.

[EKM 08] EKMEKCIOGLU E., WORRALL S.T., KONDOZ A.M., "Bit-rate adaptative down-sampling for the coding of multi-view video with depth information", *3DTV Conference: The True Vision-Capture, Transmission and Display of 3D Video*, pp. 137–140, 2008.

[FEH 02] FEHN C., KAUFF P., BEECK M., *et al.*, "An evolutionary and optimized approach on 3D-TV", *Proceedings of International Broadcast Conference*, pp. 357–365, September 2002.

[FEH 04] FEHN C., "3D-TV using depth-image-based rendering (DIBR)", *Proceedings of Picture Coding Symposium*, San Francisco, USA, December 2004.

[ITU 10] ITU-T, "Advanced video coding for generic audiovisual services", ITU-T Recommendation H.264 and ISO/IEC 14496-10 (MPEG-4 AVC), 2010.

[JAN 09] JANTET V., MORIN L., GUILLEMOT C., "Incremental-LDI for multi-view coding", *3DTV Conference: The True Vision - Capture, Transmission and Display of 3D Video*, Potsdam, Germany, 2009.

[JAN 10] JANTET V., MORIN L., GUILLEMOT C., "Génération, compression et Rendu de LDI", *COmpression et REpresentation des signaux AUdiovisuels (CORESA)*, Lyon, France, 2010.

[MAR 06] MARTINIAN E., BEHRENS A., XIN J., *et al.*, "Extensions of H.264/AVC for multiview video compression", *IEEE International Conference on Image Processing*, Atlanta, USA, 2006.

[MER 09] MERKLE P., WANG Y., MULLER K., *et al.*, "Video plus depth compression for mobile 3D services", *3DTV Conference: The True Vision – Capture, Transmission and Display of 3D Video*, Potsdam, Germany, 2009.

[NIQ 10] NIQUIN C., PRÉVOST S., REMION Y., "An occlusion approach with consistency constraint for multiscopic depth extraction", *International Journal of Digital Multimedia Broadcasting (IJDMB), Special Issue Advances in 3DTV: Theory and Practice*, vol. 2010, pp. 1–8, February 2010.

[OHM 13] OHM J.-R., SULLIVAN G., "High efficiency video coding: the next frontier in video compression", *IEEE Signal Processing Magazine*, vol. 30, no. 1, pp. 152–158, 2013

[SHA 98] SHADE J., GORTLER S., HE L., *et al.*, "Layered depth images", *Proceedings ACM SIGGRAPH*, ACM, pp. 231–242, 1998.

[SMO 08] SMOLIC A., MULLER K., DIX K., *et al.*, "Intermediate view interpolation based on multiview video plus depth for advanced 3D video systems", *15th IEEE International Conference on Image Processing, 2008, ICIP 2008*, pp. 2448–2451, October 2008.

[SMO 09] SMOLIC A., MUELLER K., MERKLE P., *et al.*, "An overview of available and emerging 3D video formats and depth enhanced stereo as efficient generic solution", *Picture Coding Symposium*, Chicago, USA, 2009.

[STE 98] STELMACH L., TAM W.J., "Stereoscopic image coding: effect of disparate image-quality in left- and right-eye views", *Signal Processing: Image Communication (Elsevier Science)*, vol. 14, pp. 111–117, 1998.

[STE 00] STELMACH L., TAM W.J., MEEGAN D., *et al.*, "Stereo image quality: Effects of mixed spatio-temporal resolution", *IEEE Transactions on Circuits and Systems for Video Technology*, vol. 10, no. 2, pp. 188–193, 2000.

[VET 10] VETRO A., "Frame compatible formats for 3D video distribution", *Proceedings of the IEEE International Conference on Image Processing*, vol. 17, pp. 2405–2408, 2010.

[VET 11] VETRO A., WIEGAND T., SULLIVAN G.J., "Overview of the stereo and multiview video coding extensions of the H.264/MPEG-4 AVC standard", *Proceedings of the IEEE*, vol. 99, no. 4, pp. 626–642, 2011.

[WIE 03] WIEGAND T., SULLIVAN G., BJØNTEGAARD G., *et al.*, "Overview of the H.264/AVC video coding standard", *IEEE Transactions on Circuits and Systems for Video Technology*, vol. 13, no. 7, pp. 560–576, 2003.

[YOO 05] YOON S.-U., KIM S.-Y., HO Y.-S., "Preprocessing of depth and color information for layered depth image coding", *Advances in Multimedia Information Processing – PCM 2004*, vol. 3333, pp. 622–629, 2005.

[YOO 07] YOON S., LEE E., KIM S., *et al.*, "A framework for representation and processing of multi-view video using the concept of layer depth image", *Journal of VLSI Signal Processing*, vol. 46, pp. 87–102, 2007.

Chapter 11

3D Mesh Compression

11.1. Introduction

With technological progress in the domains of digital acquisition, telecommunications and geometric modeling, 3D objects (both static and animated) occupy an increasingly important role. This emerging multimedia content is, for the majority, made up of 3D and 3D+t polygonal meshes, which are generally bulky due to increases in precision and the richness of the data they carry. In parallel, these 3D objects are increasingly used for Internet applications, or as part of online collaborative platforms, where they need to be transmitted rapidly across a network. Thus, the compression of these objects has become an essential scientific concern, involving multiple objectives such as the compact storage of data, interactive time transmission across networks, progressive content visualization and rapid random access.

In this chapter, we begin by considering certain basic points of information theory, the sampling and quantization notions needed to reduce the quantity of data and notions of rate-distortion. We then consider multiresolution analysis. This consists of decomposing a mesh into a series of resolution levels, for example by using wavelet transformations. This method offers an elegant and effective framework for compressing meshes and sequences of meshes, but generally requires the use of a semi-regular mesh, often obtained through an earlier remeshing process. We then present

---

Chapter written by Florent DUPONT, Guillaume LAVOUÉ and Marc ANTONINI.

monoresolution coding methods based on highly effective connectivity coders, associating prediction techniques. Other progressive compression techniques, based on iterative simplification processes or guided by geometry, may be used to obtain satisfactory levels of compression with additional functionality. Sequence compression algorithms, fewer in number, will be considered in the following section, before addressing the definition of classic and perceptual metrics which may be used to evaluate the effectiveness of compression techniques.

11.2. Compression basics: rate-distortion trade-off

Compression methods involving *information loss* enable considerable increases in compression levels, and thus reduce the signal bit rate below the Shannon limit [SHA 48]. Flows of this type are only possible in cases of non-reversible coding, generating degradations of the 3D object. The aim of compression methods with information loss is not only to eliminate signal redundancies, but also to generate losses in non-relevant information and in noise. Coding techniques of this type are required for applications requiring high levels of compression. They are designed to implement effective compression, while "controlling degradation". In these cases, we speak of rate-distortion optimization.

Quantization is the fundamental operation carried out by a compression system. Its purpose is to select, for a given real input value, the closest neighbor from a finite predetermined set of numerical values. More precisely, a scalar quantifier of size L is an application Q of $\mathbb{R}$ in a finite set C, also known as a codebook, containing L scalar symbols:

$$Q : \mathbb{R} \to C \quad \text{with } C = \{\hat{s}_1, \hat{s}_2..., \hat{s}_L\} \tag{11.1}$$

$\hat{s} = Q(s)$ is the quantization of s, where $\hat{s}$ corresponds to the centroid of the quantization interval which includes s. This corresponds to seeking the closest neighbor of s among the elements contained in the dictionary C.

In a general manner, the mean square error (MSE) of a quantizer (power of the quantization noise) is defined by the following relationship:

$$D = \sum_{m=1}^{L} \int_{s \in P_m} (s - \hat{s}_m)^2 p_S(s)\, ds \tag{11.2}$$

Here, $(s - \hat{s}_m)^2$ is the measure of distortion between a source symbol s and its quantization $\hat{s}_m$; P_m is a quantization cluster to which the source symbol belongs, represented by its centroid $\hat{s}_m$; and $p_S(x)$ is the probability density of the source signal S. For a very high number L of quantization levels (asymptotic hypothesis, or high bit rate), the MSE, expressed as a function of the bit rate $R = \log_2 L$ bits/symbol, is bounded asymptotically by [GER 92]:

$$D(R) = \frac{1}{12}2^{-2R}\left(\int_{-\infty}^{+\infty} p_S(s)^{1/3}ds\right)^3 \qquad [11.3]$$

The general behavior of this function is shown in Figure 11.1. It shows that the quantization distortion D is a strictly decreasing function of the bit rate R.

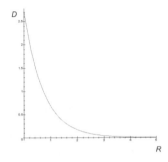

Figure 11.1. *Distortion as a function of bit rate expressed in bits per symbol*

11.3. Multiresolution coding of surface meshes

Meshes constitute a powerful tool for modeling complex 3D objects, thanks to their dual geometric and combinatory nature (positions of vertices and connectivity). While there is a number of alternatives for modeling surface forms, meshes are now omnipresent, and considerable effort has been put into developing methods for the digital processing of their geometry, essentially based on triangular meshes. If a geometry is initially described by a mesh, the mesh is considered to be an instance of the geometry. A certain degree of freedom subsists in the meshing process: we may remesh a geometric form without introducing geometric distortion in order to obtain regularity and sampling properties in the new mesh which may be used for multiresolution representation, processing and compression.

Although multiresolution techniques have long been proved to be effective in the compression of 2D images [ISO 00], their use for 3D meshes is relatively recent. This is due to the fact that it is relatively difficult to design filters for irregular subdivisions [VAL 99], and wavelet theory for meshes with regular or semi-regular subdivisions is a relatively recent idea [LOU 94]. In 1994, Lounsbery introduced a schema for the progressive compression of surfaces using a 3D wavelet transform for triangular surface meshes, obtained by regular subdivision (1:4 subdivision) of an irregular base mesh [LOU 94, LOU 97]. Schröder and Sweldens [SCH 95] and Kovacevic and Swelden [KOV 99] then developed a number of different techniques used to compress the geometry of meshes and to reduce connectivity information as much as possible. For applications where remeshing is not possible, a generalization of Lounsbery's approach for meshes with irregular subdivisions was proposed in [VAL 99]. The multiresolution approaches obtained in this way are scalable in terms of resolution, precision, quantity and even complexity (see Figure 11.2). The representation of a surface mesh with any connectivity graph (regular or irregular), representing the surface of a 3D object, using geometric wavelets is based on a subdivision inversion problem, and produces an exactly reversible compression method for the connectivity graph and the geometry of the data. Hierarchical transmission, with progressive quality and a binary bit rate, is thus possible. An effective geometrical wavelet decomposition requires prior representation of the geometric signal in a form compatible with a regular structure and uniform sampling. As perfect regularity is impossible to obtain for arbitrary topologies, the mesh is converted into a semi-regular mesh (regular by sections) [GUS 00, LEE 98] where the theory of wavelets using M-channels has been shown to be effective [KAM 12]. Thus, a geometric mesh compression schema is generally made up of three steps (see Figure 11.3):

– *semi-regular remeshing*, which transforms a given manifold mesh into a semi-regular mesh;

– *wavelet transform*, where the semi-regular mesh is decomposed into a coarse mesh of low frequencies and sets of sub-bands of high-frequency details (wavelet coefficients);

– *geometric coding*, including the coding of different low- and high-frequency sub-bands and the coding of the connectivity of the coarse mesh.

Figure 11.2. *Image of Venus: example of multiresolution meshes: a) original mesh and b) different resolutions of approximation. The information lost between two resolution levels is contained in the sub-images of wavelet coefficients*

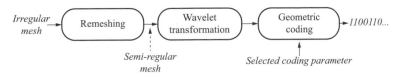

Figure 11.3. *General schema showing the coding of a surface mesh using a wavelet transform*

11.4. Topological and progressive coding

Wavelet-transform coding methods such as those presented above produce excellent performances, but require prior remeshing of the initial object to give a semi-regular structure. This remeshing stage may be undesirable in scenarios where the original connectivity of the object needs to be preserved. Other types of techniques may be used to code arbitrary geometries and connectivities. These may be grouped into two categories of algorithms:

– monoresolution algorithms, through which mesh data are compressed as a whole, and cannot be decompressed until the whole of the bit rate is decoded;

– multiresolution (or progressive) algorithms, where the mesh is compressed in the form of a coarse model (low resolution) followed by a sequence of refinement data. Thus, during decompression, as in the case of wavelet-based methods, the low-resolution model is accessible rapidly then refined iteratively through the transmission and decoding process of the compressed bit rate, until the desired resolution is obtained.

11.4.1. *Monoresolution compression*

Most monoresolution compression methods follow the same pattern: the geometry and connectivity of the object are coded separately, but not independently. The connectivity is represented by a series of symbols representing a traversal order of the mesh, while the geometry is coded by quantization, then prediction guided by this order.

The first connectivity coding method (only applicable to triangular meshes) was proposed by Deering [DEE 95]. Deering's method consisted of defining a sequence of vertices such that each vertex of the series forms a triangle with the two previous vertices. The mesh is thus decomposed into *triangle strips*, producing an efficient coding of connectivity. Working along different lines, Taubin and Rossignac [TAU 98a] proposed a method based on *spanning trees*. Their method was inspired by theoretical results obtained by [TUR 84], who established that a planar graph may be coded with a constant number of bits per vertex, using two spanning trees, one of faces and the other of vertices.

The majority of recent methods are based on a region growing approach, where a region is extended across the mesh with incremental coding of elements of the mesh and their incidence relationships with the region. This group of methods may be divided into three categories according to the type of element with the dominant role in the coding process: faces [GUM 98, KRO 01, ROS 99, SZY 01], edges [ISE 00] or vertices [ALL 01b, KHO 02, TOU 98].

Geometric information must also be coded in parallel with the connectivity coding. Generally, geometry is coded in three stages: quantization, prediction (using the parallelogram rule, for example) and coding of the difference vector. Recent algorithms propose highly efficient prediction mechanisms [COU 11].

In terms of compression rates, the best monoresolution methods generally require 2–3 bits per vertex for connectivity and 10–16 bits per vertex for geometry (depending on the method and the model in question, for a 12-bit quantization). The methods presented above are based on a connectivity-guided mesh traversal; given the volume of memory occupied by geometry, certain authors have chosen to prioritize this aspect, proposing geometry-guided methods. These include proposals by Kronrod and Gotsman [KRO 02] and Lewiner *et al.* [LEW 06].

11.4.2. *Multiresolution compression*

11.4.2.1. *Connectivity-driven approaches*

The concept of progressive mesh was first introduced by Hoppe [HOP 96]. During the coding process, the mesh is simplified by a series of edge collapsing operations, then refined during the decoding process by the opposite vertex split operator. This method, extended by Popovic and Hoppe [POP 97] for *non-manifold* arbitrary meshes, has the advantage of presenting fine granularity, but remains costly in terms of complexity and memory requirements (around 16 bits per vertex for connectivity), as the index of the vertices to split is coded explicitly on a one-by-one basis. Thus, Taubin *et al.* [TAU 98b], then Pajarola and Rossignac [PAJ 00], simplified the mesh by combinations of edge collapses applied to several edges at a time. Thus, for each iteration, a whole region of the object is simplified. The granularity produced is lower than for Hoppe's algorithm [HOP 96], but the coding cost is reduced (around 10 and 7 bits per vertex respectively for connectivity). Cohen-Or *et al.* [COH 99] used vertex deletion/insertion operators on sets of independent vertices. Alliez and Desbrun [ALL 01a] proposed an extension of monoresolution compression algorithms [ALL 01b, TOU 98] based on valency. Using their method, a set of independent vertices is removed for each iteration by two decimation conquests. This is followed by retriangulation in order to optimize the quality of the intermediate meshes by preserving regularity and connectivity as far as possible. This method requires, on average, 3.7 bits per vertex for coding connectivity.

The methods presented above are based on connectivity-driven approaches. The geometry of the objects is then coded in a similar way to that used in monoresolution approaches, by quantization then prediction (each new vertex generated by edge separation or vertex insertion may be effectively predicted using neighbors which have already been transmitted).

Several authors [AHN 11, KIM 11, LEE 12] have attempted to optimize the approach put forward by Alliez and Desbrun [ALL 01a], for example by introducing coding using bitplanes, or prediction mechanisms for valency coding.

11.4.2.2. *Geometry-driven approaches*

As with monoresolution approaches, certain authors consider that it is more efficient to use geometry to guide compression rather than connectivity. Gandoin and Devillers [GAN 02] and Devillers and Gandoin [DEV 00], for example, introduced the first geometry-driven multiresolution algorithm, based on the subdivision of space using a *kd-tree*. The compression rates obtained using this method are excellent, but the quality of the levels of detail is much lower than for connectivity-driven methods. Peng and Kuo [PEN 05] proposed a similar method based on the subdivision of space using an *octree*. Through a series of optimizations, they improved compression levels, but the quality of the levels of detail is still low in comparison with that obtained by connectivity-driven approaches.

Finally, several algorithms have been proposed in recent times which can be considered as hybrids of between geometry and connectivity-driven approaches. They notably propose variable quantization mechanisms [LEE 12, PEN 10, VAL 09].

11.5. Mesh sequence compression

Thanks to advances in the domain of camera-based 3D reconstruction, mesh sequences are becoming increasingly widespread. Using short time steps, these are extremely bulky and require particular attention for efficient compression, as in the case of video compression, which makes use of the temporal coherence of successive images. 3D animation is another domain which produces large quantities of mesh sequences. An example of a sequence is shown in Figure 11.4.

Figure 11.4. *Chicken crossing sequence*
(© Copyright 1996, Microsoft Corporation)

11.5.1. *Definitions*

A mesh sequence is a series of meshes ordered in time. Temporal coherency is required and necessitates temporal sampling at a rate which is sufficiently rapid in relation to the evolution of the object or the scene. We can distinguish at least two different types of sequences:

– the case of mesh sequences with a constant number of vertices, connectivity and topology, which we may call *constant connectivity sequences* or *dynamic meshes*;

– the case of mesh sequences with a variable number of vertices and variable connectivity over time, with a topology which may be variable or constant. These are known as *dynamic connectivity sequences*.

In the first case, the sequence consists of a geometric evolution of the vertices of the first mesh over time. In the second case, there is no natural link between a vertex at instant t and a vertex at instant $t+1$. A more precise classification of mesh sequences may be found in a thesis on the subject by R. Arcila [ARC 11].

The interest of compressing sequences with constant connectivity is evident. Connectivity is coded once at the beginning of the sequence. The geometry of each mesh is generally coded by considering the evolution of the position of each vertex over time as a trajectory. In the case of sequences with dynamic connectivity, the complexity of compression problems is considerably increased, and thus has received little attention until now [HAN 07]. In these cases, spatio-temporal redundancies are much harder to exploit without prior mapping.

We may classify sequence compression methods into three categories: methods with spatio-temporal prediction, approaches using prior segmentation and transform-based methods (principal component analysis or wavelets).

11.5.2. *Methods using spatio-temporal prediction*

These methods exploit the spatio-temporal redundancy of sequences by predicting the position of a vertex from surrounding positions (by interpolation) or from previous positions (by extrapolation). These predictors attempt to minimize the residual error between predicted positions and the real positions of vertices. Errors are then quantified and transmitted to an entropic coder.

Ibarria *et al.* [IBA 03] present two spatio-temporal predictors: the *Extended Lorenzo Predictor* (ELP) and *Replica*. The first extends the parallelogram rule used in the compression of static meshes [TOU 98], taking account of the prediction of the position of the summit at the previous instant. The *Replica* predictor makes the previous predictor robust to rigid transformations such as rotations and changes of scale, using local reference points and normalization. Ibarria *et al.* have also proposed another predictor, known as Angle Preserving (AP) [IBA 03].

Methods using this kind of geometric predictors require mapping between the vertices of meshes in the sequence, and cannot, therefore, be applied to sequences with dynamic connectivity. Moreover, they do not exploit potential spatial redundancies in cases where regions are subject to homogeneous movement.

11.5.3. *Methods with prior segmentation*

In an attempt to improve the performance of compression methods, several authors have proposed partitioning the vertices of a mesh into groups of points with similar movement. In 1999, Lengyel [LEN 99] proposed a method of this type. The basic principle involves estimating the movement of a group of vertices by a rigid transformation, which then acts as a predictor. The segmentation information, the parameters of each transformation and the prediction errors are then coded to allow reconstruction of the sequence. Alignment between successive meshes can be used to improve results, as demonstrated by Gupta *et al.* [GUP 02].

Without using direct segmentation, Zhang *et al.* [ZHA 07] placed vertex displacement vectors into a hierarchical tree structure (*octree*), which allows exploitation of spatio-temporal coherence. Once again, this method requires point-to-point mapping.

In 2007, Han *et al.* [SEU 07] proposed a first method for compressing sequences with dynamic connectivity, inspired by video compression techniques and using a division of meshes into blocks, mapping and discrete cosine transformation (DCT).

11.5.4. *Transform-based methods*

Other approaches use transformations applied to the whole of a sequence. Alexa and Muller [ALE 00], for example, propose a mesh alignment, then calculate a principal component analysis using residual errors. This method performs well if the duration of the sequence (number of frames) is long in relation to the number of vertices in the mesh.

Another transformation, the wavelet transform, which has been shown to be effective in compression, may be used on the trajectories of vertices over time. It may be applied directly or be linked to a group of correlated trajectories. After the classification phase, a motion compensation element is introduced for each cluster, and wavelet analysis is applied to their residual errors [BOU 07].

11.6. Quality evaluation: classic and perceptual metrics

The performance of a compression algorithm is evaluated in terms of the rate-distortion compromise. The goal of an algorithm is to produce a high level of compression while preserving the geometry or visual appearance of the object as far as possible. The following sections present the existing metrics used to measure the geometric and/or visual distortion between two 3D objects.

11.6.1. *Classic metrics*

A classic metric used to evaluate the distortion between two 3D objects is MSE, similar to peak signal-to-noise ratio (PSNR) which is used for images. For two meshes $\mathcal{S}_1$ and $\mathcal{S}_2$ *with the same connectivity*, each containing n vertices, this error is defined as follows:

$$MSE(\mathcal{S}_1, \mathcal{S}_2) = \frac{1}{n} \left(\sum_{i=1}^{n} \|\mathcal{S}_1[i] - \mathcal{S}_2[i]\|^2 \right) \qquad [11.4]$$

where $\mathcal{S}_1[i]$ and $\mathcal{S}_2[i]$ are the corresponding 3D vertices between the meshes in question.

We may also consider the square root of this error, the root mean square error (RMSE). Unfortunately, these distances require one-to-one correspondence between vertices of the meshes under comparison, and can therefore only be applied to meshes sharing the same connectivity.

Another widely used metric which, unlike the previous metrics, may be used to compare two objects with different connectivities and/or levels of detail is the Hausdorff distance, defined as follows: we begin by defining $e(\mathbf{p}, \mathcal{S})$ the Euclidean distance from a point $\mathbf{p}$ in the 3D space to the object $\mathcal{S}$:

$$e(\mathbf{p}, \mathcal{S}) = \min_{\mathbf{p}' \in \mathcal{S}} \|\mathbf{p} - \mathbf{p}'\| \qquad [11.5]$$

where $\mathbf{p}' \in \mathcal{S}$ means that the points $\mathbf{p}'$ are sampled in a dense manner across surface $\mathcal{S}$ (we do not only consider the vertices of $\mathcal{S}$). We may then define the asymmetric Hausdorff distance between two objects $\mathcal{S}_1$ and $\mathcal{S}_2$ as follows:

$$h_a(\mathcal{S}_1, \mathcal{S}_2) = \max_{\mathbf{p} \in \mathcal{S}_1} e(\mathbf{p}, \mathcal{S}_2) \qquad [11.6]$$

Finally, the symmetrical Hausdorff distance is defined as follows:

$$h(\mathcal{S}_1, \mathcal{S}_2) = \max\left\{h_a(\mathcal{S}_1, \mathcal{S}_2), h_a(\mathcal{S}_2, \mathcal{S}_1)\right\} \qquad [11.7]$$

Definition [11.5] also allows us to define an asymmetric mean square error:

$$mse_a(\mathcal{S}_1, \mathcal{S}_2) = \frac{1}{|\mathcal{S}_1|}\int_{\mathcal{S}_1} e(\mathbf{p}, \mathcal{S}_2)^2 \, ds \qquad [11.8]$$

The most widespread measurement is the maximum root mean square error (MRMS):

$$MRMS(\mathcal{S}_1, \mathcal{S}_2) = \max\left\{\sqrt{mse_a(\mathcal{S}_1, \mathcal{S}_2)}, \sqrt{mse_a(\mathcal{S}_2, \mathcal{S}_1)}\right\} \quad [11.9]$$

Metro[1] [CIG 98] is a program providing an efficient implementation of the Hausdorff distance and its derivatives. These geometric distances have equivalents for dynamic meshes [KAR 04].

11.6.2. *Perceptual metrics*

The drawback of the standard metrics presented above is that they are not correlated with human vision (as with PSNR for images), i.e. they do not reflect the perceived visual distortion between 3D objects. In most applications, visual quality is often more important than pure geometric distortion. More perceptual metrics have been proposed in order to solve this problem; these aim to produce scores predicting the visual impact of the distortion.

To evaluate the quality of 3D meshes compressed using their spectral compression methods, Karni and Gotsman [KAR 00] then Sorkine *et al.* [SOR 03] used a metric combining the MSE between corresponding vertices and the MSE of their Laplacians (which give an indication of the roughness of the surface). The basic idea involved states that the human eye is more sensitive to high-frequency changes in roughness than to low-frequency geometric changes. Corsini *et al.* [COR 07] then proposed perceptual metrics based on the global variation in roughness to measure the quality of meshes in the context of a watermarking application. They used two methods to calculate roughness: the variance in geometric differences between a 3D

---

1 http://vcg.isti.cnr.it/activities/surfacegrevis/simplification/metro.html.

model and its smoothed version and the variance of dihedral angles. More recently, Wang *et al.* [WAN 12] introduced another metric based on a global difference in roughness, calculated as the Laplacian of the Gaussian curvature. Like roughness, curvature is an important element in the evaluation of perceived quality. Lavoué *et al.* [LAV 06, LAV 11] have proposed metrics based on differences in curvature statistics between corresponding local windows over the objects to compare; Torkhani *et al.* [TOR 12] introduced an extension to this method to take account of directions of curvature. Finally, Vasa and Rus [VAS 12] consider the variation in dihedral angles.

We should also cite the work of Pan *et al.* [PAN 05] and Cheng *et al.* [CHE 07], who defined a quality metric for textured meshes in the context of bit allocation for transmission. They observed that quality is reduced exponentially in relation to the resolution of the geometry, and in a linear manner in relation to the texture map; from this, they deduced a simple quality estimator, based on these two pieces of resolution information.

For dynamic meshes, only Vasa and Skala [VAS 11] have introduced a perceptual metric.

11.7. Conclusion

3D meshes and mesh sequences are currently used in a number of domains, from entertainment to industry. Their presence on the Internet remains marginal, but is growing. The volume of data associated with these multimedia objects creates a need for efficient compression. In this section, we have discussed the state of the art in the domain of mesh and mesh sequence compression; the moving picture experts group (MPEG) consortium has also been involved in standardization efforts in an attempt to resolve the problem of multiple formats, which limits the acceptation of specific compression methods.

The diffusion of these 3D objects across a variety of networks and to a variety of terminals raises a need for multiresolution methods, allowing both progressive display and adaptation to the visualization terminal. The visual quality of intermediate resolution levels therefore constitutes another important point. The problem of accounting for data associated with meshes such as texture maps remains open, and solutions are needed to enlarge the field of application of compression methods.

11.8. Bibliography

[AHN 11] AHN J.-K., LEE D.-Y., AHN M., *et al.*, "R-D optimized progressive compression of 3D meshes using prioritized gate selection and curvature prediction", *The Visual Computer*, vol. 27, no. 6–8, pp. 769–779, April 2011.

[ALE 00] ALEXA M., MÜLLER W., "Representing animations by principal components", *Computer Graphics Forum*, vol. 19, pp. 411–418, 2000.

[ALL 01a] ALLIEZ P., DESBRUN M., "Progressive encoding for lossless transmission of triangular meshes", *Proceedings of the 28th annual conference on Computer graphics and interactive techniques (SIGGRAPH '01)* ACM, New York, USA, pp. 195–202, 2001.

[ALL 01b] ALLIEZ P., DESBRUN M., "Valence-driven connectivity encoding of 3D meshes", *Computer Graphics Forum*, vol. 20, no. 3, pp. 480–489, 2001.

[ARC 11] ARCILA R., Séquences de maillages: classification et méthodes de segmentation, PhD Thesis, Université Claude Bernard - Lyon I, November 2011.

[BOU 07] BOULFANI-CUISINAUD Y., ANTONINI M., "Motion-based geometry compensation for DWT compression of 3D mesh sequences", *Image Processing, 2007 (ICIP 2007) IEEE International Conference on*, IEEE, vol. 1, I-217–I-220, 16 September–19 October 2007.

[CHE 07] CHENG I., YING L., BASU A., "Packet-loss modeling for perceptually optimized 3D transmission", *Advances in Multimedia*, vol. 2007, pp. 1–10, 2007.

[CIG 98] CIGNONI P., ROCCHINI C., SCOPIGNO R., "Metro: measuring error on simplified surfaces", *Computer Graphics Forum*, IEEE Computer Society Press, vol. 17, no. 2, pp. 167–174, 1998.

[COH 99] COHEN-OR D., LEVIN D., REMEZ O., "Progressive compression of arbitrary triangular meshes", *Proceedings of the conference on Visualization '99: celebrating ten years (VIS '99)*, pp. 67–72, 1999.

[COR 07] CORSINI M., GELASCA E.D., EBRAHIMI T., *et al.*, "Watermarked 3-D mesh quality assessment", *IEEE Transactions on Multimedia*, vol. 9, no. 2, pp. 247–256, 2007.

[COU 11] COURBET C., HUDELOT C., "Taylor prediction for mesh geometry compression", *Computer Graphics Forum*, vol. 30, no. 1, pp. 139–151, 2011.

[DEE 95] DEERING M., "Geometry compression", *Proceedings of the 22nd annual conference on Computer graphics and interactive techniques (SIGGRAPH'95)*, ACM Siggraph, pp. 13–20, 1995.

[DEV 00] DEVILLERS O., GANDOIN P.-M., "Geometric compression for interactive transmission", *Proceedings of the conference on Visualization '00*, IEEE Computer Society Press, no. 8, pp. 319–326, 2000.

[GAN 02] GANDOIN P.-M., DEVILLERS O., "Progressive lossless compression of arbitrary simplicial complexes", *Proceedings of the 29th annual conference on Computer graphics and interactive techniques (SIGGRAPH '02)*, ACM, no. 8, pp. 372–379, 2002.

[GER 92] GERSHO A., GRAY R., *Vector Quantization and Signal Compression*, Kluwer Academic Publishers, 1992.

[GUM 98] GUMHOLD S., STRASSER W., "Real time compression of triangle mesh connectivity", *Proceedings of the 25th annual conference on Computer graphics and interactive techniques (SIGGRAPH '98)*, ACM, no. 8, pp. 133–140, 1998.

[GUP 02] GUPTA S., SENGUPTA K., KASSIM A., "Compression of dynamic 3D geometry data using iterative closest point algorithm", *Computer Vision and Image Understanding*, vol. 87, no. 1, pp. 116–130, 2002.

[GUS 00] GUSKOV I., VIDIMCE K., SWELDENS W., *et al.*, "Normal meshes", *Computer Graphics Proceedings*, SIGGRAPH, pp. 95–102, 2000.

[HAN 07] HAN S.-R., YAMASAKI T., AIZAWA K., "Time-varying mesh compression using an extended block matching algorithm.", *IEEE Transactions on Circuits and Systems for Video Technology*, vol. 17, no. 11, pp. 1506–1518, 2007.

[HOP 96] HOPPE H., "Progressive meshes", *Proceedings of the 23rd annual conference on Computer graphics and interactive techniques (SIGGRAPH '96)*, ACM, pp. 99–108, 1996.

[IBA 03] IBARRIA L., ROSSIGNAC J., "Dynapack: space-time compression of the 3D animations of triangle meshes with fixed connectivity", *Proceedings of the 2003 ACM SIGGRAPH/Eurographics Symposium on Computer Animation*, Eurographics Association, pp. 126–135, 2003.

[ISE 00] ISENBURG M., "Triangle fixer: edge-based connectivity compression", in *Proceedings of 16th European Workshop on Computational Geometry*, pp. 18–23, 2000.

[ISO 00] ISO/IEC 15444–1:2000, "Information technology – JPEG 2000 image coding system – Part 1: Core coding system", 2000.

[KAM 12] KAMMOUN A., PAYAN F., ANTONINI M., "Sparsity-based optimization of two lifting-based wavelet transforms for semi-regular mesh compression", *Elsevier Computers and Graphics*, vol. 36, pp. 272–282, 2012.

[KAR 00] KARNI Z., GOTSMAN C., "Spectral compression of mesh geometry", *Proceedings of the 27th annual conference on Computer graphics and interactive techniques (SIGGRAPH '00)*, ACM Press/Addison-Wesley Publishing Co., no. 8, pp. 279–286, 2000.

[KAR 04] KARNI Z., GOTSMAN C., "Compression of soft-body animation sequences", *Computers & Graphics*, vol. 28, no. 1, pp. 25–34, 2004.

[KHO 02] KHODAKOVSKY A., ALLIEZ P., DESBRUN M., *et al.*, "Near-optimal connectivity encoding of 2-manifold polygon meshes", *Journal of Graphical Models*, vol. 64, nos. 3–4, pp. 147–168, 2002.

[KIM 11] KIM J., NAM C., CHOE S., "Bayesian AD coder: mesh-aware valence coding for multiresolution meshes", *Computers & Graphics*, vol. 35, pp. 1–6, 2011.

[KOV 99] KOVACEVIC J., SWELDENS W., "Wavelet families of increasing order in arbitrary dimensions", *IEEE Transactions on Image Processing*, vol. 9, no. 3, 1999.

[KRO 01] KRONROD B., GOTSMAN C., "Efficient coding of non-triangular mesh connectivity", *Journal of Graphical Models*, vol. 63, pp. 263–275, 2001.

[KRO 02] KRONROD B., GOTSMAN C., "Optimized compression of triangle mesh geometry using prediction trees", *International Symposium on 3D Data Processing, Visualization and Transmission*, IEEE Computer Society, pp. 602–608, 2002.

[LAV 06] LAVOUÉ G., DRELIE GELASCA E., DUPONT F., *et al.*, "Perceptually driven 3D distance metrics with application to watermarking", *SPIE*, vol. 6312, SPIE, pp. 63120L–63120L–12, August 2006.

[LAV 11] LAVOUÉ G., "A multiscale metric for 3D mesh visual quality assessment", *Computer Graphics Forum*, vol. 30, no. 5, pp. 1427–1437, 2011.

[LEE 98] LEE A.W.F., SWELDENS W., SCHRÖDER P., *et al.*, "MAPS: multiresolution adaptive parameterization of surfaces", *Computer Graphics Proceedings (SIGGRAPH 98)*, ACM Siggraph, pp. 95–104, 1998.

[LEE 12] LEE H., LAVOUÉ G., DUPONT F., "Rate-distortion optimization for progressive compression of 3D mesh with color attributes", *The Visual Computer*, vol. 28, no. 2, pp. 137–153, 2012.

[LEN 99] LENGYEL J.E., "Compression of time-dependent geometry", *Proceedings of the 1999 symposium on Interactive 3D Graphics*, I3D '99, ACM, New York, pp. 89–95, 1999.

[LEW 06] LEWINER T., CRAIZER M., LOPES H., "GEncode: geometry-driven compression for general meshes", *Computer Graphics Forum*, vol. 25, no. 4, pp. 1–10, 2006.

[LOU 94] LOUNSBERY J., Multiresolution analysis for surfaces of arbitrary topological type, PhD Thesis, University of Washington, Seattle, 1994.

[LOU 97] LOUNSBERY M., DEROSE T., WARREN J., "Multiresolution analysis for surfaces of arbitrary topological type", *ACM Transactions on Graphics*, vol. 16, no. 1, pp. 34–73, 1997.

[PAJ 00] PAJAROLA R., ROSSIGNAC J., "Compressed progressive meshes", *IEEE Transactions on Visualization and Computer Graphics*, vol. 6, no. 1, pp. 79–93, 2000.

[PAN 05] PAN Y., CHENG I., BASU A., "Quality metric for approximating subjective evaluation of 3-D objects", *IEEE Transactions on Multimedia*, vol. 7, no. 2, pp. 269–279, 2005.

[PEN 05] PENG J., KUO C.-C.J., "Geometry-guided progressive lossless 3D mesh coding with octree (OT) decomposition", *ACM Transactions on Graphics (TOG)*, vol. 24, no. 3, pp. 609–616, 2005.

[PEN 10] PENG J., KUO Y., ECKSTEIN I., *et al.*, "Feature oriented progressive lossless mesh coding", *Computer Graphics Forum*, vol. 29, no. 7, pp. 2029–2038, 2010.

[POP 97] POPOVIĆ J., HOPPE H., "Progressive simplicial complexes", *Proceedings of the 24th annual conference on Computer graphics and interactive techniques (SIGGRAPH '97)*, ACM Press/Addison-Wesley Publishing Co., no. 8, pp. 217–224, 1997.

[ROS 99] ROSSIGNAC J., "Edgebreaker: connectivity compression for triangle meshes", *IEEE Transactions on Visualization and Computer Graphics*, vol. 5, no. 1, pp. 47–61, 1999.

[SCH 95] SCHRÖDER P., SWELDENS W., "Spherical wavelets: efficiently representing functions on the sphere", *Proceedings of SIGGRAPH 95*, ACM, pp. 161–172, 1995.

[SEU 07] SEUNG-RYONG HAN., YAMASAKI T., AIZAWA K., "Time-varying mesh compression using an extended block matching algorithm", *IEEE Transactions on Circuits and Systems for Video Technology*, vol. 17 , no. 11, pp. 1506–1518, 2007.

[SHA 48] SHANNON C., "A mathematical theory of communication", *Bell System Technical Journal*, vol. 27, pp. 379–423, 623–656, 1948.

[SOR 03] SORKINE O., COHEN-OR D., TOLDEO S., "High-pass quantization for mesh encoding", *Proceedings of the 2003 Eurographics/ACM SIGGRAPH symposium on Geometry processing (SGP '03)*, Eurographics Association, no. 10, pp. 42–51, 2003.

[SZY 01] SZYMCZAK A., KING D., ROSSIGNAC J., "An edgebreaker-based efficient compression scheme for regular meshes", *Computational Geometry*, vol. 20, nos. 1–2, pp. 53–68, 2001.

[TAU 98a] TAUBIN G., ROSSIGNAC J., "Geometric compression through topological surgery", *ACM Transactions on Graphics*, vol. 17, no. 2, pp. 84–115, 1998.

[TAU 98b] TAUBIN G., GUÉZIEC A., HORN W., *et al.*, "Progressive forest split compression", *SIGGRAPH*, ACM Press, New York, pp. 123–132, 1998.

[TOR 12] TORKHANI F., WANG K., CHASSERY J.-M., "A curvature tensor distance for mesh visual quality assessment", *International Conference on Computer Vision and Graphics*, 2012.

[TOU 98] TOUMA C., GOTSMAN C., "Triangle mesh compression", in DAVIS W.A., BOOTH K.S., FOURNIER A. (eds), *Graphics Interface*, Canadian Human-Computer Communications Society, pp. 26–34, 1998.

[TUR 84] TURAN G., "Succinct representations of graphs", *Discrete Applied Mathematics*, vol. 8, pp. 289–294, 1984.

[VAL 99] VALETTE S., KIM Y., JUNG H., *et al.*, "A multiresolution wavelet scheme for irregularly subdivided 3D triangular mesh", *IEEE International Conference on Image Processing*, Kobe, Japon, vol. 1, pp. 171–174, October 1999.

[VAL 09] VALETTE S., CHAINE R., PROST R., "Progressive lossless mesh compression via incremental parametric refinement", *Computer Graphics Forum*, vol. 28, no. 5, pp. 1301–1310, 2009.

[VAS 11] VASA L., SKALA V., "A perception correlated comparison method for dynamic meshes", *IEEE Transactions on Visualization and Computer Graphics*, vol. 17, no. 2, pp. 220–230, 2011.

[VAS 12] VASA L., RUS J., "Dihedral angle mesh error: a fast perception correlated distortion measure for fixed connectivity triangle meshes", *Computer Graphics Forum*, vol. 31, no. 5, pp. 1715–1724, 2012.

[WAN 12] WANG K., TORKHANI F., MONTANVERT A., "A fast roughness-based approach to the assessment of 3D mesh visual quality", *Computers & Graphics*, vol. 36, no. 7, pp. 808–818, 2012.

[ZHA 07] ZHANG J., OWEN C.B., "Octree-based animated geometry compression", *Computers and Graphics*, vol. 31, no. 3, pp. 463–479, 2007.

Chapter 12

Coding Methods for Depth Videos

12.1. Introduction

Recent advances in the domain of three-dimensional (3D) cinema and the emergence of new multimedia services such as free viewpoint television (FTV) or 3D television (3DTV) have created a need for new, lighter and less costly 3D video formats. This is a current research focus in both the academic and industrial domains.

The first proposal for 3D video representation was the classic stereoscopic format (stereoscopic 3D (S3D)). This consists of two views of the same scene, recorded by two cameras separated by a certain distance (the baseline). The association of a different view with each of the spectator's eyes produces a 3D effect by stereoscopy.

Mixed resolution stereo (MRS), followed S3D. This format exploits the theory of binocular suppression, which stipulates that if the two views have a different image quality, the perceived quality of the stereoscopic video will be closer to that of the view with the highest quality [BRU 09]. In practice, this means that if one of the two views is encoded at a lower resolution, the overall quality perceived by the spectator will remain good. Therefore, MRS allows us to reduce the cost (in terms of bit rate) of encoding and transmitting 3D information.

Chapter written by Elie Gabriel MORA, Joël JUNG, Béatrice PESQUET-POPESCU and Marco CAGNAZZO.

However, two views are not sufficient to deliver wide-range fluid 3D content. This fluidity is necessary for applications such as FTV, where the user has the ability to navigate between different 3D scenes using a controller. Furthermore, for 3DTV, two views are not sufficient to ensure that a spectator not situated directly in front of the screen will experience 3D. The multi-view video (MVV) format was introduced to respond to these issues. The approach involves N views representing the same scene, captured by N cameras arranged and spaced in a specific manner. However, this format is generally costly due to the bit rate required to code and transmit N views.

Depth-based formats are the most recently proposed type of 3D video formats. They are still the subject of study and research, both in the academic sphere and in industry. First, the video + depth (V+D) format consists of a texture view and an associated depth view. Using these two views, a second texture view can be extrapolated at a different position in space, using a technique known as depth image based rendering (DIBR). The newly created texture view and the original view allow stereoscopic viewing of the 3D video. If more views are required, the multiview video + depth (MVD) format may be used. This involves N texture views and their associated depth views. Several intermediate views may then be interpolated using the DIBR technique.

The main advantage of depth-based formats is their low cost, as the bit rate required to code and transmit depth information is lower than that needed for texture information. However, the acquisition of depth videos is not simple. The algorithms for depth estimation, such as stereo matching, are not perfect; the estimated depth often presents artifacts. Obtaining depth directly using time-of-flight cameras is an interesting alternative, but the technology has not yet reached a sufficient level of maturity. Another possibility would involve estimating depth, then *cleaning* each depth map manually, but this is only an option for major cinema studios with the necessary financial resources. Nevertheless, depth-based formats present considerable potential and are currently the object of standardization efforts, allowing these formats to gradually take their place in industry.

The delivery of 3D content is costly in terms of bit rate. High compression gains are required in order to stay within the bandwidth authorized by service providers. In cases where depth-based formats are used, depth videos must be compressed (as in the case of texture videos). However, depth videos have different characteristics from classic texture videos, and, therefore, specific depth compression algorithms must be used. In this chapter, we will present the different methods or tools that have been developed to code depth videos.

This chapter is organized as follows: section 12.2 presents an analysis of depth maps and their specific characteristics. It also includes a discussion of different types of redundancies, which may be exploited in order to maximize compression levels of depth videos. Three main categories of depth coding tools can thus be defined. These tools are presented in section 12.3, with examples. Finally, section 12.4 concludes the chapter.

12.2. Analyzing the characteristics of a depth map

A depth map associates a depth value (a distance to the camera) with each pixel in the texture picture. It is therefore a single-component image (Luma), essentially made up of smooth regions separated by contours, as shown in Figure 12.1. The depth map is not affected by illumination and contains neither textures nor shadows. Depth compression may be carried out in three different ways. First, the intrinsic characteristics of depth maps may be exploited, either at the block level in the coding loop, or in higher level data structures, such as slices or frames. Second, correlations with the associated texture may be exploited. Texture and depth are highly correlated, notably in the contours of objects. Specific depth transforms or certain coding decisions may also depend on texture information. Finally, as depth videos are not shown on screen but rather used to synthesize intermediate views, depth coding may be optimized for the quality of these synthesized views. This means that new distortion models, considering the effects of depth coding on synthesized views, may be established.

Figure 12.1. *Texture frame and associated depth map*

As a result, we may identify three broad categories of depth coding methods: methods using the intrinsic characteristics of depth maps, those using correlations with associated textures and, finally, those which optimize depth coding for the quality of synthesized views. The following section provides a detailed overview of each category, giving examples of tools proposed either in the literature on the subject or in response to the Moving Pictures Expert Group (MPEG) Call for Proposals (CfP) for 3D video in November 2011 [IEC 11].

12.3. Depth coding methods

In this section, we present the three categories of methods used in depth coding, with examples of tools used in each category.

12.3.1. *Methods using the intrinsic characteristics of depth maps*

These methods may be grouped into two subcategories: tools that operate at the block level and those that operate at a higher level. The first subcategory uses the block coding structure found in classic video encoding, with specific modifications through the introduction of new coding modes, better suited for depth coding. The second subcategory considers depth images as a whole (without the use of blocks).

12.3.1.1. *High-level coding tools*

The first example of a coding tool in this category is the reduction in resolution of depth maps before coding, which may be considered as a depth data compression technique [RUS 11]. The artifacts associated with up-sampling the depth map after decoding are minimal, and we may thus obtain considerable compression gains by simply reducing the resolution of the depth videos to be encoded. The test model for advanced video coding (AVC) based 3DV solutions [HAN 12], defined in March 2012, proposes down-sampling depth videos to a quarter of their initial resolution (half in each direction) before encoding.

Another tool involves reducing the resolution of motion vectors used in depth coding. The eight-tap filters used for motion-compensated interpolations, as defined in the sixth High Efficiency Video Coding (HEVC) working draft [BRO 11], produce artifacts on sharp edges in depth. As a result, motion (or disparity) compensated predictions may be modified to avoid interpolations, meaning that only full-pel motion vectors and disparity vectors will be used. The reduction in the precision of motion vectors also reduces the bit rate required to transmit the motion vector differences [SCH 11].

View synthesis prediction is another efficient coding tool, used for both texture and depth [RUS 11]. Using a reconstructed depth map D_0^* at view 0, a depth map D_1' may be synthesized for view 1 and used as a predictor to code D_1. Synthesizing an image involves making the pixels of a source image $s(x, y)$ correspond to a destination image $t(x, y)$, situated at a required position (view), as follows:

$$t\left(\lfloor x + D(x,y)\rfloor, y\right) = s(x, y)$$

$$D(x, y) = \frac{f * B}{z(x, y)}$$ [12.1]

$$z(x, y) = \left(\frac{d(x, y)}{255}\left(\frac{1}{Z_{near}} - \frac{1}{Z_{far}}\right) + \frac{1}{Z_{far}}\right)^{-1}$$

where f is the focal distance, B is the distance between cameras (for views 0 and 1 in our case), $D(x, y)$ is the value of the disparity at position (x, y) and $d(x, y)$ is the depth value at position (x, y). As, in this case, we are synthesizing a depth map, and we obtain $s(x, y) = d(x, y)$.

Another useful tool is the Z_{near} Z_{far} compensation [DOM 11] for weighted prediction. Different frames of the same view or different views at the same instant in time, for a depth video sequence, may have different extremal depth values (denoted as Z_{near} and Z_{far}). As depth maps are rescaled in the interval $[0, 255]$, different depth images may be scaled differently. In this case, the use of a depth map as the reference image for another depth map would lead to poor predictions and inefficient compression. To solve this problem, coherent scaling needs to be applied for all depth maps in question. This is easy to achieve using extremal values. For example, if a depth map with extremal values Z_{near}^s and Z_{far}^s is used as the reference image for a current depth map with extremal values Z_{near}^t and Z_{far}^t, the scaling for the current depth map is carried out as follows:

$$L_T = L_S * \frac{Z_{far}^s - Z_{near}^s}{Z_{far}^t - Z_{near}^t} + 255 * \frac{Z_{near}^s - Z_{near}^t}{Z_{far}^t - Z_{near}^t}$$ [12.2]

where L_S is the original depth value and L_T is the rescaled value.

12.3.1.2. *Block-based coding tools*

The first tool for consideration in this category is the approximation of depth blocks using modeling functions, presented in [MER 08]. As the depth map is essentially made up of smooth regions separated by contours, a depth block may be approximated by four different types of functions: a constant function, a linear function, a constant piecewise function or a linear piecewise function. If no approximation can be found for the current block, then the block is divided into four blocks in a quad-tree manner. The process is repeated for each block until an approximation function is found for each leaf of the quad-tree.

The same principle was used and improved in a contribution to the MPEG CfP for 3D video [SCH 11]. This contribution proposes four new Intra modes, two of which (mode 3 and mode 4) use texture information and will be discussed further in section 12.3.2.2. These modes are known as depth modeling modes (DMM). In mode 1, the current depth block is approximated by two constant regions, denoted by R_1 and R_2, separated by a line segment DF, as shown in Figure 12.2. The predictor signal (or block) is thus made up of two regions, and the value P_i, where $i = \{1; 2\}$, of each pixel in a region R_i is equal to the mean value of the pixels in the original block covered by R_i. Thus, P_1 and P_2 are transmitted to the decoder, along with the partition information, which consists of the beginning and end points (D and F) of the segment DF separating the two regions. In mode 2, P_1 and P_2 are transmitted without the partitioning information, which, in this case, is deduced from the neighboring blocks. However, the resulting partition may not be adequate for the current block. Thus, an offset F_{off}, correcting the arrival point of the straight line, as shown in Figure 12.2, is also transmitted in the bitstream.

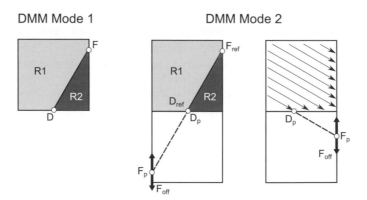

Figure 12.2. *Depth modeling modes 1 and 2*

While the first two tools involve Intra prediction, the adaptive 2D block matching (2D-BM)/3D block matching (3D-BM) selection tool concerns inter-prediction [KAM 10]. 2D-BM is the classic way in which an encoder, during a motion estimation stage, finds the temporal reference block that best corresponds to the current block. The correspondence is measured by the sum of squared errors (SSE), defined as follows:

$$SSE(i, j) = \sum_{m=0}^{M-1} \sum_{n=0}^{N-1} (f(m, n) - g(m + i, n + j))^2 \qquad [12.3]$$

where f is the original current block of size $M \times N$ to code, and g is the temporal reference block, also of size $M \times N$. In this case, the search is carried out in two dimensions: horizontal and vertical. However, in depth videos, we also have motion in the depth direction. Search precision may thus be increased by extension to a third dimension. This idea is implemented in 3D-BM. Thus, the best temporal correspondence is the one that minimizes the new formulation of the SSE, as follows:

$$SSE(i, j, k) = \sum_{m=0}^{M-1} \sum_{n=0}^{N-1} (f(m, n) - g(m + i, n + j) + k)^2 \qquad [12.4]$$

3D-BM is more efficient than 2D-BM for high bit rates, where the gains produced by the reduction of distortion are higher than the costs associated with the addition of a new component for motion vectors. Inversely, at low bit rates, 2D-BM is more efficient than 3D-BM. Adaptive 2D-BM/3D-BM selection evaluates the R-D cost for the coding of each block with 2D-BM and 3D-BM and chooses the method that minimizes cost. The choice is then sent in the bitstream for decodability. This adaptive selection is more efficient than pure 2D-BM and 3D-BM for middle bit rates.

Transforms may also be adapted for depth content. Edge-adaptive transforms (EAT) are proposed in [SHE 10]. These transforms directly take into account the contours of a depth block, and consequently enable a reduction in the number of non-zero coefficients after filtering. An initial contour detection stage is carried out on the residual block, and a binary contour map is established. An adjacency matrix A is then calculated based on this binary contour map, as follows:

$$A(i, j) = \begin{cases} 1 & \text{if } i \text{ and } j \text{ are direct neighbors} \\ 0 & \text{otherwise} \end{cases} \qquad [12.5]$$

Using the adjacency matrix, a degree matrix D may be calculated as follows:

$$D(i,j) = \begin{cases} 0 & \text{if } i \neq j \\ \sum_{k=0}^{N-1} A(i,k) & \text{otherwise} \end{cases} \qquad [12.6]$$

The Laplacian matrix L may thus be calculated as the difference between D and A: $L = D - A$. Finally, from L, the transform matrix E^t may be calculated using the cyclical Jacobi method. Take a simple example of a residual of size 2×2, as shown in Figure 12.3. In this example, matrices A, D, L and E^t will be equal to:

$$A = \begin{bmatrix} 0 & 1 & 0 & 0 \\ 1 & 0 & 0 & 0 \\ 0 & 0 & 0 & 1 \\ 0 & 0 & 1 & 0 \end{bmatrix} \qquad D = \begin{bmatrix} 1 & 0 & 0 & 0 \\ 0 & 1 & 0 & 0 \\ 0 & 0 & 1 & 0 \\ 0 & 0 & 0 & 1 \end{bmatrix}$$

$$L = \begin{bmatrix} 1 & -1 & 0 & 0 \\ -1 & 1 & 0 & 0 \\ 0 & 0 & 1 & -1 \\ 0 & 0 & -1 & 1 \end{bmatrix} \qquad E^t = \begin{bmatrix} \frac{1}{\sqrt{2}} & \frac{1}{\sqrt{2}} & 0 & 0 \\ 0 & 0 & \frac{1}{\sqrt{2}} & \frac{1}{\sqrt{2}} \\ \frac{-1}{\sqrt{2}} & \frac{1}{\sqrt{2}} & 0 & 0 \\ 0 & 0 & \frac{-1}{\sqrt{2}} & \frac{1}{\sqrt{2}} \end{bmatrix}$$

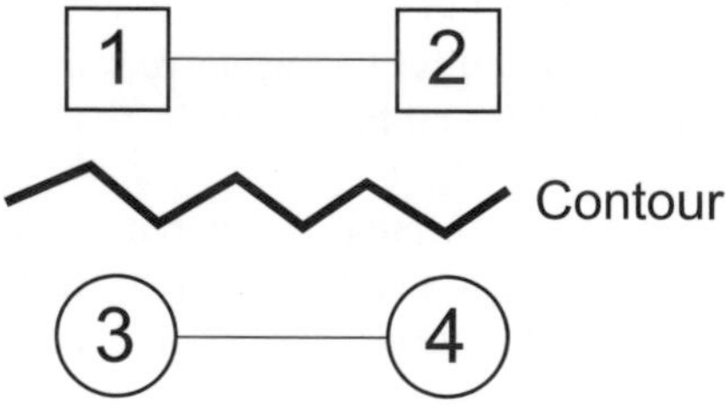

Figure 12.3. *Residual of a 2 × 2 block*

We can prove that for a residual of size $N \times N$, piecewise constant with M constant regions, the EAT coefficients will be composed of at most M non-zero coefficients and $N^2 - M$ zero coefficients. As a result, coding efficiency may be improved by using EAT instead of the classic discrete cosine transform (DCT). However, the use of EATs requires transmission of the binary contour map, which will not otherwise be available to the decoder. Thus, EATs may be less effective than the DCT. An R-D criterion may be specified in order to identify the best transform, but this choice must also be transmitted to the decoder.

12.3.2. *Methods exploiting correlation with associated textures*

There is a strong correlation between the texture and depth components, and this correlation can be exploited to produce coding gains at depth level. First, the prediction modes used for different depth blocks may be selected according to the information present in texture. Second, the prediction information for depth blocks, such as motion vectors, reference picture indices or Intra modes, may be inherited from texture blocks. Finally, texture information may be used to construct spatial transforms better suited to depth coding in order to increase coding efficiency.

12.3.2.1. *Inheritance/selection of prediction modes*

Depth block skip (DBS), is a tool presented in a contribution [LEE 11b] to the MPEG CfP for 3DV. It allows coding of a depth macroblock in Skip if the temporal correlation in texture is judged to be sufficient. If there is a high temporal correlation in texture, there is a high chance that this correlation will also be strong with depth, and, in this case, the best mode (in terms of R-D) for a depth block will be Skip. In DBS, this block is forced to be coded using Skip, meaning that the Skip mode does not need to be signaled. Formally, for a current depth block X to be coded in the depth frame D_i at instant i, we calculate the sum of squared differences (SSD) between A, the colocated block in the reconstructed texture frame T_i, and B, the colocated block in the reconstructed texture frame T_{i-1}. If this SSD is below a certain threshold, we consider that the temporal correlation in texture – and also in depth – is high, and we code X in Skip mode with C, the colocated block in D_{i-1}, as a candidate, meaning that the prediction information for C will be inherited by X, as shown in Figure 12.4. This tool has certain drawbacks. First, it uses two texture frames to code a depth block. This represents a relatively large amount of additional information to use, only justifiable if the method produces significant gains in terms of depth compression, which is not the case. Moreover, using this tool, depth coding is subject to errors in the texture, which affect an access unit different to that of the current depth frame, thus reducing the robustness of the coding. Finally, the tool violates a general codec design principle, which stipulates that the parsing of the bitstream should be independent of signal processing applied to already decoded pictures.

In a more direct manner, Skip mode in depth may also be forced each time that the colocated texture block is coded using Skip [KIM 09]. The main idea behind this tool is that the spatial uniformity of movement in texture is likely to also be present in depth. Skip mode may not necessarily be chosen under normal circumstances due to the presence of certain artifacts inherent in the original depth, which create false movements. Imposing Skip mode on a depth block if it is chosen for the colocated texture block eliminates these artifacts, thus increasing the quality of views that will be synthesized with the current depth frame. The tool also produces compression gains due to the fact that no signaling is required for Skip mode. Finally, the tool also reduces the complexity of the encoder as it reduces the number of complex motion estimations and compensations to perform.

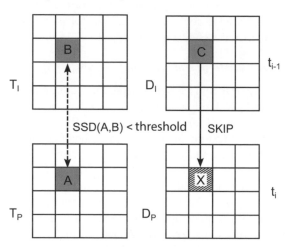

Figure 12.4. *The DBS coding tool*

12.3.2.2. *Inheritance of prediction information*

Motion information (motion vectors + reference picture indices) is highly correlated between texture and depth, notably around object contours. In [SEO 10], a new mode, known as motion sharing (MS), is added to the list of existing modes (Intra, Inter, Skip). In this new mode, the motion information for a depth block is directly inherited from the corresponding texture block. No motion estimation is carried out. MS mode is not systematically imposed for depth blocks: an R-D criterion verifies the cost and compares it to those entailed by other prediction modes. MS is then selected only if it offers the lowest cost.

Motion parameter inheritance (MPI), introduced in [SCH 11], is another tool that allows inheritance of motion information from texture for depth, but the technique used is somewhat different. In MPI, there is no new prediction mode; the texture motion information and its partitions are considered as a new candidate for Merge mode for the current depth block. Merge is a coding mode, introduced in HEVC, in which the current block inherits prediction information from a given candidate, selected from a group of candidates (we thus speak of "competition" between different candidates). The selection is based on an R-D criterion, and only the index of the selected candidate is transmitted to the decoder in order to construct the predictor signal.

The Intra mode of a texture may also be inherited, as proposed in [BAN 11], for the colocated depth block. As we see in Figure 12.5, if the texture block is partitioned more finely than the depth block, the Intra mode of the top left block will be inherited. If, on the other hand, the depth block is more finely partitioned than the texture block, the Intra mode of the texture block will be inherited for all depth partitions. Here, inheriting the texture of the Intra mode means that the mode will be added to the list of most probable Intra candidates most probable mode (MPM) for the current depth block, where it will act as a predictor for the depth Intra mode.

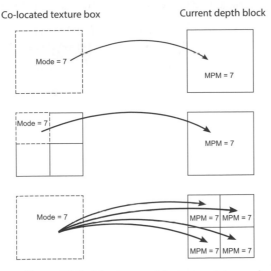

Co-located texture box   Current depth block

Figure 12.5. *Inheritance of the texture Intra mode*

The method described above involved inheritance of the texture Intra mode for all depth blocks. However, the Intra modes of the two components do not always match. In cases where there is no matching, inheritance of the texture Intra mode can lead to coding losses, as the inherited mode may replace a good spatial predictor in the list of MPM candidates. In [MOR 12], a study shows that texture and depth Intra modes are particularly correlated in areas where contours are clearly defined in texture. Thus, inheritance is only applied if a criterion measuring the sharpness of the contours present in the colocated texture block is above a certain threshold.

Texture information may also be used to define new Intra modes for modeling depth blocks. We have already discussed modes 1 and 2 in section 12.3.1.2. Mode 3 still seeks to approximate the depth block using two constant regions separated by a straight line, but in this case, the partition information is inherited directly from the texture. A simple thresholding of Luma texture samples may be used to separate the texture block into two distinct regions. This partition is used directly for the depth block. Thus, no partition information is transmitted as the same procedure may be carried out by the decoder. Constants P_1 and P_2, on the other hand, must always be transmitted. Mode 4 operates in the same way as mode 3, except the depth block is divided into three constant regions, as shown in Figure 12.6 (this is known as "Contour" type partitioning, as opposed to "Wedgelet" partitioning, where the block is divided into two constant regions, as in modes 1, 2 and 3).

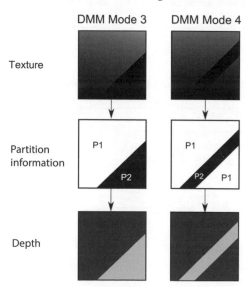

Figure 12.6. *Depth modeling modes 3 and 4*

12.3.2.3. *Spatial transforms*

Texture information may also be used to construct spatial transforms suited to depth coding. Daribo *et al.* [DAR 08] propose an adaptive wavelet lifting scheme, in which short filters are applied to depth areas containing contours and long filters are applied to homogeneous depth areas. Contour detection is carried out in texture for decodability, and is based on observed correlations between texture and depth contours.

12.3.3. *Methods optimizing depth coding for the quality of synthesized views*

As depth maps are not shown on screen but rather used to synthesize views, the R-D model used in depth coding must consider distortion directly on synthesized views in order to optimize coding for the truly important aspects, i.e. the intermediate synthesized views. There are different methods, some of which effectively synthesize views (or parts of views) during depth coding, while others estimate this correlation using texture information. We will compare these different techniques in this section.

12.3.3.1. *View synthesis optimization*

This method, known as view synthesis optimization (VSO), was proposed in [SCH 11]. It calculates the change in the distortion of the synthesized views caused by a change at the level of the current depth block. This metric, called synthesized view distortion change (SVDC), is calculated by synthesizing, in-loop, part of the view, which will be affected by a change in the depth block. Figure 12.7 shows how this metric is calculated. We consider a frame s_D, where the depth block is in its original form. All succeeding blocks (in coding order) are in their original forms and all previous blocks have already been coded and reconstructed. Another depth frame $\tilde{s}_D$ is formed in the same way as s_D, except the depth block is replaced by a reconstructed version as a function of the test (mode) being carried out. s_D and $\tilde{s}_D$ are used to synthesize two texture views, s'_T and $\tilde{s}'_T$. Another view $s'_{T,ref}$ may also be synthesized using the original depth and texture frames. The calculation of the SSE between s'_T and $s'_{T,ref}$ gives E, and between $\tilde{s}'_T$ and $s'_{T,ref}$ gives $\tilde{E}$. The SVDC metric is simply equal to the difference $\tilde{E} - E$.

This method is useful for three main reasons. First, it offers an exact measure of distortion, taking account of occlusions and disocclusions in the synthesized views. Second, the measure is related to a block. Third, using this metric, partial distortions are additive. If a block is divided into four sub-blocks, as is often the case when using quad-tree coders such as HEVC, the sum of the distortions caused by a change in each sub-block must be equal to the distortion caused by the corresponding change at the level of the whole block. SVDC guarantees this, as only the change in distortion in the synthesized view caused by a change in the current depth block is considered, and not the total distortion itself in the synthesized view. The main drawback of this method is the increase in complexity in terms of coding, as the metric is calculated (and thus a partial view synthesis is produced) for each possible depth partition. This view synthesis, although partial, requires a considerable number of operations, which may lead to a rapid increase in complexity.

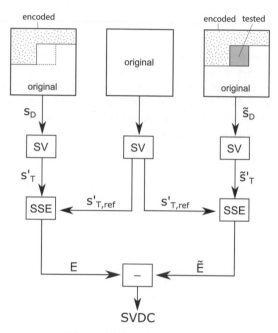

Figure 12.7. *Calculating SVDC*

12.3.3.2. *Distortion models*

Other distortion models do not effectively synthesize views in the depth coding loop; the synthesized view distortion is simply estimated. In [LEE 11a], the view synthesis distortion (VSD) metric is calculated as a distortion in the depth block, weighted by one-pixel texture translation differences:

$$VSD = \sum_{(x,y)} \left[\frac{\alpha}{2} \cdot \left| D_{x,y} - \tilde{D}_{x,y} \right| \cdot \left(\left| \tilde{C}_{x,y} - \tilde{C}_{x-1,y} \right| \right.\right.$$

$$\left.\left. + \left| \tilde{C}_{x,y} - \tilde{C}_{x+1,y} \right| \right) \right]^2 \qquad [12.7]$$

where $D_{x,y}$, $\tilde{D}_{x,y}$ and $\tilde{C}_{x,y}$ are, respectively, the original depth value, the reconstructed depth value and the reconstructed texture value at position (x, y). The value of α is defined as follows:

$$\alpha = \frac{f \cdot B}{255} \cdot \left(\frac{1}{Z_{near}} - \frac{1}{Z_{far}} \right) \tag{12.8}$$

where f is the focal distance, B is the distance between the camera for the current view and the one for the synthesized view and Z_{near} Z_{far} are the extremal depth values.

In equation [12.7], we suppose that two adjacent pixels will remain adjacent after the warping operation, which is not always the case as occlusions or disocclusions may occur in synthesized views. To rectify this, $\tilde{C}_{x-1,y}$ and $\tilde{C}_{x+1,y}$ in equation [12.7] may be replaced by $\tilde{C}_{x_L,y}$ and $\tilde{C}_{x_R,y}$, respectively, with x_L and x_R defined as follows:

$$\begin{aligned} x_L &= x + \arg\max\nolimits_{l \geq 1} \lfloor \alpha \cdot (D_{x-l,y} - D_{x,y}) - l \rfloor \\ x_R &= x + \arg\min\nolimits_{r \geq 1} [\max(\alpha \cdot (D_{x+r,y} - D_{x,y}) + r, 0)] \end{aligned} \tag{12.9}$$

Equation [12.7] is then corrected to take into account the regions of occlusion and disocclusion as follows:

$$VSD = \begin{cases} \sum\limits_{(x,y)} \left[\frac{\alpha}{2} \cdot \left| D_{x,y} - \tilde{D}_{x,y} \right| \cdot P_C \right]^2 & \text{si } x_L < x \\ 0 & \text{si } x_L \geq x \end{cases} \tag{12.10}$$

where P_C, the weight in texture, is defined as follows:

$$P_C = \left| \tilde{C}_{x,y} - \tilde{C}_{x_L,y} \right| + \left| \tilde{C}_{x,y} - \tilde{C}_{x_R,y} \right| \tag{12.11}$$

An even simpler distortion model, which estimates the distortion in synthesized views, was proposed in [KIM 09]. In fact, a distortion ΔD_{prof} in a depth value at position (x, y) produces a translation error ΔP in the synthesized view. A linear relationship exists between the two quantities, defined by:

$$\Delta P = \alpha \cdot \Delta D_{prof} \tag{12.12}$$

where α is defined by equation [12.8]. Moreover, there is a linear relationship between a global pixel translation t_x and the distortion measured in terms of the SSD of the original video signal translated by t_x. This relationship is defined by:

$$d_{SSD}(t_x) = \sum_x \sum_y \left(V_{(x,y)} - V_{(x-t_x,y)} \right)^2 \qquad [12.13]$$

where $V_{(x,y)}$ is the value of the original video signal at position (x,y). If several values of t_x are used and we calculate the corresponding distortions, we may define a scale factor s in the following way:

$$s = \frac{\mathbf{d}_{\mathbf{SSD}}^{\mathbf{T}} \cdot \mathbf{t}_{\mathbf{x}}}{\mathbf{t}_{\mathbf{x}}^{\mathbf{T}} \cdot \mathbf{t}_{\mathbf{x}}} \qquad [12.14]$$

where $\mathbf{d}_{\mathbf{SSD}}$ and $\mathbf{t}_{\mathbf{x}}$ are the vectors formed by the aggregation of multiple values of $d_{SSD}(t_x)$ and t_x, respectively, and $\mathbf{T}$ is the transpose operator. For a certain position error ΔP, this parameter s provides an estimation of the resulting distortion in the synthesized view. If the views are synthesized from a left view and a right view, each contribution generally has a weight, p for the left and $(1-p)$ for the right where $p \in [0;1]$:

$$V_{synth} = p \cdot V_{left} + (1-p) \cdot V_{right} \qquad [12.15]$$

A scale factor representing the global characteristics of V_{left} may be defined by:

$$k = p \cdot s \qquad [12.16]$$

Using the two parameters above, a new distortion metric may be calculated in the following manner:

$$\Delta D_{synth}^2 = k \cdot \Delta P = k \cdot \alpha \cdot |\Delta D_{prof}| \qquad [12.17]$$

Finally, the Lagrangian cost function used in R-D optimization may be written as:

$$J = \sum_x \sum_y \Delta D_{synth}^2(x,y) + \lambda \cdot R_{prof}$$

$$J = k \cdot \alpha \sum_x \sum_y |\Delta D_{prof}| + \lambda \cdot R_{prof}$$

The main advantage of these distortion models, based on the disparity, is that they are less complex than solutions that effectively synthesize views (or parts of views) in the depth coding loop itself. However, the distortion evaluated in the second case is more precise than that obtained in the first case, a fact that may influence R-D decisions and increase coding efficiency.

Other depth coding tools aim to increase the sparsity of depth in the transform domain in order to increase coding efficiency. This is carried out by modifying the original values of the depth map before coding, as long as the change only affects synthesized views below a certain, predefined threshold. Depth map coding using *Don't Care Regions* (DCR) [CHE 10] falls into this category.

12.4. Conclusion

In this chapter, we have introduced a number of different depth coding tools found in the literature and in contributions to the MPEG CfP for 3DV. These tools are divided into three broad categories. Certain approaches exploit the intrinsic characteristics of depth maps, such as representations in smooth regions separated by contours, their role in view synthesis or their capacity to represent motion in a depth direction. Other tools exploit the correlation between depth and associated textures. Choices may be made concerning the prediction modes used in coding depth maps, depth block prediction information may be inherited from texture blocks and spatial transforms may be designed specifically for depth using texture information. Finally, certain tools do not use depth coding algorithms; instead, they introduce new distortion models used in R-D optimization during depth coding to evaluate distortion in the most critical areas, i.e. directly on synthesized views. Several 3DV standardization meetings are planned for the near future with the aim of defining a new standard for 3D video coding. In the course of these meetings, inefficient depth coding tools will be removed, others will be studied and potentially integrated until depth compression rates reach an acceptable level. In any case, it seems that depth-based 3D formats have finally begun to attract the interest of the industrial sector.

12.5. Bibliography

[BAN 11] BANG G., YOO S., NAM J., "Description of 3D video coding technology proposal by ETRI and Kwangwoon University", ISO/IEC JTC1/SC29/WG11 MPEG2011/M22625, November 2011.

[BRO 11] BROSS B., HAN W.-J., OHM J.-R., *et al.*, "Working draft 5 of High Efficiency Video Coding", ITU-T SG16 WP3 & ISO/IEC JTC1/SC29/WG11 JCTVC-G1103, November 2011.

[BRU 09] BRUST H., SMOLIC A., MUELLER K., *et al.*, "Mixed resolution coding of stereoscopic video for mobile devices", *3DTV Conference: The True Vision – Capture, Transmission and Display of 3D Video, 2009*, Potsdam, Germany, pp. 1–4, May 2009.

[CHE 10] CHEUNG G., KUBOTA A., ORTEGA A., "Sparse representation of depth maps for efficient transform coding", *Picture Coding Symposium (PCS)*, IEEE, pp. 298–301, December 2010.

[DAR 08] DARIBO I., TILLIER C., PESQUET-POPESCU B., "Adaptive wavelet coding of the depth map for stereoscopic view synthesis", *IEEE 10th Workshop on Multimedia Signal Processing (MMSP)*, Cairns, Queensland, pp. 413–417, 8–10 October 2008.

[DOM 11] DOMANSKI M., GRAJEK T., KARWOWSKI D., *et al.*, "Technical description of Poznan University of Technology proposal for call on 3D video coding technology", ISO/IEC JTC1/SC29/WG11 MPEG2011/ M22697, November 2011.

[HAN 12] HANNUKSELA M., "Test model for AVC based 3D video coding", ISO/IEC JTC1/SC29/WG11 MPEG2012/N12558, March 2012.

[IEC 11] IEC, "Call for proposals on 3D video coding technology", ISO/IEC JTC1/SC29/WG11 N12036, March 2011.

[KAM 10] KAMOLRAT B., FERNANDO W., MRAK M., "Adaptive motion-estimation-mode selection for depth video coding", *IEEE International Conference on Acoustics Speech and Signal Processing (ICASSP)*, IEEE, Dallas, TX, pp. 702–705, 14–19 March 2010.

[KIM 09] KIM W.-S., ORTEGA A., LAI P., *et al.*, "Depth map distortion analysis for view rendering and depth coding", *16th IEEE International Conference on Image Processing (ICIP)*, Cairo, pp. 721–724, 7–10 November 2009.

[LEE 11a] LEE J., OH B.T., LIM I., "Description of HEVC compatible 3D video coding technology by Samsung", ISO/IEC JTC1/SC29/WG11 MPEG2011/M22633, November 2011.

[LEE 11b] LEE J.Y., WEY H.-C., PARK D.-S., "A fast and efficient multi-view depth image coding method based on temporal and inter-view correlations of texture Images", *IEEE Transactions on Circuits and Systems for Video Technology*, vol. 21, no. 12, pp. 1859–1868, December 2011.

[MER 08] MERKLE P., MORVAN Y., SMOLIC A., *et al.*, "The effect of depth compression on multiview rendering quality", *3DTV Conference: The True Vision – Capture, Transmission and Display of 3D Video*, IEEE, Istanbul, Turkey, pp. 245–248, May 2008.

[MOR 12] MORA E., JUNG J., PESQUET-POPESCU B., *et al.*, "Codage de vidéos de profondeur basé sur l'héritage des modes Intra de texture", *COmpression et REprésentation des Signaux Audiovisuels (CORESA)*, Lille, France, April 2012.

[RUS 11] RUSANOVSKY D., HANNUKSELA M., "Description of Nokia's response to MPEG 3DV call for proposals on 3DV video coding technologies", ISO/IEC JTC1/SC29/WG11 MPEG2011/M22552, November 2011.

[SCH 11] SCHWARZ H., BARTNIK C., BOSSE S., "Description of 3D video technology proposal by Fraunhofer HHI", ISO/IEC JTC1/SC29/WG11 MPEG2011/M22571, November 2011.

[SEO 10] SEO J., PARK D., WEY H.-C., *et al.*, "Motion information sharing mode for depth video coding", *3DTV-Conference: The True Vision – Capture, Transmission and Display of 3D Video (3DTV-CON)*, Tampere, Finland, pp. 1–4, 7–9 June 2010.

[SHE 10] SHEN G., KIM W.-S., NARANG S., *et al.*, "Edge-adaptive transforms for efficient depth map coding", *Picture Coding Symposium (PCS)*, Nagoya, Japan, pp. 566–569, December 2010.

Chapter 13

Stereoscopic Watermarking

13.1. Introduction

The success of movie releases such as *Avatar* or *Alice in Wonderland* marks, for audiences, the coming of age of three-dimensional (3D) stereoscopic cinematography. The launch of 3D television systems allowing direct retransmission of major events has reinforced the trend. This new content presents high added value and increases the risk of illegal copying; therefore, we need to find the means of preserving its intellectual property rights, without imposing additional operational constraints on the underlying chain of production, distribution and consumption.

Watermarking technology presents a response to this need: digital works can be authenticated and traced by inserting additional information that is invisible (transparent) to the user but persistent (robust) against attacks. In practice, the main watermarking challenge is to find a point of operational balance between transparency, robustness and the amount of inserted information. Computational cost is also a key factor in the case of real-time applications.

The state of the art in stereoscopic watermarking is broadly made up of a variety of approaches inherited from 2D/2D+t watermarking. These approaches are based either on spread spectrum (SS) communications theory [COX 02] or on side information (SI) theory [COS 83, SHA 58].

Chapter written by Mihai MITREA, Afef CHAMMEM and Françoise PRÊTEUX.

SS techniques offer excellent levels of transparency while guaranteeing robustness for a mark of small size. SI techniques allow the insertion of a larger quantity of information, often to the detriment of transparency and robustness levels. Hybrid methods, which make use of both SS and SI techniques, aim to ensure robustness and transparency while inserting a large quantity of information [MIT 07].

While the capture, production and projection of stereoscopic content are based solely on two views, left and right, the reduction of bandwidth during transmission and/or storage requires the use of intermediary representations, such as disparity maps. This raises questions for the watermarking of stereoscopic video content concerning the optimal domain for insertion.

This issue is the subject of this chapter that is structured into four sections. In section   13.2, we present the constraints of stereoscopic video watermarking. Section 13.3 contains an analysis of different watermarking techniques. Section 13.4 presents experimental results obtained using these techniques, and section 13.5 concludes the chapter, offering perspectives for future research.

13.2. Constraints of stereoscopic video watermarking

13.2.1. *Theoretical framework*

Watermarking techniques [COX 02] consist of inserting additional information (a mark) into multimedia content. This mark must be imperceptible (transparent) and persistent (robust). The information must be found in all replicas of the watermarked content, even following more or less severe transformations, whether malicious or mundane. From a theoretical standpoint, all watermarking processes are modeled by a noisy transmission channel (see Figure 13.1). The mark represents a sample of the information source, which must be transmitted through a channel of this type. The host content and malicious attacks represent noise sources. In this context, the properties of transparency and robustness are contradictory. Our aim is to optimize transmission via the subjacent information channel, i.e. to maximize the quantity of information while respecting transparency and robustness constraints. While the right and left views of stereoscopic content constitute a natural domain for the insertion of a mark, other modes of representation may be considered: the representations defined for optimal transmission in terms of bandwidth for efficient storage of stereoscopic content, for example, also constitute possible domains of insertion. The disparity map, which is clearly interesting as a representation derived from stereoscopic content, may also be considered as original content in Figure 13.1.

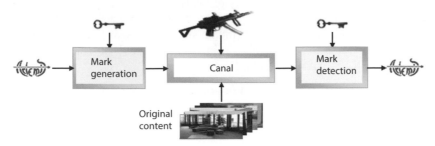

Figure 13.1. *Theoretical model of watermarking*

13.2.2. *Properties*

13.2.2.1. *Transparency*

The property of transparency is linked to the (non-)perception of artifacts introduced during the watermarking procedure. It relates to the notions of watermark fidelity and quality. A watermarking procedure is said to demonstrate fidelity if there is no perceivable difference between the original and marked media. However, it is said to be of good quality if the artifacts do not disturb the human user [COX 02].

The assessment of transparency is essentially subjective, and depends on the observer (who will possess personal characteristics such as age, professional experience and visual acuity) and other factors such as the environment, test conditions and evaluation methodology. The International Telecommunications Union (ITU) defines recommendations for evaluating image quality in documents ITU-R BT 500-12 [ITU 02] and BT 1438 [ITU 00]. These recommendations include general test methods, conditions linked to the evaluators, evaluation scales and observation conditions. Evaluators must possess a prescribed level of visual acuity and normal color vision. The number of evaluators depends on the sensitivity and reliability of the selected test procedure, with a lower limit of 15 individuals. Laboratory observation conditions are also specified, including observation distances, screen dimensions and lighting conditions.

In spite of these recommendations, a wide variety of test methods are used. In practice, each evaluation problem requires a particular method. Nevertheless, a five-level perception quality scale has been widely adopted: bad, poor, fair, good and excellent.

As this type of procedure is long and costly, approaches using objective evaluation measurements have been developed. While these measurements cannot guarantee the absolute quality of visual data, they are widely used due to the practical advantages they present: the measurements are automatic, precise, rapid and highlight trends. In this study, the visual quality of content has been assessed objectively using two types of measurements [CHI 11, ESK 95, WAN 04]:

– Pixel difference: peak signal-to-noise ratio (PSNR) and image fidelity (IF).

– Correlation: normalized cross correlation (NCC) and structural similarity (SSIM).

These measurements are calculated individually for each frame of the video sequence, and then averaged across the whole sequence. In an ideal case, for two identical images, the PSNR tends toward infinity and the IF, NCC and SSIM values are equal to 1.

13.2.2.2. Robustness

Robustness is the ability of a watermarking process to resist mundane or malicious attacks, i.e. to recover inserted information after an attack. It can be measured by the bit error rate in the detected mark after an attack. Attacks on video content may be classified according to their effects on the original content. The first attack type (noise, quantification and collusion) aims to make the mark unreadable at detection level. Geometric attacks (rotations, random geometric transformations, etc.) destroy the synchronization of the mark. After an attack of this type, the mark remains present in the video, but is difficult to locate. Protocol attacks aim to make the mark unusable by creating ambiguities. Finally, cryptographic attacks aim to detect the mark without possessing the security key. As the last two attack classes are broadly generic, this study will focus on the first two classes, which are specific to watermarking.

13.2.2.3. Data payload

The data payload is defined by the size of the inserted mark (expressed in bits). It depends on the target application and may vary from 64 bits to approximately 100 kbits per video sequence. The data payload is directly linked to the robustness and imperceptibility of the watermark. The larger the mark, the more perceptible it will be, and its robustness will decrease. The capacity of a watermark is defined as the theoretical upper limit of the quantity of information inserted into original content, which satisfies the imposed constraints of robustness and transparency.

13.2.2.4. *Computation cost*

The computation cost of a watermarking technique takes into account the execution time and the complexity of the algorithm. The execution time corresponds to the time needed for the mark insertion and detection phases. It is expressed in milliseconds (ms) and must be compatible with real-time services, such as video on demand (VOD) or live television. Knowledge of the complexity of an algorithm allows us to allocate the necessary resources in terms of memory. Analyzing the complexity of candidate algorithms for a given problem allows us to identify the most efficient candidate.

13.2.3. Corpus

13.2.3.1. *Constitution criteria*

For all types of application and all types of content (text, image, video, etc.), a number of criteria are involved in organizing a corpus [PIN 99]. The main criteria are as follows:

– Significance conditions: as a corpus is generally constructed for a specific study relating to a specific object, a reality as perceived from a certain viewpoint, the study corpus must be relevant and coherent. In our experiments, we will consider both high-definition television sequences and content intended for use by computers or cell phones.

– Acceptability conditions: the corpus must be representative and comprehensive. It must present a level of detail suited to the degree of precision and the desired richness of the results of analysis. For the sequences analyzed in this study, particular attention will be paid to the generation source, the variety of sizes, shape factors and visual/semantic content.

– Exploitation conditions: the corpus must be homogeneous and of a large size. The data that form the corpus must be commensurable and provide enough elements to allow identification of significant behaviors, in the statistical sense of the term. These aspects will be expressed via statistical analysis of 95% confidence intervals

13.2.3.2. *Considered corpus*

Experimental tests were carried out on two corpora: 3DLive and moving picture experts group (MPEG). Each includes indoor and outdoor scenes, both static and dynamic, in a variety of lighting conditions (see Figures 13.2 and 13.3).

Figure 13.2. *Examples of left and right views sampled from the 3DLive corpus: rugby, volleyball, dance, theater play and rock concert sequences*

Figure 13.3. *Examples of left and right views sampled from the MPEG corpus: roller, office, city tour and cartoon sequences. The frames are represented at the same size, although their original sizes differ*

The 3DLive corpus was generated as part of the 3DLive project [3DL 12], and is made up of 2 h 11 min of stereoscopic video sequences, or 197, 000 stereoscopic pairs encoded at a rate of 25 frames per second. It includes sequences from rugby matches (*Rugby*), volleyball (*Volley*), a dance performance (*Dancing*), theater (*Theater*) and a rock concert (*Rock*). The sequences are coded in full high definition (HD) (1, 920 × 1, 080 pixels).

The MPEG corpus [MOB 12] is made up of 41 sequences, corresponding to 17 min 29 s, or 29,908 stereoscopic pairs, with several different frame scanning frequencies. A variety of resolutions are present, from 320×192 to 640×480 pixels. The content includes street-based sports (roller skating or bicycle races), indoor scenes (office), city tours and cartoons.

13.2.4. *Conclusion*

The principle of watermarking consists of associating additional information (copyright information) with original multimedia content in a transparent and robust manner. In specifying an operational watermarking technique, the aim is to establish a method satisfying the operational balance between the data payload, transparency, robustness and computation time.

13.3. State of the art for stereoscopic content watermarking

A number of recent studies concerning stereoscopic content watermarking demonstrate the state of the art in the domain [BHA 09, HWA 03a, HWA 03b, KUM 09, YU 11, ZHA 07]. The insertion method is inherited from $2D/2D + t$ watermarking. These studies may be grouped into two categories by domain of insertion. This domain may be one view (left or right) from the stereoscopic pair, or a representation derived from this pair (the disparity map). Watermarking approaches involving inserting a mark in one of the views of the stereoscopic pair are more widespread than other techniques, and are mostly used for still stereoscopic images [BHA 09, HWA 03a, HWA 03b, KUM 09, YU 11].

In their successive studies [HWA 03a, HWA 03b], Dong-Choon *et al.* consider problems associated with SS watermarking of stereoscopic images. In [HWA 03a], a 1,024 bit mark (representing a visual logo) was inserted in the discrete cosine transform (DCT) of each right view of a sequence of 25 stereoscopic pairs, with a resolution of 256×256 pixels. Transparency performances were evaluated in terms of the PSNR between the marked and original images, with a mean PSNR of 34.89 dB. Robustness was also evaluated in terms of PSNR calculated between the detected and inserted logos; this produced a mean value of 19.04 dB. However, the study includes no information on the applied attacks. In [HWA 03b], mark insertion was carried out in the discrete wavelet transform (DWT). The study used a sequence of three image pairs of 512×512 pixels with a logo of 64×64 pixels for the mark. The mean transparency value in this case was increased by 3.92 dB, reaching a limit of 38.81 dB. The robustness decreased by 2.36 dB (to 16.68 dB).

Kumar et al. [KUM 09] also considered an SS watermarking technique. The disparity map was first estimated in the DWT domain. Insertion was carried out in the singular value decomposition (SVD) of the DWT of the left view and optimized using a genetic algorithm. Validation tests were carried out using five grayscale stereo images with a resolution of 512×512 pixels. The transparency was expressed by a PSNR of 42.88 dB. The robustness against average filtering, rotations, redimensioning and the addition of Gaussian noise was evaluated by calculating the NCC between original messages and those detected after the attacks, producing results of 0.91, 0.91, 0.90 and 0.89, respectively.

In the study carried out by Bhatnager et al. [BHA 09], the inserted mark is represented by the disparity map, calculated directly in the pixel domain. Additionally, this disparity map was inserted using an SS technique applied in the SVD of the discrete fractional Fourier transform (DFrFT) of the left view. The experimental validation process considered a corpus of three stereo images (two image pairs of 256×256 pixels and one pair of 512×512 pixels). The transparency evaluation gave a mean PSNR of 45.92 dB. This value is excellent, but was obtained at the expense of robustness. The robustness, expressed by the NCC between the original and detected marks, was calculated to be 0.64, 0.71, 0.69, 0.98, 0.57, 0.63 and 0.46 for average filtering, median filtering, redimensioning, Joint Photographic Experts Group (JPEG) compression, noise addition, cropping and rotation, respectively.

Campisi [CAM 08] developed a semi-fragile stereoscopic watermarking approach based on the quantization index modulation (QIM) insertion method applied in the DWT domain. The quantity of information inserted was $2,000$ bits per image. Experimental tests showed the proposed method to be robust to JPEG and JPEG 2000 compression, but fragile in relation to simple signal manipulations. The fragility property was evaluated by calculating BER (bit error rate) values. The following numerical results were obtained: 0.07 after Gaussian filtering, 0.11 after median filtering, 0.38 after removal of a line/column, 0.30 after a rotation of 0.25 degree and 0.39 after a rotation of 0.50 degree. No information is supplied concerning transparency and the experimental database.

Yu et al. [YU 11] inserted the mark into the left and right views of the stereoscopic pair, taking account of statistical relationships within and between blocks, established by combining the DWT and the DCT. A parity quantization was also designed to handle the cases not verifying a relationship of this type. During the experiments, a binary pixel logo was inserted into a stereo image of 640×480 pixels. Transparency was evaluated in terms of PSNR between the original and watermarked images, giving values of 52.14 and 51.99 dB, respectively, for the left and right views. Robustness was evaluated in terms of watermarking recovering ratio (WRR) (also called HC in [YU 11]) given by $WRR = 1 - \sum w \oplus w' / m \times m$, where $\oplus$ is exclusive-OR, w is the original binary mark of size $m \times m$ and w' is the detected mark. WRR values of 0.94, 0.90, 0.94 and 0.81 were obtained after application of JPEG compression, salt-and-pepper noise, median filtering and cropping, respectively. Note that $WRR = 1 - BER$.

Watermarking techniques where the mark is inserted into a disparity map are represented by the study described in [ZHA 07]. The insertion technique combines the SS principle with the low-density parity control (LDPC) error correction code. Experimental tests were carried out using four video sequences: three of these sequences were made up of 22 stereoscopic images, while the fourth included 192 frames of 720×576 pixels. The transparency was calculated using the mean image degradation index, defined as $\bar{\delta} = \bar{\alpha} - \bar{\beta}$, where $\bar{\alpha}$ and $\bar{\beta}$ designate the PSNRs of the unmarked and marked video images, respectively (no details are given on the way in which the mean PSNR of the non-marked images was calculated). In the study, the authors showed that $\bar{\delta} = 0.06$. The robustness was evaluated using the *watermarking detection ratio* (WDR), evaluated for each product and defined by WDR = $e_0 e_1 / m_0 m_1$, where m_0 and m_1 are, respectively, the numbers of "0" and "1" in the original mark, while e_0 and e_1 denote, respectively, the numbers of "0" and "1" in the detected mark. The reported WDR values were more than 0.78 after recoding and 0.74 after recoding with the addition of noise.

Table 13.1 provides an overview of different approaches to watermarking stereoscopic data. From an application perspective (e.g. protection of 3D HD TV content in real time), there are a number of limitations:

– The domain of insertion is still chosen by the experimenter, and has not been the subject of objective studies to determine optimality.

– Transparency is only evaluated in terms of PSNR and not using subjective approaches, limiting the legitimacy and the relevance of results.

– Robustness to geometric attacks has yet to be attained.

– Computation cost has not been studied.

– The structure and size of the corpus considered are too small to guarantee the generality and relevance of results.

	Insertion technique	Corpus	Transparency Watermarked views versus originals	Robustness detected mark versus original	Data payload (per frame)	Computation cost
View from stereoscopic pair	SS on DCT [HWA 03a]	25 frames of 256 × 256 pixels	Mean PSNR 34.89 dB	Mean PSNR 19.04 dB	Visual logo 32 × 32 pixels binary image	Unavailable
	SS on DWT [HWA 03b]	3 frames of 512 × 512 pixels	Mean PSNR 38.81	Mean PSNR 16.68 dB	Visual logo 64 × 64 pixels binary image	Unavailable
	SS on SVD-DWT [KUM 09]	5 frames of 512 × 512 pixels	Mean PSNR 44.05 dB	Mean normalized cross-correlation (NCC) Average filtering: 0.9 Rotation: 0.91 Redimensioning: 0.90 Gaussian noise: 0.89	Disparity map Image of 128 × 128 pixels	Disparity map calculated on DWT: reduction in computation cost. Genetic algorithm: drastic augmentation in computation cost
	SS on FrFT [BHA 09]	2 frames of 256 × 256 pixels 1 frame of 512 × 512 pixels	Mean PSNR 48.5 dB	Mean normalized cross-correlation (NCC)) JPEG compression: 0.98 Gaussian noise: 0.46 Median filtering: 0.71 Average filtering: 0.64 Rotation: 0.63 Redimensioning: 0.69 Cropping: 0.57	Disparity map Image of 64 × 64 pixels Image of 128 × 128 pixels	Unavailable
	QIM on DWT [CAM 08]	Unavailable	Unavailable	Bit error rate (BER) Gaussian noise: 0.07 Median filtering: 0.11 Column/line removal: 0.38 Rotation 0.25: 0.30 Rotation 0.5: 0.39	Randomly generated bits 2, 000 bits	Unavailable
	SS on DWT-DCT combination [YU 11]	1 frame of 640 × 480 pixels	Mean PSNR 52.07 dB	Watermark recovering rate (WRR) JPEG compression: 0.94 Salt and pepper noise: 0.90 Median filtering: 0.94 Cropping: 0.81	Visual logo Binary image of 64 × 64	Unavailable
Disparity map	SS on disparity map [ZHA 07]	3 video sequences of 22 frames of 512 × 512 pixels 1 video sequence of 192 frames of 720 × 576 pixels	Mean degradation index 0.06 dB	WRR Recoding: 0.86 Recoding with addition of noise: 0.81	Unavailable	Unavailable

Table 13.1. *Synthesis of performances of stereoscopic image/video watermarking methods*

In the remainder of this chapter, we will consider the issues mentioned above, by analyzing the following:

– The extension of 2D/2D+t watermarking methods: SS [COX 97], binary QIM, 5-QIM [BEL 10, HAS 11] and IProtect [CHA 10].

– The potential offered by the 3DV-NTSS disparity map [CHA 12] for watermarking in terms of optimality as a domain of insertion.

– Transparency, evaluated using (1) a subjective protocol considering the recommendations contained in UIT-R BT 500-12 [ITU 02] and BT 1438 [ITU 00], and (2) an objective protocol using four measurements: PSNR, IF, NCC and SSIM.

– Robustness, objectively expressed in terms of BER in mark detection after attacks such as linear and nonlinear filtering, compression and geometric transformations (StirMark random bending).

– Computation cost, estimated for each stage in the watermarking process.

13.4. Comparative study

A general view of our watermarking evaluation procedure is shown in Figure 13.4. The insertion domains under consideration include the left view of video sequences and the disparity map, calculated using three algorithms: NTSS [LI 94], FS-MPEG [STA 08] and 3DV-NTSS [CHA 12]. These algorithms use NCC as their block matching criterion. Mark insertion is carried out using the DWT representation of each of these four domains of insertion. To simplify the notation, these domains will be referred to as left view, NTSS, FS-MPEG and 3DV-NTSS, respectively. Four insertion methods will be considered, taken from the SS, binary QIM, 5-QIM and IProtect method families. The performances of the watermarking method will be evaluated in terms of transparency (subjective and objective), robustness and computation cost. The quantity of information inserted will remain constant at 1 bit per frame (i.e. 25 bits per second).

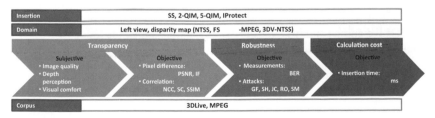

Figure 13.4. *Watermarking procedure evaluation protocol*

13.4.1. *Transparency*

13.4.1.1. *Subjective protocol*

The evaluation was carried out (1) in professional test conditions, in the 3D cinema room at Cesson Sévigné and (2) in laboratory conditions in the ARTEMIS department, using a 47" LG LCD monitor with full HD 3D $(1,920 \times 1,080$ pixels) and maximal luminosity 400 cd/m^2. Twenty-five non-expert observers participated in the tests. This is above the lower limit, set at 15 by the ITU-R. The participants were aged between 20 and 33 years, with a mean age of 25. They all presented visual acuity and color vision tested using a Snellen chart and the Ishihara test, respectively [ITU 00]. The tests involved two subjects per session. The subjects sat in a line in front of the center of the screen at a distance equal to three times the height of the screen. Each observer randomly evaluated 34 video excerpts, each with a duration of 40s. These excerpts represented the two corpora and all of the possibilities under study: original videos and watermarked videos, obtained using the four insertion methods applied in the four domains of insertion. The double stimulus continuous quality scale (DSCQS) method was used. Visual comfort, image quality and depth perception were noted independently on a quality scale from 1 to 5 (1 = bad and 5 = excellent).

The results were analyzed using the mean opinion score (MOS), taking the mean of the individual scores for each test condition. These results are summarized in Figure 13.5, where MOS values are indicated with a confidence interval of 95% [WAL 72].

In a synthetic manner, these observations provide us with the following preliminary conclusions:

– The visual experience was better for marked 3DLive content than for that sampled from the MPEG database. The average difference in MOS was 0.16. This result can be linked to differences in quality in the original material itself, which may influence observers in their evaluation of the quality of marked content.

– IProtect offers the best visual quality, with a mean MOS value of 0.18 for the 3DLive corpus and 0.11 for the MPEG corpus above the values for the SS method, which took second place. This result is unexpected, as when using 2D video content, SS methods give the best visual quality [COX 97, COX 02].

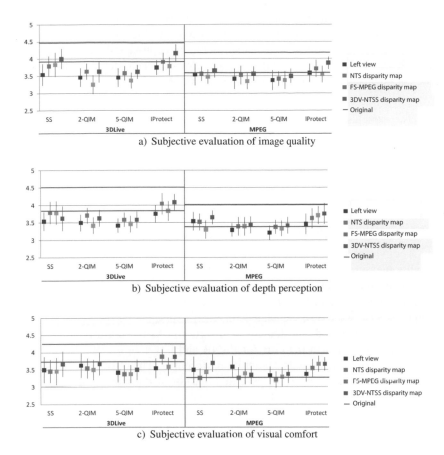

Figure 13.5. *Subjective evaluation: a) image quality, b) depth perception and c) visual comfort. The squares represent MOS values for each insertion technique/domain (from left to right: left view, NTSS, FS-MPEG and 3DV-NTSS), and the vertical lines centered on the squares show the 95% confidence intervals for watermarked content. The bold horizontal lines show the 95% confidence intervals for subjective evaluation of the original content*

– In terms of domains of insertion, 3DV-NTSS offers the best visual quality, with an average MOS value of 0.06 for the 3DLive corpus and 0.12 for MPEG above NTSS, the second best domain. This result highlights the advantages produced by managing the interdependence of stereoscopic views in calculating the disparity map, and thus shows the relevance of 3DV-NTSS for watermarking applications.

– The IProtect method, applied to the 3DV-NTSS domain, was the only solution allowing imperceptible mark insertion. For the two corpora and the three evaluation criteria (image quality, depth perception and visual comfort), IProtect/3DV-NTSS was the only solution providing confidence limits within the confidence intervals of the original context.

The conclusions presented above are of a general nature and may be enriched by noting that several method/insertion domain combinations may be taken into account for specific applications. For example, with depth perception constraints, low-resolution sequences from the MPEG corpus may be protected by three types of solutions: SS/NTSS, IProtect/FS-MPEG or IProtect/3DV-NTSS (see Figure 13.5).

13.4.1.2. *Objective protocol*

The visual quality of watermarked content was evaluated using four quality measurements: PSNR, IF, NCC and SSIM. These measurements were calculated individually for each watermarked frame of the video sequence, and then averaged out for the whole sequence (see Figure 13.6). Confidence limits at 95% were also calculated for each experimental test (each corpus, watermarking method and domain of insertion). As the corresponding error was always lower than 0.45 dB for PSNR and 0.001 in the other cases, the associated confidence limits cannot be represented. The only exception is for the 3DLive corpus using IProtect with insertion of the mark in a view from the stereoscopic pair, which allows this visualization.

From the obtained experimental data, we find the following:

– For a given similarity measurement and domain of insertion, the 3DLive and MPEG corpora behaved in a very similar manner. This is a reflection of the balanced conditions of the comparative analysis.

– The mean PSNR values were above 30 dB (except for the 5-QIM method applied to the left view of the 3DLive corpus). All of the method/domain of insertion combinations were able to ensure fundamental transparency properties. Note the transparency of more than 35 dB obtained using SS methods for all domains of insertion and for the two corpora. The same results can be achieved by using the IProtect method combined with 3DV-NTSS for 3DLive, and for any of the four domains of insertion in the case of MPEG. According to the PSNR values, the SS methods represent the best watermarking methods, followed by IProtect.

a) Mean PSNR values

b) Mean IF values

c) Mean NCC values

d) Mean SSIM values

Figure 13.6. *Objective evaluation of visual quality of watermarked content using different objective quality metrics*

– The IF, NCC and SSIM measurement values show that transparency values between 0.95 and 1.05 can be virtually guaranteed using all of the watermarking methods under consideration (with certain constraints in the choice of a domain of insertion). Only the SS method was refuted by the IF values, estimated for the MPEG corpus. In general, the SS and IProtect methods appear to offer the best solutions.

– For each watermarking method and for each corpus, the four objective quality indicators selected the 3DV-NTSS disparity map as the optimal domain of insertion, except for NCC values calculated for the 5-QIM insertion method applied to the 3DLive corpus.

13.4.2. *Robustness*

Robustness was evaluated in terms of BER during detection of the mark following the application of three types of attacks: filtering, compression and geometric attacks. The Gaussian and enhancement filtering techniques considered a 3×3 convolution kernel. JPEG compression was applied with an average quality factor of $Q = 60$. Rotations were generated at random with angles less than ± 0.5 degree. The StirMark random geometric attack was applied using default parameters [PET 98, PET 00]. Each attack was applied to a view of a stereo image. The corresponding BER values were averaged for the corpus and the corresponding 95% confidence intervals were calculated (see Figure 13.7).

Analysis of the results shows that IProtect/3DV-NTSS is the only combination presenting robustness, expressed by a reduction in BER of 0.05 after filtering and JPEG compression and a reduction of less than 0.1 after geometric attacks for either corpus. The 2-QIM method, applied to the 3DV-NTSS domain, gives the same BER values after Gaussian filtering, JPEG compression and geometric attacks, but does not respond to robustness requirements after contour enhancement (for both the 3DLive corpus and the MPEG corpus) or after StirMark random bending attacks in the case of the MPEG corpus. The SS method did not fulfill robustness requirements for any domain of insertion for either corpus.

13.4.3. *Computation cost*

The results presented above show that the IProtect watermarking method, applied to the 3DV-NTSS disparity map, is the most effective method for protecting stereoscopic videos, considering constraints of transparency and robustness, and for a fixed quantity of information.

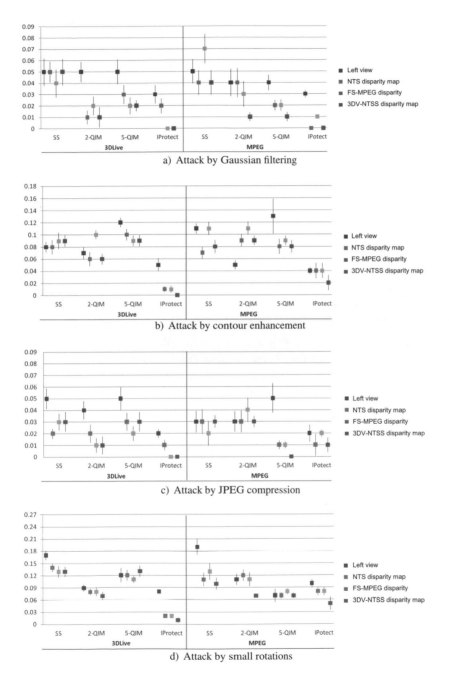

a) Attack by Gaussian filtering

b) Attack by contour enhancement

c) Attack by JPEG compression

d) Attack by small rotations

Figure 13.7. *Robustness of watermarking methods to different attacks: mean BER values and associated 95% confidence intervals*

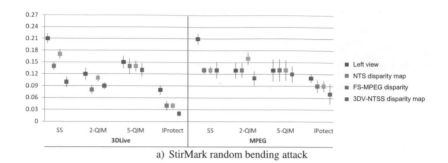

a) StirMark random bending attack

Figure 13.7. *(Continued) Robustness of watermarking methods to different attacks:*
mean BER values and associated 95% confidence intervals

Computation cost, which is crucial in the context of real-time applications (for example the protection of live broadcasting HD 3DTV content), should also be considered.

Thus, for each of the five main steps in the watermarking process (calculation of the 3DV-NTSS disparity map, DWT, insertion of the mark using IProtect, calculation of the inverse DWT and reconstruction of the image), the computation time is estimated at 137, 547, 38, 579 and 47 ms, respectively. The values are calculated at frame level and averaged for the corpus. They were obtained using a Core 2 CPU@ 2.13 GHz PC with 2 GB RAM.

The computation times for the 3DV-NTSS disparity map, IProtect insertion and image reconstruction are, respectively, 4, 15 and 12 times lower than those for the DWT. For solutions where DWT/IDWTs are calculated in real time [GAL 11], IProtect/3DV-NTSS watermarking can therefore be carried out in real time.

13.4.4. *Conclusion*

To summarize, methods inherited from 2D/2D+t watermarking offer very good performances in protecting stereoscopic videos. Our experimental results show that from a selection of these approaches, the hybrid IProtect method, which combines the aspects of SS and SI approaches, applied to a 3DV-NTSS disparity map offers a generic solution responding to a wide variety of applications:

– This is the only one of the studied insertion methods that ensures imperceptibility of the watermark according to subjective tests (ITU-R BT 1438 and BT 500-12 recommendations) and criteria of image quality, depth perception and visual comfort.

– This subjective evaluation of transparency is reinforced by PSNR values larger than 35 dB and IF, NCC and SSIM approaching the ideal value of 1 with an error lower than 0.05.

– It is the only one of the studied methods that ensures robustness expressed in BER with a value less than 0.05 after JPEG compression and linear/nonlinear filtering, and less than 0.1 after geometric transformations and StirMark random bending attacks.

– The computation cost is compatible with real-time applications (in the sense discussed in section 13.4.3).

The size and composition of the two corpora (almost 2 h 30 min of heterogeneous stereoscopic video content) and statistical error monitoring (confidence limits of 95% with relative error less than 0.1) guarantee the validity of our results and conclusions from a statistical perspective.

13.5. Conclusions

The multiplication of stereoscopic content, and their high added values increase the risks of digital piracy. Therefore, we need to find the means of preserving intellectual property rights, without imposing additional operational constraints in the chain of production, distribution and consumption of this content. Digital watermarking offers a solution to this issue, and consists of inserting additional information in a way that is imperceptible to the user and resistant to attacks.

In practice, the main challenge in watermarking is to find a point of operational balance between transparency, robustness and the amount of inserted information. Computation cost is also a key factor in the case of real-time applications. The optimality of the domain of insertion (one of the two views, or another derived representation, such as a disparity map) is another essential consideration in terms of the performance of stereoscopic watermarking approaches.

The research work presented in this chapter can be divided into three broad parts:

– An introduction concerning issues in the application of digital watermarking for stereoscopic content.

– An overview of the state of the art of stereoscopic video watermarking.

– A comparative analysis of the main classes of watermarking methods, considering their properties of transparency, robustness and the computation costs.

The experimental results obtained from almost 2 h 30 min of stereoscopic videos provide objective proof concerning the concept of stereoscopic watermarking: the IProtect method, applied to 3DV-NTSS, ensures operational equilibrium between constraints concerning the quantity of inserted information, transparency, robustness and computation cost. Future research will aim to establish a theoretical viewpoint of the limitations of stereoscopic video watermarking. As a result, watermarking may be integrated in a relevant manner into applications from content protection to adaptive enrichment for collaborative virtual environments.

13.6. Bibliography

[3DL 12] 3DLIVE, "Projet français 3D Live", available at http://3dlive-project.com, 2012.

[BEL 10] BELHAJ M., MITREA M., PRÊTEUX F., et al., "MPEG-4 AVC robust video watermarking based on QIM and perceptual masking", 8th International Conference on Communications (COMM), Bucharest, Romania, pp. 477–480, 10–12 June 2010.

[BHA 09] BHATNAGAR G., KUMAR S., RAMAN B., et al., "Stereo image coding via digital watermarking", Journal of Electronic Imaging, vol. 18, no. 3, pp. 033012–033012, 2009.

[CAM 08] CAMPISI P., "Object-oriented stereo-image digital watermarking", Journal of Electronic Imaging, vol. 17, no. 4, pp. 043024–043024, 2008.

[CHA 10] CHAMMEM A., MITREA M., PRÊTEUX F., "Speeding-up the hybrid video watermarking techniques in the DWT domain", Society of Photo-Optical Instrumentation Engineers (SPIE) Conference Series, vol. 7535, p. 75350E, 2010.

[CHA 12] CHAMMEM A., MITREA M., PRÊTEUX F., "High-definition three-dimentional television diapsrity map computation", Journal of Electronic Imaging, vol. 21, no. 4, pp. 043024–043024, 2012.

[CHI 11] CHIKKERUR S., SUNDARAM V., REISSLEIN M., *et al.*, "Objective video quality assessment method: a classification, review, and performance comparison", *IEEE transaction on broadcasting*, vol. 57, no.2, pp 165–182, 2011.

[COS 83] COSTA M., "Writing on dirty paper (corresp.)", *IEEE Transactions on Information Theory*, vol. 29, no. 3, pp. 439–441, 1983.

[COX 97] COX I., KILIAN J., LEIGHTON F., *et al.*, "Secure spread spectrum watermarking for multimedia", *IEEE Transactions on Image Processing*, vol. 6, no. 12, pp. 1673–1687, 1997.

[COX 02] COX I., MILLER M., BLOOM J., *Digital Watermarking*, Morgan Kaufmann, San Francisco, New york, 2002.

[ESK 95] ESKICIOGLU M.A., FISHER P.S., "Image quality measures and their performance", *IEEE Transaction on Communications*, vol. 43, no. 12, pp 2959–2965, 1995

[GAL 11] GALIANO V., LÓPEZ O., MALUMBRES M., *et al.*, "Improving the discrete wavelet transform computation from multicore to gpu-based algorithms", *Proceedings of International Conference on Computational and Mathematical Methods in Science and Engineering*, Benidorm, Alicante, Spain, 26–30 June 2011.

[HAS 11] HASNAOUI M., BELHAJ M., MITREA M., *et al.*, "mQIM principles for MPEG-4 AVC watermarking", *Society of Photo-Optical Instrumentation Engineers (SPIE) Conference Series*, vol. 7881, p. 78810L, 2011.

[HWA 03a] HWANG D., BAE K., KIM E., "Stereo image watermarking scheme based on discrete wavelet transform and adaptive disparity estimation", *Proceedings of SPIE*, vol. 5208, pp. 196–205, 2003.

[HWA 03b] HWANG D., BAE K., LEE M., *et al.*, "Real-time stereo image watermarking using discrete cosine transform and adaptive disparity maps", *Proceedings of SPIE*, vol. 5241, pp. 233–242, 2003.

[ITU 00] ITU-R BT.1438, "Subjective assessment of stereoscopic television pictures ITU-R BT.1438", Recommendation, 2000.

[ITU 02] ITU-R BT.500-12, "Methodology for the subjective assessment of the quality of television pictures", Recommendation, 2002.

[KUM 09] KUMAR S., RAMAN B., THAKUR M., "Real coded genetic algorithm based stereo image watermarking", *IJSDIA International Journal of Secure Digital Information Age*, vol. 1, no. 1, pp. 23–33, 2009.

[LI 94] LI R., ZENG B., LIOU M., "A new three-step search algorithm for block motion estimation", *IEEE Transactions on Circuits and Systems for Video Technology*, vol. 4, no. 4, pp. 438–442, 1994.

[MIT 07] MITREA M., PRÊTEUX F., "Tatouage robuste des contenus multimédias", in CHAOUCHI H., LAURENT-MAKNAVICIUS M., *La sécurité dans les réseaux sans fil et mobiles 1: Concepts fondamentaux*, Hermes-Lavoisier, Paris pp. 169–224, 2007

[MOB 12] MOBILE3DTV, "MPEG data base", available at http://sp.cs.tut.fi/mobile3dtv/stereo-video/, 2012.

[PET 98] PETITCOLAS F., ANDERSON R., KUHN M., "Attacks on copyright marking systems", *Information Hiding*, Springer, Berlin, Heidelberg, pp. 218–238, 1998.

[PET 00] PETITCOLAS F., "Watermarking schemes evaluation", *IEEE Signal Processing Magazine*, vol. 17, no. 5, pp. 58–64, 2000.

[PIN 99] PINCEMIN B., "Construire et utiliser un corpus: le point de vue d'une sémantique textuelle interprétative", in CONDAMINES A., PÉRY-WOODLEY M.-P. & FABRE C. (eds), *Atelier Corpus et TAL: pour une réflexion méthodologique*, pp. 26–36, 1999.

[SHA 58] SHANNON C., "Channels with side information at the transmitter", *IBM journal of Research and Development*, vol. 2, no. 4, pp. 289–293, 1958.

[STA 08] STANKIEWICZ O., WEGNER K., "Depth map estimation software version 2", *ISO/IEC MPEG meeting M*, vol. 15338, p. 2008, 2008.

[WAL 72] WALPOLE R., MYERS R., MYERS S., *et al.*, *Probability and Statistics for Engineers and Scientists*, vol. 5, Macmillan, New York, 1972.

[WAN 04] WANG Z., BOVIK A.C., SHEIKH H.R., *et al.*, "Image quality assessment: from error visibility to structural similarity", *Image Processing, IEEE Transactions on*, 2004.

[YU 11] YU M., WANG A., LUO T., *et al.*, "New block-relationships based stereo image watermarking algorithm", *ICSNC 2011, The Sixth International Conference on Systems and Networks Communications*, pp. 171–174, 2011.

[ZHA 07] ZHANG Z., ZHU Z., XI L., "Novel scheme for watermarking stereo video", *International Journal of Nonlinear Science*, vol. 3, no. 1, pp. 74–80, 2007.

PART 4

Rendering and 3D Display

Chapter 14

HD 3DTV and Autostereoscopy

The ultimate display would, of course, be a room within which the computer can control the existence of matter.
Ivan SUTHERLAND, 1965

14.1. Introduction

The difference between 3D and 2D displays is not always clearly defined, in spite of the seemingly clear 2D/3D dichotomy. With the notable exception of volumetric devices, most of the so-called 3D displays currently available are, in fact, simple 2D displays. The images projected onto these displays may be assimilated to 2D surfaces, using psychovisual cues to create an illusion of depth and increase its perception. With these limitations in mind, we may define 3D displays as devices able to reproduce dynamic depth signals on the basis of psychological (motion parallax and kinetic depth) and/or physiological cues (stereoscopy, accommodation and convergence).

A broad range of technologies currently allow 3D display [HOL 11, LUE 11, MAT 04]. In this chapter, we will only consider those based on apparent depth with the objective of separating information destined for the right and left eyes using the same surface (the screen). The methods used to guide optical beams exiting the screen have permitted the

Chapter written by Venceslas BIRI and Laurent LUCAS.

development of a number of different 3D display models, which are generally classified as stereoscopic or autostereoscopic.

A classification of these methods is shown in Figure 14.1. The proposed taxonomy consists of arranging these methods so that the number of views transmitted by each type of display increases from left to right, from 2-view stereoscopic displays to multiview horizontal parallax displays and multiview volumetric displays. These systems fall into the following categories:

– Helmet-mounted displays (HMDs): often used in virtual reality, these devices allow distinct images to be sent to the user, one for each eye. The principle involved is similar to that used in head-up displays, but with a separate miniature screen for each eye, often integrated into a helmet.

– Stereoscopic displays: these devices require users to wear glasses that filter incident light into separate image signals for the right and left eyes.

– Autostereoscopic displays: unlike the previous categories, these devices do not require the user to wear glasses. An optical technique applied to the screen directs the light so that each view n (where $n \geq 2$) is correctly transmitted to different observers.

– Holographic displays: this last category is based on technology, still *confidential*, able to recreate virtual holographic images [LUC 95], mostly static for the moment. This category is essentially composed of prototypes designed around specific optical elements with the ability to *dynamize* a hologram[1]. Other systems using similar principles exist, under the name of holoscopy [BOG 89]; these often include full-parallax (or integral imaging) autostereoscopic displays.

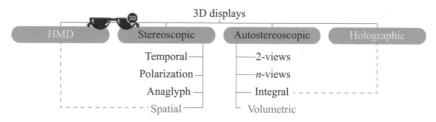

Figure 14.1. *Taxonomy of 3D displays (see [HOL 11] for further details)*

---

1 www.imec.be/ScientificReport/SR2010/2010/1159126.html.

All, or almost all, of these technologies are already in use in a number of domains of application, both in the civilian and military sectors, in academia and in industry, in connection with virtual reality [KOO 07] (see Chapter 15), biomedical imaging (see Chapter 20) or multimedia creation [SMO 11] and many other applications.

We will begin by discussing the subjacent technological elements involved in these techniques, before describing the principles of multiplexing multiview images, including filter design and use. We will conclude the chapter by considering the generation of multiview images and offering perspectives for further research.

14.2. Technological principles

The projection of 3D images, created using stereoscopic techniques, involves a number of processes to allow these images to be displayed on a flat surface. In this section, we will present the technological principles used to recreate the sensation of depth, which, we should remember, is simply an illusion.

14.2.1. *Stereoscopic systems using glasses*

Four types of projection are generally used:

– Alternating: these "active" devices display left and right views, in turn, on the screen (or projector). The impression of 3D is recreated via goggles using liquid crystals, and each pair must be perfectly synchronized with an image emitter. The emitter alternately obscures one of the two lenses (frequencies of ≥ 60 Hz per eye to avoid a "shimmer" effect) so that only the other eye receives the corresponding image. The retinal persistence effect allows the brain to recreate an illusion of depth by temporal mixing of the stereo pairs.

– Polarizing: these systems exploit the orientation property of light. This is known as polarization (see Figure 14.2). Screens using this principle are mainly based on linear polarization, which gives the best optical performance. However, systems using projection onto a metallic screen, as in cinemas, use circular polarization that allows spectators to sit in a wider variety of locations. In all the cases, the filters used effectively *sieve* the light, leading to a loss in resolution.

– Anaglyph: these systems use complementary color filters (different wavelengths) to transpose images forming a stereo pair. They do not generally allow correct recreation of the colors of images, but are simple to use and cheap to produce.

– Simultaneous: in this case, the collocation of left and right images is not guaranteed. These systems, similar to Wheatstone and Brewster's stereoscopes (see Chapter 1), are generally used in HMDs, the modern equivalent of the stereoscopes mentioned above.

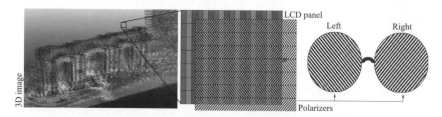

Figure 14.2. *Light polarization principle*

14.2.2. *Autostereoscopic displays*

Unlike the systems described above, autostereoscopic devices [DOD 05, HAL 05] do not require users to wear optical equipment. Angular view separation, notably in the case of displays, is carried out by an optical element, the operation and properties of which are discussed in detail below. We may distinguish four types of systems:

– 2-views: these displays simultaneously show two views (one for each eye). The position of the user is essential for correct relief perception. Certain models include an optical tracking mechanism that allows free movement of the head while controlling image distortion[2].

– n-views: these displays extend the horizontal field of vision by simultaneously recreating more than two views (generally between five and nine at the time of writing). This gives a wide range of preferential positions from which the spectator may observe different stereo pairs. Moreover, this technology enables collaborative 3D vision, allowing several individuals to observe the same scene simultaneously from slightly different angles.

---

2 *Fraunhofer Heinrich Hertz Institute*: www.hhi.fraunhofer.de/en/departments/interactive-media-human-factors/department-overview/.

– Integral imagery: the optical elements in this equipment allow a double angular separation of views, vertically as well as horizontally. Devices using this technology thus offer a visual experience close to real life, as a scene may be observed from several angles (around, above and below) [MAR 09]. These systems reproduce a 4D light field (plenoptic function [GOR 96, LEV 96]), creating double parallax stereoscopic images when the observer moves.

– Volumetric: unlike the three previous system types (where the optical image exists in the plane of the screen), these systems [JON 07, STA 10] produce a genuine 3D display by generating images at different positions in space. Different techniques are used to do this, such as the use of a rotating projection display to produce a spherical image volume [FAV 05] or the use of variable focus lenses to position several "slices" at different optical depths (see Figure 14.8).

14.2.3. *Optical elements*

In recent years, a number of university and industrial laboratories have developed autostereoscopic 3D displays. While certain attempts remain at an experimental stage, others have resulted in genuine commercial products. Most of these devices currently use conventional liquid crystal display (LCD) tiles, with the addition of an optical element that serves to redirect the incoming image (combination of lower resolution images) (see section 14.4) in priority viewing directions (see Figures 14.3 and 14.5). The number of views that these screens can handle and their angular separation (parallax) also characterize critical factors which designers, content producers and users must take into account, as they affect the whole chain of production of 3D images, from capture to diffusion. These optical elements, seen as an extension of work by Lippmann, who established the foundations of integral photography at the start of the 20th Century, are based on the use of parallax barriers or lens filters (see Figure 14.4). Several variants of these filters are currently used: strip barriers or lenticular sheets for horizontal parallax systems, and pinhole barriers or micro-lenses for full-parallax systems. Diffraction optics may also vary depending on the system (linear or circular) and may be mixed over several layers. Other solutions use colored barriers that allow selective filtering based on the wavelengths emitted by the LCD tile. These technologies have enabled the creation of a number of different display models, the main characteristics of which are shown in Table 14.1.

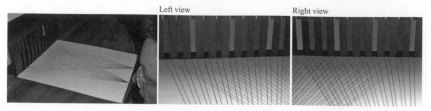

Figure 14.3. *Operating principle of a parallax barrier*

14.2.4. *Measurement of autostereoscopic display*

The accurate quality of perceived depth is an essential element in the use and practice of autostereoscopic techniques. Several factors are involved, in addition to physiological aspects concerning the observer (see Chapter 4). These include:

– The reproduction device itself. The autostereoscopic displays currently available are characterized by (1) the number of viewpoints they reproduce ($n \in (5, 7, 8, 9)$ for the most common devices), (2) the resolution (generally a full high-definition LCD tile), (3) the distance range offering high-quality 3D restitution and (4) the optical equipment ensuring angular separation of the n views.

– The nature of the displayed media, i.e. the conditions (real or virtual) in which the images were created. Chapter 4 gives an overview of this issue.

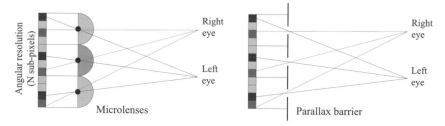

Figure 14.4. *Optical filters*

	Parallax barrier	Lenticular
2D/3D commutation	✓	✓
Portrait/landscape orientation	✓	
View separation	+++	++
Undesirable effects (3D cross-talk, moiré)	++	+++
Screen luminosity	+	+++
2-view display	✓	

Table 14.1. *Main characteristics of autostereoscopic 3D displays*

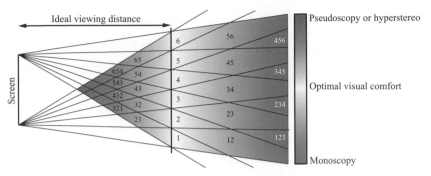

Figure 14.5. *Observation windows produced by different optical beams. The optimal comfort zones for good 3D vision (stereopsy) correspond to zones 1–6*

That said, the user may move laterally, advance or reverse in relation to the display without leaving the zone of acceptability, a condition which ensures that the quality of stereoscopic visualization will be maintained. Figure 14.5 illustrates this principle for a 6-view 3D display. For each zone of the viewing space, it shows the image numbers visible from left to right. The optimal viewing distance corresponds to regions 1–6. If the observer is placed so that one of his/her eyes is in zone 1 and the other is in zone 2, he/she will receive the full stereoscopic effect on screen. However, if the observer moves to a position where his or her left eye is in zone 23 and the right eye is in zone 34, he or she will still receive a stereoscopic view of the content, but there may be perceptible visual discomfort in the transition zone of views 2, 3 and 4. This artifact is known as cross-talk, and produces ghosting effects; these effects may be attenuated using specific software and/or material resources as described in [CHU 11]. The color gradation zone in Figure 14.5 represents valid positions for both eyes for an observer, excluding the issue of cross-talk. Therefore, the user has lateral freedom of movement in front of the screen across a distance known as the lobe, defined by the relationship $(n - 1) \times b$ (where n is the number of views and b is the interocular distance), but also has the possibility of moving toward or away from the screen. This allows several individuals to simultaneously perceive 3D images using different stereoscopic pairs. If the same observer is located in front of the plane representing ideal viewing positions, he or she will be subject to a hyperstereo or pseudoscopy phenomenon. The latter phenomenon corresponds to a permutation of left and right views of a stereo pair that produces an inversed relief effect, giving a confusing image that is difficult to interpret.

14.3. Design of mixing filters

The matrix representation of a 2D digital image $\mathcal{I}$ associates each position $(x, y) \in [0, M[\times[0, N[$ with an intensity $c \in [c_{min}, c_{max}]^p$ (generally $[0.255]^3$ in the case of color images). This arrangement facilitates not only access to and processing of data (the image is defined as a matrix of integer values), but also their display on an *ordinary* display device. If we then consider a volumetric device, a third parameter coding the depth of a (voxel) point needs to be added to the 2D coordinates. In the case of certain stereoscopic displays, access to this third dimension depends on another parameter: time. This clearly shows the interdependence of these dimensions (3D + time), in particular when it comes to properly addressing a 3D multiview visualization device in a unified manner ($n \geq 2$). Grasnick [GRA 10] and Ju-Seog *et al.* [JUS 04] discuss this issue, and we will use the first of these references as a basis for discussion of multiview image multiplexing in the following section. The multiplexing algorithm presented below allows us to produce arrangements of multiview images for different display devices, both real and virtual, volumetric and stereoscopic; we will illustrate the principle for, and using, autostereoscopic displays. While this algorithm is generic, it is not suitable for specifying all multiplexing schemas.

For a sub-pixel $i = f(x)$, the identification of a view V in a sequence of images (n) in the case of a one-dimensional display may be simply defined by the relationship $V = i \bmod n$. Taking $n = 3$ and $i \in [0.5]$, we obtain the interleaving sequence $(0, 1, 2, 0, 1, 2)$, which corresponds to the mixing of the three reference views. In 2D, this extended relationship is shown as follows:

$$V_{i,j} = \left(\lfloor \frac{j}{q_x} \rfloor \times q_a + \lfloor \frac{i}{q_y} \rfloor \times q_b \right) \bmod n \qquad [14.1]$$

where q_x and q_y correspond to repetition factors and q_a and q_b represent position modulation parameters. The matrix form $(V_{i,j})_{i=0,...,M;j=0,...,N}$ of this relationship then allows us, specifying the number of views and the different parameters mentioned above, to determine the masks to use in order to mix different views before display. Examples of the use of this algorithm will be given below. In Figure 14.7, we see that view interleaving is carried out not using successive pixels in the LCD screen, but directly in the red, green and blue (RGB) channels. The notion of position must therefore be clearly assimilated to one of the sub-pixels (see Figure 14.6).

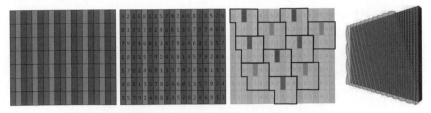

Figure 14.6. *Multiview representation of a pixel (nine views)*

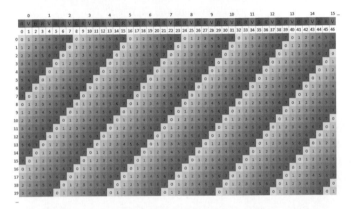

Figure 14.7. *Mixing filters for 4D-view screens with eight views*
$(i \in [0.46], j \in [0.19], n = 8, q_a = 1, q_b = 1, q_x = 1, q_y = 1)$

The generalization of equation [14.1] is presented in a very similar manner, as shown by the following relationship:

$$V_{e_1, e_2, \ldots, e_n} = \left(\sum_{i=1}^{n} \left(\lfloor \frac{e_i}{q_{R_i}} \rfloor q_{D_i} \right) \right) \bmod n \qquad [14.2]$$

knowing that for $n = 2$, we return to:

$$e_1 \rightleftharpoons i \qquad q_{D_1} \rightleftharpoons q_a \qquad q_{R_1} \rightleftharpoons q_x$$
$$e_2 \rightleftharpoons j \qquad q_{D_2} \rightleftharpoons q_b \qquad q_{R_2} \rightleftharpoons q_y$$

This equation is also suitable for displays using several layers of liquid crystals, such as the *DepthCube*[3], with $z = n = 20$, or the *Perspecta* [FAV 02], where $z = n = 198$ with a value of z expressed as an angle.

---

3 www.lightspacetech.com/Specifications.html.

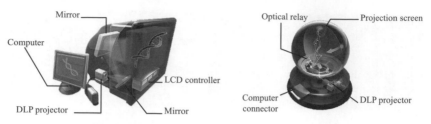

Figure 14.8. *DepthCube and Perspecta volumetric screens,*
Actuatily Systems Inc. [FAV 02]

14.4. View generation and interleaving

14.4.1. *Virtual view generation*

When autostereoscopic devices are coupled with a 3D rendering engine rather than multiple video flows, it becomes necessary to request n synthesized images. As the value of n may be high (5, 7, 8, 9 etc.), it becomes difficult to render these images within a reasonable interaction time ($\geq$30 Hz), and this may have detrimental effects on image quality. However, during the rendering of these n-views, a variety of information is shared, not only including geometric information (positions, normals of synthesized objects, etc.) but also radiometric information (diffuse color, texture, etc.). While proprietary techniques (Nvidia [DEV 06]) exist for stereoscopy, they are poorly suited to autostereoscopy and do not solve the issue of rendering time. Certain optimizations have been developed for specific rendering algorithms: ray tracing [ADE 93] and point splatting or volumetric rendering [HUB 06, HUB 07]. We will concentrate on approaches that improve rendering time by *rasterization*, the technique most commonly used for rendering synthesized images. One approach that aims to optimize the rendering time of n-views of the same scene, illustrated in Figure 14.9, exploits geometry shaders[4] in order to automatically duplicate each triangle as many times as there are views. Each of these triangles must then be sent to a *buffer* associated with each camera before final composition (see section 14.4.2).

---

4 Geometry shaders are programmable units that, broadly speaking, replace primitive generation in the graphics pipeline. Using input data (such as a vertex triplet with characteristics for each triangle), the program can delete, move, or duplicate a primitive or even create new ones. First used in late 2006, geometry shaders were included in the OpenGL 3.2 standard in December 2009. They are the successors of the *vertex shader* and preceded the *fragment shader*.

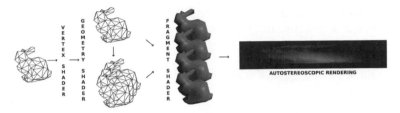

Figure 14.9. *Overview of our multiview stereoscopic GPU rendering method*

During the first stage, the graphics pipeline duplicates the 3D scene for each view; there is therefore no need to transfer data to the pipeline more than once, a transfer which can be very costly for bulky scenes. In the *vertex shader* stage, there is no need for projection into the camera space, as this will be carried out by the *geometry shader* for each rendered view. The *vertex shader* is responsible for all calculations relating to mesh vertices, which are carried out only once (diffuse color, calculation of normals, texture coordinates, etc.). The bulk of the work is then carried out by the geometry shader, where each primitive is duplicated and projected onto each viewpoint (see algorithm 14.1). The geometry shader has the capacity to duplicate each primitive (triangle) and to position it as desired. The final stage involves explicit generation of views, for which two possibilities exits: either the n views are stored as n distinct images (or *buffers*) or they are directly generated into a vast texture made up of the n viewpoints.

The first technique requires the use of *frame buffer objects*, which are simply rendering buffers, associated with the *multiple render target* technique, which allows all of these buffers to be filled in a single step. This technique, however, has significant limitations relating to the depth buffer, which is shared by all views, generating undesirable artifacts on the edges of objects.

The second technique consists of correctly positioning each primitive in each subpart of the image corresponding to the view indicated by the primitive (see Figure 14.10). In this case, we need to be attentive to *clipping* problems between each sub-image; this problem may be solved using explicit clipping in the geometry shader (see [DE 10][5]). The simplest solution, however, is to

---

5 Note that there is an error in this article in listing 1, where $coeff$ should take a value of $2.0 * tmp.w/NV$ and not $2.0 * tmp.w * NV$.

use the *viewport array* extension in OpenGL[6], shown in algorithm 14.1, where each generated primitive is sent to a specific viewport, thus managing clipping implicitly.

Algorithm 14.1. Example of a geometry shader for geometry cloning using an extension of *viewport arrays*

```
#extension GL_ARB_viewport_array : enable;
layout(triangles) in;
layout(triangle_strip, max_vertices=48) out;
uniform int numView;                          // Number of views
uniform mat4 projMatrix[MAXVIEW];             // Projection matrix
void main() {
    int i=0,k=0;
    for k < numView do
        for i < gl_VerticesIn do
            /* Projection onto image i                        */
            gl_Position = projMatrix[k]*gl_PositionIn[i];

            /* transmit all input data to fragment shader     */

            /* Set viewportfor vertex                          */
            gl_ViewportIndex = k;
            EmitVertex();
            i++;
        end
        EndPrimitive();
        k++;
    end
}
```

14.4.2. *View interleaving*

Once the n views have been generated, an image acceptable to the autostereoscopic device must be produced. To do this, we have n views, which are either stored in separate textures or combined in the same texture. Each view passes through a filter, which distributes pixels in the final image

---

6 The specifications of this extension are available at
http://developer.download.nvidia.com/opengl/specs/GL_ARB_viewport_array.txt.

in a way suitable for the autostereoscopic device (see section 14.3). To render the final image in graphic processing units (GPU), a final rendering stage is necessary, where a triangle is drawn to cover the whole of the image[7]. Interleaving must be carried out in the fragment shader, which fills each pixel of the final image using equation [14.1] and the n textures corresponding to the n views, as shown in Figure 14.11.

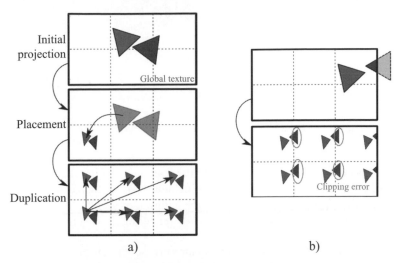

Figure 14.10. *a) Use of a texture for multiview rendering;*
b) clipping issue for this technique

During the final stage, we may consider *anti-aliasing*, for which several techniques have been proposed [MOL 05, VET 07, ZWI 07]. However, we recommend *morphological anti-aliasing* type approaches (MLAA), such as [JIM 11], which operate in *postprocessing*, applied during this final stage.

14.5. Future developments

While it is currently possible to create our own 3D displays [HIR 10], several factors, such as the intrinsic resolution of the selected LCD tiles or, to a lesser extent, the exclusive consideration of horizontal disparity, limit the expansion of autostereoscopic techniques.

---

7 www.altdevblogaday.com/2011/08/08/interesting-vertex-shader-trick/.

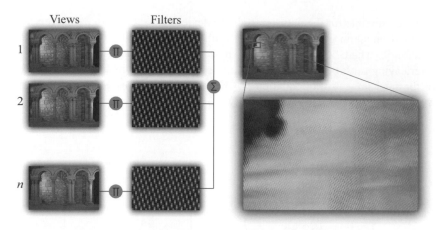

Figure 14.11. *View interleaving principle with mixing filters*

Several studies are currently underway in an attempt to limit these effects. First, studies based on current technology include work on characterizing 3D displays [LER 09]. These elements, with the addition of specific anti-aliasing strategies for autostereoscopy, improve the 3D rendering of existing content.

Another pathway for improving autostereoscopic use consists of considerably increasing display resolution, without necessarily increasing the number of views. The recent development of a new generation of 4K (ultra HD) screens is promising in this respect, and industrial actors have announced that 4K-based 3D solutions will be released in 2013.

Finally, other approaches propose holoscopic systems, allowing diffusion of integral imagery in the form of discretized plenoptic functions [FUC 08, LAN 10, WET 11] (lumigraphs [GOR 96] or *lightfields* [LEV 96]). These systems use multilayer 3D displays or pico-projectors [JUR 11].

14.6. Conclusion

A wide variety of devices currently allow 3D image display. While they mostly remain associated with specific domains of application, they clearly show a long-term trend toward the democratization of these technologies and a genuine, permanent spread of 3D content. In this context, this chapter has essentially been devoted to autostereoscopic techniques, which present a number of advantages along with certain limitations. More specifically, we discussed the way in which these systems operate and the approaches used, either using calculated images or real images, before presenting a number of recent developments providing considerable improvements in the quality of perceived images.

14.7. Bibliography

[ADE 93] ADELSON S.J., HODGES L.F., "Stereoscopic ray-tracing", *The Visual Computer*, vol. 10, pp. 127–144, 1993.

[BOG 89] BOGUSZ A., "Holoscopy and holoscopic principles", *Journal of Optics*, vol. 20, no. 6, pp. 281–284, 1989.

[CHU 11] CHULHEE L., GUIWON S., JONGHWA L., et al., "Auto-stereoscopic 3D displays with reduced crosstalk", *Optics Express*, vol. 19, no. 24, pp. 24762–24774, 2011.

[DE 10] DE SORBIER F., NOZICK V., SAITO H., "GPU-based multi-view rendering", *Computer Games, Multimedia and Allied Technology (CGAT 2010)*, Singapore, pp. 7–13, April 2010.

[DEV 06] DEVELOPPER TEAM N., "Nvidia: GPU Programming Guide version 2.5.0 (GeForce 7 and earlier GPUs)", electronic document, available at http://developer.nvidia.com/object/gpu_programming_guide.html, 2006.

[DOD 05] DODGSON N.A., "Autostereoscopic 3D displays", *Computer*, vol. 38, no. 8, pp. 31–36, 2005.

[FAV 02] FAVALORA G.E., NAPOLI J., HALL D.M., et al., "100-million-voxel volumetric display", *Proceedings of SPIE*, vol. 4712, pp. 300–312, 2002.

[FAV 05] FAVALORA G.E., "Volumetric 3D displays and application infrastructure", *Computer*, vol. 38, no. 8, pp. 37–44, 2005.

[FUC 08] FUCHS M., RASKAR R., SEIDEL H.-P., et al., "Towards passive 6D reflectance field displays", *ACM SIGGRAPH 2008 Papers, SIGGRAPH '08*, ACM, New York, NY, pp. 58:1–58:8, 2008.

[GOR 96] GORTLER S.J., GRZESZCZUK R., SZELISKI R., et al., "The lumigraph", *Proceedings of the 23rd Annual Conference on Computer Graphics and Interactive Techniques, SIGGRAPH '96*, ACM, New York, NY, pp. 43–54, 1996.

[GRA 10] GRASNICK A., "Universal 4D multiplexing of layered disparity image sequences for pixel and voxel based display devices", vol. 7526, pp. 75260V–75260V-12, 2010.

[HAL 05] HALLE M., "Autostereoscopic displays and computer graphics", *ACM SIGGRAPH 2005 Courses, SIGGRAPH '05*, ACM, New York, NY, 2005.

[HIR 10] HIRSCH M., LANMAN D., "Build your own 3D display", *ACM SIGGRAPH 2010 Courses, SIGGRAPH '10*, ACM, New York, NY, pp. 4:1–4:106, 2010.

[HOL 11] HOLLIMAN N.S., DODGSON N.A., FAVALORA G.E., *et al.*, "Three-dimensional displays: a review and applications analysis", *IEEE Transactions on Broadcasting*, vol. 57, pp. 362–371, 2011.

[HUB 06] HUBNER T., ZHANG Y., PAJAROLA R., "Multi-view point splatting", *Proceedings of the 4th International Conference on Computer Graphics and Interactive Techniques in Australasia and Southeast Asia, GRAPHITE '06*, ACM, New York, NY, pp. 285–294, 2006.

[HUB 07] HUBNER T., ZHANG Y., PAJAROLA R., "Single-pass multi-view rendering", *IADIS International Journal on Computer Science and Information Systems*, vol. 2, no. 2, pp. 122–140, October 2007.

[JIM 11] JIMENEZ J., MASIA B., ECHEVARRIA J.I., *et al.*, " Practical morphological anti-aliasing", in ENGEL W. (ed.), *GPU Pro 2*, AK Peters Ltd., Natick, MA, USA, pp. 95–113, 2011.

[JON 07] JONES A., MCDOWALL I., YAMADA H., *et al.*, "Rendering for an interactive 360 degree light field display", *ACM SIGGRAPH 2007 Papers, SIGGRAPH '07*, ACM, New York, NY, 2007.

[JUR 11] JURIK J., JONES A., BOLAS M., *et al.*, "Prototyping a light field display involving direct observation of a video projector array", *IEEE International Workshop on Projector-Camera Systems*, Colorado Springs, CO, 2011.

[JUS 04] JU-SEOG J., YONG-SEOK O., BAHRAM J., "Spatiotemporally multiplexed integral imaging projector for large-scale high-resolution three-dimensional display", *Optics Express*, vol. 12, no. 4, pp. 557–563, February 2004.

[KOO 07] KOOIMA R., PETERKA T., GIRADO J., *et al.*, "A GPU sub-pixel algorithm for autostereoscopic virtual reality", *Proceedings VR, IEEE Virtual Reality Conference*, Charlotte, NC, USA, pp. 131–137, 2007.

[LAN 10] LANMAN D., HIRSCH M., KIM Y., *et al.*, "Content-adaptive parallax barriers: optimizing dual-layer 3D displays using low-rank light field factorization", *ACM SIGGRAPH Asia 2010 Papers, SIGGRAPH ASIA '10*, ACM, New York, NY, pp. 163:1–163:10, 2010.

[LER 09] LEROUX T., BOHER P., BIGNON T., *et al.*, "VCMaster3D: a new fourier optics viewing angle instrument for characterization of autostereoscopic 3D displays", *SID Symposium Digest of Technical Papers*, vol. 40, no. 1, pp. 115–118, 2009.

[LEV 96] LEVOY M., HANRAHAN P., "Light field rendering", *Proceedings of the 23rd Annual Conference on Computer Graphics and Interactive Techniques, SIGGRAPH '96*, ACM, New York, NY, pp. 31–42, 1996.

[LUC 95] LUCENTE M., GALYEAN T.A., "Rendering interactive holographic images", *Proceedings of the 22nd Annual Conference on Computer Graphics and Interactive Techniques, SIGGRAPH '95*, ACM, New York, NY, pp. 387–394, 1995.

[LUE 11] LUEDER E., *3D Displays*, Wiley Series in Display Technology, Wiley, 2011.

[MAR 09] MARTINEZ-CUENCA R., SAAVEDRA G., MARTINEZ-CORRAL M., *et al.*, "Progress in 3-D multiperspective display by integral imaging", *Proceedings of the IEEE*, vol. 97, no. 6, pp. 1067–1077, June 2009.

[MAT 04] MATUSIK W., PFISTER H., "3D TV: a scalable system for real-time acquisition, transmission, and autostereoscopic display of dynamic scenes", *ACM SIGGRAPH 2004 Papers, SIGGRAPH '04*, ACM, New York, NY, pp. 814–824, 2004.

[MOL 05] MOLLER C.N., TRAVIS A. R.L., "Correcting interperspective aliasing in autostereoscopic displays", *IEEE Transactions on Visualization and Computer Graphics*, vol. 11, no. 2, pp. 228–236, March 2005.

[SMO 11] SMOLIC A., "3D video and free viewpoint video, from capture to display", *Pattern Recognition*, vol. 44, no. 9, pp. 1958–1968, 2011.

[STA 10] STAVNESS I., LAM B., FELS S., "pCubee: a perspective-corrected handheld cubic display", *Proceeding CHI, ACM Computer Human Interaction*, Atlanta, GA, Etats-Unis, pp. 1381–1390, 2010.

[VET 07] VETRO A., YEA S., ZWICKER M., *et al.*, "Overview of multiview video coding and anti-aliasing for 3D displays", *Proceedings of the International Conference on Image Processing (ICIP 2007)*, IEEE, San Antonio, TX, pp. 17–20, 2007.

[WET 11] WETZSTEIN G., LANMAN D., HEIDRICH W., *et al.*, "Layered 3D: tomographic image synthesis for attenuation-based light field and high dynamic range displays", *ACM SIGGRAPH 2011 Papers, SIGGRAPH '11*, ACM, New York, NY, pp. 95:1–95:12, 2011.

[ZWI 07] ZWICKER M., VETRO A., YEA S., *et al.*, "Resampling, antialiasing, and compression in multiview 3-D displays", *IEEE Signal Processing Magazine*, vol. 24, pp. 88–96, 2007.

Chapter 15

Augmented and/or Mixed Reality

15.1. Introduction

The term "augmented reality" (AR) appeared in the early 1990s. The aim of AR is to increase user perception by adding information, such as sound, textual notations or virtual objects to a perceived scene. By its very nature, augmented reality is interactive and three-dimensional (3D), meaning that at any time, added elements must be correctly placed in relation to the real world as seen by the user. This concept has numerous applications [AZU 01] in fields such as medical imaging, maintenance assistance, collaborative working, architecture, cultural heritage and gaming. From a practical perspective, scene visualization is carried out using a specific helmet or goggles, or, more simply, using a portable device (telephone, tablet computer, etc.).

To achieve coherent integration, virtual objects must be rendered at any given time using the viewpoint or pose of the portable camera carried by the user. Reliable, real-time position calculation methods are therefore needed. Moreover, a model of the scene is essential, first because most pose computation methods are based on image–model mapping, and second in order to generate interactions between real-world and virtual objects, such as occlusion and mutual shadowing. Point cloud models are used for pose computation, but the management of interactions between real and virtual elements requires the use of surface models.

Chapter written by Gilles SIMON and Marie-Odile BERGER.

In this chapter, we will discuss the state of the art of two key points in augmented reality: pose computation and the acquisition of a scene model, in the common framework of monocular vision.

15.2. Real-time pose computation

15.2.1. *Pose computation requirements*

In this section, we will only consider the widespread case where a 3D model of the fixed scene is available, usually in the form of a set of 3D points. Pose computation thus consists of identifying a set of n correspondences $(m_i, M_i)_{1 \leq i \leq n}$ between points in the image and points in the model, and finding the best possible solution, generally using a least squares type approach, to the following set of equations:

$$\mathbf{PM}_i \approx \tilde{\mathbf{m}}_i, \ 1 \leq i \leq n \qquad\qquad [15.1]$$

where $\mathbf{P} = \mathbf{K}[R, T]$ represents the camera projection matrix, $\mathbf{M}_i$ is the vector associated with the 3D coordinates of point M_i and $\tilde{\mathbf{m}}_i$ is the homogeneous vector associated with the two-dimensional (2D) coordinates of point m_i. The intrinsic parameters of the camera, $\mathbf{K}$, are generally known, and the unknown values in the system of equations [15.1] are the extrinsic properties of the camera, $[R, T]$. In summary, the difficulties involved in the process may be grouped into two categories:

– identification of a set of image/model correspondences (m_i, M_i);

– pose computation by solving equations [15.1].

Work on viewpoint computation has been underway since the early 1990s, and it would be impossible to give a full overview of the state of the art in the domain in a few pages. Lepetit and Fua [LEP 05], published in 2005, give a state of the art of pose computation based on rigid objects. In this chapter, we will focus on certain basic problems involved in augmented reality: the (re)initialization of pose algorithms, the reliability of pose computation over time and the need for efficient methods suitable for use with mobile architectures (telephones and tablets) with limited memory and computational power.

Due to space constraints, we will not consider robust estimation problems involved in tackling the presence of abnormal data in kernel matching. Spectacular progress has been made in this domain over the last 15 years.

Robust M-estimator [HUB 81] or consensus search-type approaches (random sample consensus (RANSAC)) [FIS 81] and their improvements (progressive sample consensus (PROSAC), maximum likelihood sample consensus (MLESAC), etc.) are now widespread techniques used to take account of imprecise or abnormal data. References [LEP 05, STE 99, ZHA 97] provide descriptions of these techniques and their application in computer vision. In this chapter, we will concentrate on the problem of correspondence identification and the resolution of equation [15.1].

15.2.2. *Image/model mapping*

15.2.2.1. *Iterative tracking methods*

The difficulty of the 2D/3D mapping problem varies depending on the situation, whether in the tracking phase, with relatively smooth movement, or the initialization or reinitialization phase. Let (m_i^t, M_i) be the correspondences between the point in the model M_i and the image established at instant t. In the case of relatively smooth movement, if P_t is the position calculated at instant t, then $P_t(M_i)$ should be close to m_i^{t+1}. m_i^{t+1} is then sought in the neighborhood of the predicted point using photometric resemblance criteria, such as correlation or the proximity of contour points. This gives a first estimation of P_{t+1}, which is then refined in an iterative manner. A number of systems operate in this way, such as [DRU 99, KOL 92, SIM 98], to cite just a few examples.

These iterative methods are strongly based on the availability of a correct initial estimation of the pose, often the calculated pose for the previous instant. They cannot, therefore, be used at the start of the application, or in the case of brusque camera movements during use. The mapping problem then becomes more complex, as the number of points to search in the image is no longer limited. While the use of easily identifiable markers placed in the environment, such as those made popular by ARtoolkit, allows easy identification of image/model correspondences [KAT 99], the problem is much more complex for non-equipped scenes, and recognition techniques are generally required in order to establish mapping.

15.2.2.2. *Recognition techniques*

Using recognition methods, the mapping problem is converted into a classification problem [LEP 04]. The subjacent idea involves the use of a learning phase, using several images of a scene to "learn" the photometric appearance of the same 3D point from different viewpoints. This appearance database, is then used in the operational phase to map a point of interest onto a 3D point by seeking the elements in the database, which are most similar to that point [OZU 06]. Figure 15.1 illustrates this procedure: several images of a scene, from different viewpoints, have been taken. Points of interest are extracted and mapped in the images, and the set of corresponding 3D points is determined using a *structure from motion* type algorithm [AGA 09]. A point is therefore represented by a class containing the collection of patches taken from the 2D images in which it was detected. During the application phase, for each acquired image and each extracted point of interest x, we seek a class C such that P(x|C) is maximized, allowing the detected point to be associated with a 3D point.

Figure 15.1. *Representation of a point by a collection of its aspects: the eye of the dinosaur is represented by the various appearances which it may take in a set of learning images*

These methods have enjoyed considerable success in recent years. Unlike iterative tracking-type methods, they do not block in cases of failure, even if the recognition method fails for one of the images. As with all learning-based methods, however, difficulties arise if the viewpoints adopted in the operational phase are very different from those used in the learning phase. For this reason, other techniques, which aim to synthesize a large number of additional viewpoints, are often used to complement the learned database, via the application of affine or homographic distortions to learning images [HSI 10]. This method generates realistic views if the object is locally planar; in other cases, it only generates an approximation of reality. If a textured model of the object is available, it is also possible to generate realistic views using a rendering engine.

15.2.2.3. *Choice of points of interest and real-time constraints*

Methods may differ according to the type of points of interest and the classification methods used, and are driven by the need for real-time computation of the mapping phase. Point of interest detectors, which are not subject to certain transformation groups (similarity and affine) and of which scale-invariant feature transform (SIFT) and speeded-up robust features (SURF) are the best known examples, allow mapping of points seen from relatively different aspects. However, the descriptor is relatively large (128) and the calculation and mapping involved are slow, meaning it is hard to use with mobile architectures, although improvements to reduce descriptor size have been proposed, such as PCA-SIFT [KE 04]. More recently, faster computing detectors with smaller descriptors have emerged, including BRIEF [CAL 12]; BRIEF uses a binary descriptor, where each element is the result of an intensity comparison between two points of the patch. This descriptor may be implemented rapidly and compares favorably with SIFT and SURF in performance terms. However, it is subject to problems of rotation invariance. A new detector, Oriented FAST and Rotated BRIEF (ORB), has recently been released to correct these invariance faults [RUB 11].

15.2.3. *Pose computation: principal PnP algorithms*

In this section, we will focus on pose computation using perspective-n-point camera pose computation (PnP) methods, which calculate poses using n 2D/3D point correspondences (m_i, M_i), with known intrinsic parameters.

A number of articles have been published that consider pose computation based on a small number of exact correspondences (often from three to five). There is not always a single solution: the perspective-3-point camera pose computation (P3P) problem, with three corresponding points, generally possesses four solutions [FIS 81]. In cases with four points, some cases produce two or more solutions. However, the solution will be unique if the points are coplanar without co-linear point triplets [FIS 81].

15.2.3.1. *Pose computation by minimization of reprojection error*

In practice, we have access to a large number of mapped points and these points are subject to noise. It is not, therefore, useful to seek an exact solution; instead, we aim to estimate a solution using the least squares method to minimize the reprojection error for the point set [LOW 87]

$$[R, T] = argmin_{R,T} \sum dist^2(z(\mathbf{PM}_i), m_i)$$

where z is the application that transforms a homogeneous vector $[x, y, z]^t, z \neq 0$ into a point with Cartesian coordinates $(x/z, y/z)$ and $dist$ is the application that calculates the Euclidean distance between two points. In this case, there is no direct solution to the minimization problem. For this reason, iterative minimization methods are generally applied, starting from an initial estimate. In practice, we often choose the calculated pose for the previous instant as the estimate, presuming, implicitly, that the movement of the camera is sufficiently smooth. For the initial solution, we use an exact solution obtained from a small number of points. This clearly does not guarantee that the method will converge, particularly if the camera is moving rapidly, something that occurs frequently in augmented reality when the user turns his or her head. For all of these algorithms, sensor data greatly facilitate pose computation by providing an initial estimate close to the optimum [ARO 06].

Certain methods also use approximations of the projection model [DEM 95] or abandon the orthogonality constraint of matrix R in order to facilitate convergence. However, these techniques still do not guarantee convergence. Other methods use multiple initializations, but are very costly. In a similar vein, the expansion of sampling techniques and particle filtering methods [ISA 96] has brought significant improvements in terms of the robustness of pose computation. The basic idea is to sample the distribution of the pose in the vicinity of the current estimation. In this way, we generate a set of potential poses, weighted in relation to reprojection error (the higher the weight, the lower the projection error). In this way, the pose can be calculated as a weighted mean of these particles. The sampling process is based on an *a priori* distribution (often a Gaussian distribution centered on the last estimation) or linked to knowledge of movement dynamics that allow us to predict pose in the current image [CHA 02, PUP 05]. The advantage of these methods is that they allow us to generate potential poses that are relatively distant from the current pose, thus accounting for non-smooth camera trajectories. The precision of these techniques clearly depends on the number of particles used, but they provide a considerable improvement in the robustness of pose computation in cases of sudden movement or occlusions in the sequence.

15.2.3.2. *Direct pose computation methods*

In recent years, the convergence difficulties present in iterative methods and the impossibility of mastering processing time have led to research concerning non-iterative pose computation solutions. The efficient perspective-n-point camera pose estimation (EPnP) [LEP 09] and direct least squares (DLS) [HES 11] algorithms are good examples of this.

In most non-iterative methods, the objective is to express coordinates of 3D points in the camera frame by calculating the depth associated with each image point. A relationship exists linking the distances d_{ij} between points in the model, the depth of the points x_i in the image frame and the angle θ_{ij} between lines of sight. Using trigonometry (see Figure 15.2), we obtain $x_i^2 + x_j^2 - 2cos(\theta_{ij}) - d_{ij} = 0$. The d_{ij} values are known from the model and, as the camera is calibrated, θ_{ij} may be calculated using the image points m_i and m_j. We thus obtain second-degree polynomial equations of which the unknown values are depth values in the camera frame [ANS 03]. Pose computation thus involves finding the rigid transformation that best aligns the points in the model expressed in the camera frame and those expressed in the general plane, for example using [UME 91]. Unfortunately, these methods are complex and do not take account of errors in extracted points.

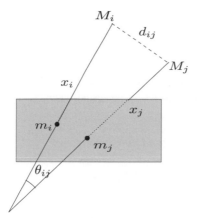

Figure 15.2. *Basic geometry involved in pose computation using corresponding points*

The recent EPnP method [LEP 09] is based on the idea of expressing each 3D point as the weighted sum of four virtual points *via* barycentric coordinates. This weighting is the same whether the point is expressed in the general plane or the camera frame. For each pair (m_i, M_i), we may thus write two equations depending on the image point and the weightings, for which the unknown values are the coordinates of the four reference points expressed in the image frame. The concatenation of the equations for n observations allows us to estimate reference points using least squares, then the model points in the camera frame and finally the pose by aligning the two sets of points. The core of the algorithm involves the calculation of the coordinates of the four reference points, obtained by least squares resolution of a system of size $2n \times 12$ ($2n \times 9$ if the points are coplanar), which may be carried out with a complexity in $O(n)$.

One drawback of this type of method is that it does not consider noise in extracted points. The DLS approach [HES 11] tackles this problem and offers a method for finding all solutions of a least squares formulation of the PnP problem. In this approach, the rigid constraint that exists between the points expressed in the general plane and the camera frame is used to express the depth of points and the translation as a function of the rotation R. The least squares criterion minimizes the difference between the measured line of sight and the line of sight deduced from the model by the transformation $[R, T]$, and is expressed as a fourth-degree polynomial in rotation parameters, providing third-degree optimality conditions with calculable roots. A non-optimized implantation produced using Matlab provided a time of around 15 ms with precision levels close to those produced by iterative methods (in cases of convergence).

The development of efficient initialization-free methods has enabled pose computation methods suited to augmented reality applications in which the need for re-initialization arises on a regular basis.

15.2.4. *Pose computation and planar surfaces*

Many applications of AR take place in urban or indoor environments that are often piecewise planar. In these cases, the scene is often described in terms of a computer-aided-design (CAD) model of the environment using a set of planar surfaces to construct the scene. Pose computation must then be based on correspondences between 3D and 2D faces. The basic notion used in this case is that of planar homography (see Figure 15.3). Let X be any point in a plane π of which the equation is given by $n^t X + d = 0$. The plane is observed by two cameras, of which the projection matrices are denoted by $\mathbf{P}$ and $\mathbf{P}'$. We may presume that $\mathbf{P} = \mathbf{K}[\mathbf{I}, \mathbf{0}]$ and $\mathbf{P}' = \mathbf{K}[R, T]$, within a change of frame. Let $\mathbf{x}$ and $\mathbf{x}'$ be the homogeneous coordinates of the projections of point X by $\mathbf{P}$ and $\mathbf{P}'$. We can then easily demonstrate [HAR 00] that a linear transformation exists, 3×3 $\mathbf{H}$, known as homography, such that $\mathbf{x}' = \mathbf{H}\mathbf{x}$. This transformation is given by:

$$\mathbf{H} \approx \mathbf{K}(R - Tn^t/d)\mathbf{K}^{-1} \qquad\qquad [15.2]$$

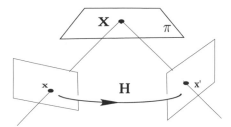

Figure 15.3. *Plane-induced homography*

The linear transformation **H** may be calculated using at least four points in correspondence belonging to the same plane using a least squares type estimation. If the matrix of intrinsic parameters **K** is known, then knowledge of **H** allows us to write nine proportional equations, i.e. eight equations including the unknown values R and T. Representing a spatial rotation using three parameters (Euler's angle, quaternions, etc.), we thus obtain eight equations with six unknown values, allowing us to calculate the relative movement between three positions. A linear solution to this problem can be found supposing (which is always possible) that the observed plane has the equation $z = 0$ [SIM 00]. The precision of the pose computation may be improved by using several planes [VIG 03]. A number of projects are currently underway with the aim of automatic acquisition of the plane equation in a scene for pose computation (see section 15.3). Certain approaches use the notion of dominant planes in urban environments and the extraction of points at infinity for the main directions in the scene [KOS 05]. Others use interactivity for precise reconstruction of planes in a scene [SIM 11].

15.3. Model acquisition

An overview of techniques used to reconstruct objects is given in Chapter 8 In this section, we will consider the generation of 3D models that may be used for real-time pose computation and the management of spatial (occlusions and collisions) and photometric interactions (shadows and light reflection) between the real and virtual worlds. We will consider scenes of varying sizes, from manufactured objects to urban landscapes.

15.3.1. *Offline modeling*

A point cloud associated with descriptors may be obtained using a structure from motion (SFM) algorithm. To do this, the scene for reconstruction must be filmed using a moving camera. Points of interest are detected and mapped between sequence images, and then recreated in 3D using binocular or multi-ocular stereovision techniques (see Chapter 7). Non-linear optimization is generally used to refine this initial solution, taking account of the global coherence of the 3D reconstruction in relation to the 2D points measured throughout the sequence. This step is known as *bundle adjustment*. Much has been written concerning SFM methods, the basics of which are covered by Hartley and Zisserman [HAR 00] and in an overview by Triggs *et al.* [TRI 00].

Denser point clouds can generally be obtained by combining image data with laser data. In [FRU 01], for example, two inexpensive 2D lasers and a camera, with precision-calibrated positions and relative orientations, were mounted onto a vehicle traveling through the streets of a town. One laser scanned buildings horizontally, the other laser scanned vertically. The point map produced by the horizontal scan was compared to aerial photographs and digital road maps to improve the precision of the obtained geometry.

The interest of point cloud models for AR, however, is limited for several reasons. First, positioning a 3D object in a point cloud requires calibration to within six degrees of freedom, which may be difficult in point clouds where the structural elements of a scene are difficult to identify. It is, however, easy to place an object on a surface of a model by forcing the vertical of the object to remain aligned with the normal of the surface, for example. Second, a surface model allows us to obtain a visibility map that is useful both in managing occlusions between the real and virtual world and for obtaining the primitives of the model that are visible at a given instant and may be used for pose computation. Finally, the calculation of shading and light reflections between real and virtual elements depends on the use of surface models.

Certain authors have considered the automatic detection of coplanar point groups in a point cloud [KOS 05], from which planar surfaces may be obtained. However, the proposed methods are generally unreliable for a number of reasons, discussed in [SIM 11]. Their use notably requires calibration of a number of parameters of which the optimal values are linked to the size of the scene and the context of application. A surface mesh may also be obtained by triangulating a point cloud. In [FRU 01], for example, a cloud is triangulated by connecting neighboring points under certain regularity constraints, and the images captured by a camera were used to texture the generated triangles. However, the meshes obtained using this type of procedures generally contain a number of holes and/or phantom triangles.

For this reason, it is more reliable to involve the user in segmenting point clouds to convert them into structured and/or textured surface models. The authors of [VAN 07] propose using point clouds as a basis for an interactive modeling of the scene. A mouse is used to trace polygons directly onto video images, which are then reconstructed in 3D for optimal adjustment to SFM-generated point groups. Straight lines or curves may then be traced onto these polygons for division or subdivision purposes. Autodesk ImageModeler is another example of a user interface that allows scene modeling based on a point cloud. The scene is represented by a set of elementary blocks (boxes, spheres, etc.) assembled by the user; the structural elements of these blocks (vertices, axes, etc.) are aligned with reconstructed points. In these two examples, textures taken from video images may be pasted onto the reconstructed model.

Other commercially available tools allow scenes to be modeled directly from images without passing through a point cloud. This allows a scene to be reconstructed from a single photograph, with the possibility of adding further photographs to provide a complete view of the scene. These programs use manually inserted parallelism and orthogonality constraints to obtain intrinsic parameters and the camera pose [WIL 02]. The 3D scene may then be modeled incrementally by the user, using an image as a visual support [OH 05]. Google SketchUp is one such program that uses this technique.

15.3.2. *Online modeling*

As AR spreads, it has become increasingly difficult to require inexperienced users to model their own environment before using an AR application. Currently, artificial markers [KAT 99] are generally used to avoid this stage, but the markers still need to be printed onto paper and the solution would not be appropriate for complex or large-scale environments (generally anything more than 10 times the size of the marker). Recent and current work has focused on circumventing the modeling phase, with environment acquisition taking place at the same time as positioning, or in a rapid initial phase requiring little user expertise. These approaches also enable what you see is what you get (WYSIWYG) environment modeling, avoiding postprocessing problems, and presents an interest for other contexts of use such as special effects in cinema, with high economic stakes at play.

For small-scale scenes (objects, work surfaces, etc.), visual simultaneous localization and mapping (SLAM)-type methods may be used [DAV 02, EAD 06, KLE 07]. These methods are based on the hypothesis of the camera being moved, allowing points identified in several images to be reconstructed using stereovision techniques. A Kalman filter [DAV 02] or a particle filter [EAD 06] is generally used to refine the 3D position of these points during the process. The 3D map may also be refined as a whole using bundle adjustment, an operation that may be carried out at a frequency lower than that of positioning, and using a different processor in order to respect the real-time constraint [KLE 07].

Positioning virtual objects in a 3D map is even harder in this context than in an offline situation. In [KLE 07], a dominant plane is calculated for placement of the virtual object, but this procedure is not suited to complex or multiplane environments. The authors of [PAN 09] propose "on-the-fly" transformation of a point cloud into a textured surface mesh. A Delaunay tetrahedrization is applied to the point cloud, followed by a probabilistic *sculpture* phase used to find the hollows in convex solids. This procedure is designed to reconstruct handheld objects, presented to the camera from different angles. It is not suited to use with non-manipulable scenes. Moreover, the reconstructed surface often contains faults (unfilled holes or parasite triangles) that may lead to strange visual effects when the model is used to manage occlusions.

The appearance of time-of-flight (TOF) cameras, particularly the relatively inexpensive Microsoft Kinect, has attracted strong interest from the AR community. The Kinect uses a structured light technique [FRE 08] to generate depth maps of a scene in real time, from which a discrete point cloud may be obtained. In [IZA 11], the iterative closest point (ICP) algorithm is used for real-time alignment of this initial cloud with new clouds obtained when the camera moves. This allows us to obtain a camera pose for each instant, and to supplement the initial point cloud with points that become visible as the camera moves. Unlike the 3D maps obtained using SLAM or by SFM, the point clouds generated in this way are very dense. Representing each 3D point by a voxel, we obtain a relatively precise volumetric model that can be used to manage occlusions and collisions between the real scene and virtual objects. This technique appears promising, but is currently limited to indoor scenes with a size limited by the camera range (around 3 m). Moreover, a volumetric representation in voxel form gives an understanding of the scene no better than that obtained using a point cloud.

In situ modeling methods have been proposed in response to this issue, allowing us to obtain a structured representation of the scene by modeling directly over the flow of video images obtained in real time. In [BUN 08], a poster is used to obtain camera poses until the reconstructed model is sufficiently rich to take over. While positioning is taking place, the user may designate a point in the current image (generally the apex of an object) using a mouse. When the camera moves, the epipolar line corresponding to this point is tracked, and the user may move the point along this line until it reaches the correct position for the current image. The polygons modeled in this way may be extruded to produce volumes. Once a sufficient number of edges have been modeled, contour-based tracking is used in the place of the poster for pose computation and the remaining modeling operations. This method is relatively inconvenient for the user, due to the use of the poster in the activation phase and the complexity of the interactions required to obtain a complete model.

Figure 15.4. *In situ environment modeling using a method developed by the MAGRIT team. a) The walls and floor are targeted and captured through the camera, then reconstructed in real time using simple interactions. After a few seconds, a virtual scene (of simple palm trees, in this case) may be integrated into the modeled environment. A video showing these operations is available at www.loria.fr/˜gsimon/vc/blobs_outdoor.avi. b) Example of* blobs *obtained in the corridors at the Loria. These* blobs *can be used to easily position virtual objects in relation to the real scene*

In [SIM 11], we proposed a *paintbrush*-type technique, more intuitive and less inconvenient, for *in situ* modeling of piecewise planar environments (objects, indoor scenes, urban environments, etc). The user "paints" visible faces of the model through a disk displayed in the middle of the video image (see Figure 15.4(a)). When the user presses a key, the disk is placed onto the face, then tracked in real time by mapping points of interest between consecutive images and robust homography calculation. Disks posed on the same phase are concatenated to form convex polygons known as *blobs*. When two *blobs* are tracked on secant planes, the image of the intersection of the two planes is determined using a particle filter with geometric and photometric constraints. The user may then visually validate the obtained line, or restart the filter if the trace does not converge or converges to an imprecise solution. Once the image of the intersection has been obtained, the equations of the planes corresponding to the two *blobs* are calculated, taking into account this information and the geometric constraints linked to the movement of the camera. Using this method, we do not obtain an exhaustive representation of the environment or the object, but textured 3D *blobs* that allow pose computation and facilitate the positioning of virtual objects (see Figure 15.4(b)).

15.4. Conclusion

AR has made significant progress in recent years, allowing us to envisage increasingly complex applications. This progress is due to the combination of several factors:

– the spread of mobile devices including video cameras and captors that allow approximate pose estimation;

– increased robustness of model/image mapping methods, particularly due to recognition technology and the implementation of direct viewpoint calculation methods;

– the increased availability of methods for environmental reconstruction, either using devices (Kinect) or automatic or interactive visual reconstruction techniques.

However, few applications are currently able to operate in large-scale environments, particularly outdoors. In these contexts, changes may occur between the moment the scene is acquired and the time the application is used, presenting major difficulties. These modifications may be spatial (changes in an urban environment, traffic and pedestrians), and may concern the ambient lighting or climatic conditions (rain, snow, etc.) creating considerable differences between the model and the current scene and making the model/image mapping stage much more complex. Another significant difficulty in applications requiring high levels of realism lies in the photometric coherence of the real/virtual mix. The problem of shading or relighting is on the border between the domains of computer vision and graphical information technology; it is particularly complex in the case of natural light, and is the subject of current research.

15.5. Bibliography

[AGA 09] AGARWAL S., SNAVELY N., SIMON I., *et al.*, "Building rome in a day", *Proceedings of the 9th International Conference on Computer Vision*, Kyoto, Japon, October 2009.

[ANS 03] ANSAR A., DANIILIDIS K., "Linear pose estimation from points or lines", *IEEE Transactions Pattern Analysis and Machine Intelligence*, vol. 25, pp. 578–589, 2003.

[ARO 06] ARON M., SIMON G., BERGER M.-O., "Use of inertial sensors to support video tracking", *Computer Animation and Virtual Worlds*, vol. 18, pp. 57–68, 2006.

[AZU 01] AZUMA R.T., BAILLOT Y., BEHRINGER R., *et al.*, "Recent advances in augmented reality", *Computer Grahics Applications*, vol. 21, pp. 34–47, December 2001.

[BUN 08] BUNNUN P., MAYOL-CUEVAS W.W., "OutlinAR: an assisted interactive model building system with reduced computational effort", *IEEE/ACM International Symposium on Mixed and Augmented Reality*, IEEE Computer Society, Los Alamitos, CA, pp. 61–64, 2008.

[CAL 12] CALONDER M., LEPETIT V., OZUYSAL M., *et al.*, "BRIEF: computing a local binary descriptor very fast", *IEEE Transactions on Pattern Analysis and Machine Intelligence*, vol. 34, pp. 1281–1298, 2012.

[CHA 02] CHANG P., HEBERT M., "Robust tracking and structure from motion through sampling based uncertainty representation", *Proceedings of ICRA '02*, Washington, May 2002.

[DAV 02] DAVISON A.J., MURRAY D.W., "Simultaneous localization and map-building using active vision", *IEEE Transactions on PAMI*, vol. 24, pp. 865–880, 2002.

[DEM 95] DEMENTHON D., DAVIS L., "Model based object pose in 25 lines of code", *International Journal of Computer Vision*, vol. 15, pp. 123–141, 1995.

[DRU 99] DRUMMOND T., CIPOLLA R., "Real-time tracking of complex structures with on-line camera calibration", *Proceedings of the British Machine Vision Conference, BMVC 99*, Nottingham, 1999.

[EAD 06] EADE E., DRUMMOND T., "Scalable monocular SLAM", *Proceedings of the 2006 IEEE Computer Society Conference on Computer Vision and Pattern Recognition*, IEEE Computer Society, Washington, DC, pp. 469–476, 2006.

[FIS 81] FISCHLER M.A., BOLLES R.C., "Random sample consensus: a paradigm for model fitting with applications to image analysis and automated cartography", *Commununications of the ACM*, vol. 24, no. 6, pp. 381–395, June 1981.

[FRE 08] FREEDMAN B., SHPUNT A., MACHLINE M., *et al.*, "Depth mapping using projected patterns", Patent Application, 10 2008. WO 2008/120217 A2, 2008.

[FRU 01] FRUH C., ZAKHOR A., "3D model generation for cities using aerial photographs and ground level laser scans", *Proceedings of the 2001 IEEE Computer Society Conference on Computer Vision and Pattern Recognition, 2001, CVPR 2001*, vol. 2, pp. II-31–II-38, 2001.

[HAR 00] HARTLEY R.I., ZISSERMAN A., *Multiple View Geometry in Computer Vision*, Cambridge University Press, 2000.

[HES 11] HESCH J.A., ROUMELIOTIS S.I., "A direct least-squares (DLS) method for PnP", *International Conference on Computer Vision*, Barcelona, pp. 383–390, 2011.

[HSI 10] HSIAO E., COLLET ROMEA A., HEBERT M., "Making specific features less discriminative to improve point-based 3D object recognition", *IEEE Conference on Computer Vision and Pattern Recognition (CVPR)*, San Francisco, June 2010.

[HUB 81] HUBER P.J., *Robust Statistics*, Wiley, New York, 1981.

[ISA 96] ISARD M., BLAKE A., "Contour tracking by stochastic propagation of conditional density", *Proceedings of the 4th European Conference on Computer Vision*, Cambridge, UK, vol. 1064, pp. 343–356, 1996.

[IZA 11] IZADI S., KIM D., HILLIGES O., *et al.*, "Kinect-fusion: real-time 3D reconstruction and interaction using a moving depth camera", *ACM Symposium on User Interface Software and Technology*, Santa Barbara, USA, pp. 559–568, 2011.

[KAT 99] KATO H., BILLINGHURST M., "Marker tracking and HMD calibration for a video-based augmented reality conferencing system", *Proceedings of the 2nd International Workshop on Augmented Reality*, San Francisco, CA, 1999.

[KE 04] KE Y., SUKTHANKAR R., "PCA-SIFT: A more distinctive representation for local image descriptors", *2012 IEEE Conference on Computer Vision and Pattern Recognition*, IEEE Computer Society, vol. 2, pp. 506–513, 2004.

[KLE 07] KLEIN G., MURRAY D., "Parallel tracking and mapping for small AR workspaces", *Proceeding of the 6th IEEE and ACM International Symposium on Mixed and Augmented Reality (ISMAR'07)*, Nara, Japan, November 2007.

[KOL 92] KOLLER D., DANIILIDIS K., NAGEL H.H., "Model-based object tracking in traffic scenes", *Proceedings of the 2nd European Conference on Computer Vision*, Santa Margherita Ligure, Italy, vol. 588, pp. 437–452, 7 October, 1992.

[KOS 05] KOSECKA J., ZHANG W., "Extraction, matching and pose recovery based on dominant rectangular structures", *Computer Vision and Image Understanding*, vol. 100, no. 3, pp. 274–293, 2005.

[LEP 04] LEPETIT V., PILET J., FUA P., "Point matching as a classification problem for fast and robust object pose estimation", *Proceedings of IEEE Conference on Computer Vision and Pattern Recognition*, Washington, DC, pp. 244–250, 2004.

[LEP 05] LEPETIT V., FUA P., "Monocular model-based 3D tracking of rigid objects: a survey", *Foundations and Trends in Computer Graphics and Vision*, vol. 1, no. 1, pp. 1–89, 2005.

[LEP 09] LEPETIT V., MORENO-NOGUER F., FUA P., "EPnP: an accurate O(n) solution to the PnP problem", *International Journal of Computer Vision*, vol. 81, no. 2, pp. 155–166, February 2009.

[LOW 87] LOWE D.G., "Three-dimensional object recognition from single two-dimensional images", *Artificial Intelligence*, vol. 31, pp. 355–395, 1987.

[OH 05] OH J., STUERZLINGER W., DANAHY J., "Comparing SESAME and sketching on paper for conceptual 3D Design", *EUROGRAPHICS Workshop on Sketch-Based Interfaces and Modeling*, Konstanz, Germany, 2005.

[OZU 06] OZUYSAL M., LEPETIT V., FLEURET F., *et al.*, "Feature harvesting for tracking-by-detection", *Proceedings of European Conference on Computer Vision*, Graz, Austria, 2006.

[PAN 09] PAN Q., REITMAYR G., DRUMMOND T., "ProFORMA: probabilistic feature-based on-line rapid model acquisition", *Proceedings of the 20th British Machine Vision Conference (BMVC)*, London, September 2009.

[PUP 05] PUPILLI M., CALWAY A., "Real-time camera tracking using a particle filter", *Proceedings of British Machine Vision Conference*, British Machine Vision Association 2009, pp. 519–528, 2005.

[RUB 11] RUBLEE E., RABAUD V., KONOLIGE K., *et al.*, "ORB: an efficient alternative to SIFT or SURF", *International Conference on Computer Vision (ICCV)*, Barcelona, Spain, pp. 2564–2571, 2011.

[SIM 98] SIMON G., BERGER M.-O., "A two-stage robust statistical method for temporal registration from features of various type", *Proceedings of the 6th International Conference on Computer Vision*, Bombay, India, pp. 261–266, 1998.

[SIM 00] SIMON G., FITZGIBBON A., ZISSERMAN A., "Markerless tracking using planar structures in the scene", *Proceedings of International Symposium on Augmented Reality*, Darmstadt, Germany, pp. 120–128, October 2000.

[SIM 11] SIMON G., BERGER M.-O., "Interactive building and augmentation of piecewise planar environments using the intersection lines", *The Visual Computer*, vol. 27, no. 9, pp. 827–841, February 2011.

[STE 99] STEWART C.V., "Robust parameter estimation in computer vision", *SIAM Reviews*, vol. 41, pp. 513–537, 1999.

[TRI 00] TRIGGS B., MCLAUCHLAN P., HARTLEY R., *et al.*, "Bundle adjustment – a modern synthesis", *Vision Algorithms: Theory and Practice*, LNCS, vol. 1883, pp. 298–375, 2000.

[UME 91] UMEYAMA S., "Least-squares estimation of transformation parameters between two point patterns", *IEEE Transactions on Pattern Analysis and Machine Intelligence*, vol. 13, pp. 376–380, 1991.

[VAN 07] VAN DEN HENGEL A., DICK A., THORMÄHLEN T., *et al.*, "VideoTrace: rapid interactive scene modelling from video", *ACM SIGGRAPH 2007 Papers*, ACM New York, NY, USA, p. 86–91, 2007.

[VIG 03] VIGUERAS F., BERGER M.-O., SIMON G., "Iterative multi-planar camera calibration: improving stability using model selection", *Vision, Video and Graphics (VVG '03)*, Bath, UK, July 2003.

[WIL 02] WILCZKOWIAK M., BOYER E., STURM P., "3D modeling using geometric constraints: a parallelepiped based approach", *7th European Conference on Computer Vision (ECCV '02)*, vol. 4, Springer-Verlag, Copenhagen, Denmark, pp. 221–237, 2002.

[ZHA 97] ZHANG Z., "Parameter estimation techniques; a tutorial with application to conic fitting", *Image and Vision Computing Journal*, vol. 15, no. 1, pp. 59–76, 1997.

Chapter 16

Visual Comfort and Fatigue in Stereoscopy

16.1. Introduction

The development of three-dimensional (3D) video has impacted each part of the transmission chain, from acquisition to reproduction. It also strongly impacts user experience through the sensation of perceived depth. In this context, the Quality of Experience (QoE) [QUA 12] may be defined using three major factors [LAM 11]: *visual quality*, which refers to the quality of each view taken separately and corresponds to quality evaluated in terms of two-dimensional (2D) viewing; *quality of the depth effect*, qualified by a variety of indicators such as the reproduction quality of the 3D scene and the sense of presence induced by the 3D effect; and *visual fatigue* and *visual (dis)comfort*, often induced by technical constraints related to viewing stereoscopic content.

While recent progress in terms of capture and reproduction has enabled the use of high-quality 3D content according to the first two criteria (visual quality and quality of the depth effect), fatigue and discomfort continue to present significant limitations in the deployment of 3D video. Certain studies have shown that users tend to prefer the 2D version of a video to its 3D equivalent in cases where the latter may induce fatigue or discomfort [BAR 10].

Chapter written by Matthieu URVOY, Marcus BARKOWSKY, Jing LI and Patrick LE CALLET.

In this chapter, we will provide an overview of visual fatigue and discomfort, both in terms of the mechanisms involved in the human visual system (HVS) and in terms of 3D content and processing technologies. We will begin by defining the notions of fatigue and discomfort (see section 16.2). To facilitate understanding of the rest of the chapter, section 16.3 will provide an overview of cognitive and optical manifestations of fatigue and discomfort. Section 16.4 is devoted to different sources of visual fatigue and discomfort, in connection with binocular vision. In section 16.5, we will consider the problems posed by 3D technologies and content.

16.2. Visual comfort and fatigue: definitions and indications

Few studies provide definitions of the notions of fatigue and discomfort and the associated terminology [LAM 07]. Our definitions in this section are inspired by the dominant medical technology, extended to a broader context. Figure 16.1 shows the way in which the analysis of objective and/or subjective data may be used to identify the presence of fatigue or discomfort. Recently, a considerable number of studies have focused on the construction of objective models for predicting fatigue or discomfort based on the nature of 3D content or related to processing and visualization conditions.

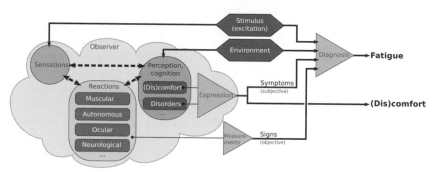

Figure 16.1. *Visual fatigue and discomfort: context and terminology*

16.2.1. *Visual fatigue*

We consider that visual fatigue is manifested and may be diagnosed by the presence of objective indications (*signs*) and subjective indications (*symptoms*). *Signs* correspond to the experimental measurement of one or more physiological reactions by the observer, and a measurement protocol is required in order to measure them. *Symptoms* are expressed by the observer and constitute an indication of his or her perceived mental or physical state. Visual fatigue is generally caused by the repetition of excessive visual effort, and can build up as these efforts are repeated. It disappears after a sufficient rest period. The nature, intensity and temporal characteristics (time of appearance, duration, rising slopes and falling edges) of signs and symptoms constitute parameters that may be used to evaluate the severity of visual fatigue. The terms *asthenopia* and *eye strain* may also be used in this context. Their differentiation from visual fatigue varies according to the context, and the terms are sometimes used interchangeably; precise definitions are generally based on the severity of signs and symptoms. In this chapter, the three terms will be considered as equivalent.

16.2.2. *Visual comfort and discomfort*

According to definitions used in the field of psychology [ROO 00], visual discomfort is made up of the integration of one or more sensations (multimodal construction) and depends on perception. Visual discomfort is, therefore, a percept constructed and evaluated by the user in reaction to an environment and a presented stimulus. It corresponds to the degree of immediate physical or mental unpleasantness. Discomfort may thus be linked to the experience of one or more symptoms, to difficulties experienced in the accomplishment of a visual task or to any other negative sensation associated with the visual treatment of presented information. Visual discomfort appears and disappears with the presence of these negative associations. Unlike visual fatigue, the rise and fall time of visual discomfort is comparably short. Comfort and discomfort are sometimes defined in separate, distinct ways (for example, in design) [VIN 07].

In these studies, comfort is not defined solely as the absence of discomfort, but also requires the presence of *well-being*.

In this context, we will consider the presence of comfort as the absence of discomfort.

16.2.3. *Detection and evaluation of visual fatigue and discomfort*

Symptoms linked to visual fatigue and discomfort are generally evaluated using questionnaires. The signs of fatigue, on the other hand, are commonly evaluated using appropriate examinations. Visual fatigue may be diagnosed by the presence of characteristic signs and/or symptoms, including the degree of fatigue reported, subjectively, by observers [YAN 02]. It may also be voluntarily provoked when an observer is required to make an excessive visual effort over a long period [WAN 02]. For observers, the only means of detecting discomfort is to subjectively evaluate and/or quantify it. In [YAN 02], for example, discomfort is evaluated continuously using the *Single Stimulus Continuous Quality Evaluation* (SSCQE) protocol [ITU 04]. Some studies have connected the sensation of discomfort to other characteristics involved in visual tasks, using symptoms or objective signs.

16.3. Signs and symptoms of fatigue and discomfort

Manifestations of visual fatigue may take different forms, most of which are ocular or cognitive. Other manifestations, such as shoulder stiffness, modifications in respiratory or cardiac rhythms [AND 02, NAG 08] or the concentration of cortisol in saliva [NAG 08] may also be taken into consideration.

16.3.1. *Ocular and ocularmotor fatigue*

There are a number of signs and symptoms of visual fatigue [CAI 92], including the presence of mucus or tear formation around the eyelids, modifications in blinking frequency and the sensation of ocular dryness. The near vision triad (accommodation, vergence and pupillary response) has been the subject of particular attention.

16.3.1.1. *Accommodation disturbances*

Accommodation (A) allows us to focus on a presented visual stimulus and is measured in accommodation diopters. In practice, the accommodation distance is not the exact distance from the target object, but generally shifts in the direction of the tonic accommodation distance, given by the state of refraction of the human eye at rest ($\approx 1.0\ \delta$ -diopter-).

This *accommodation adaptation* thus introduces an accommodation lead or lag, depending on the context (see Figure 16.2(a)). The tonic accommodation distance varies from person to person. Jaschinski-Kruza [JAS 98] showed that individuals with high tonic accommodation are more susceptible to visual fatigue. In [HAS 01], the visual fatigue induced by monocular or binocular accommodation and vergence efforts led to a reduction in tonic accommodation (of up to $0.5\ \delta$).

In [YAN 02, YAN 04], subjective scores (1–5) for perceived visual fatigue were correlated with a reduction in accommodation amplitude (for an accommodation demand of $\approx$ +5 δ) after viewing 2D or stereoscopic 3D video sequences. The reduction was higher in 3D ($>$0.5 δ) than in 2D ($<$0.5 δ).

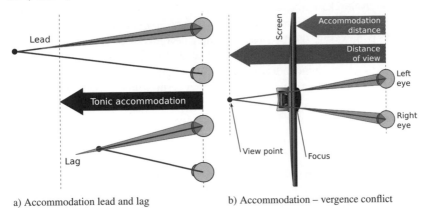

a) Accommodation lead and lag   b) Accommodation – vergence conflict

Figure 16.2. *Adaptation and conflicts between accommodation and vergence*

16.3.1.2. *Vergence disorders*

(Con)vergence (C) involves the orientation of the eyes toward a target object and is measured in prismatic diopters. As in the case of accommodation, a vergence adaptation, or prismatic adaptation, mechanism exists. Vergence errors are known as fixation *disparities* or *stability* and introduce a shift toward the value at rest, or *dark vergence*. In [JAS 91], the author shows that in near vision, visual fatigue increases in line with the distance between the viewpoint and the dark vergence. In [NOR 81], the authors found that within a group of 15 patients presenting asthenopia or binocular vision problems, the majority showed deficient adaptation in the optometric prism facility test, both to base-in and base-out prisms. Similar results were observed after viewing stereoscopic 3D content [LAM 09].

16.3.1.3. *Oculomotor control models*

In real-world situations, the point of focus (accommodation) and the gaze point (vergence) are synchronized; within the oculomotor system, accommodation and vergence are mutually interdependent [HUN 01]. *Convergence accommodation* (CA) occurs when convergence induces an accommodative response, and *accommodative convergence* (AC) occurs when an accommodative change triggers vergence. Fusion convergence (C) is faster than accommodative convergence (AC): the former is based on the estimation of retinal disparity and the latter is based on retinal blurring [HOF 08]. This interdependence may be measured using the ratios CA/C and AC/A.

In [WAN 02], a reduction in the two ratios AC/A and CA/C was measured after visualization of depthwise movements of a 3D stimulus. In [SCH 87], however, the two relationships developed in opposite directions after a visual effort: the relationship associated with the fastest mechanisms (vergence or accommodation) increased, while the other decreased. The modification of these relationships may also be interpreted as the result of an adaptation process: vergence and accommodation may respond to the newly experienced conflicting demands in a more efficient manner.

16.3.1.4. *Pupil diameter changes*

In [MUR 01], perceived fatigue is correlated with a reduction in pupil diameter. The authors of [UKA 97] showed that in 30% of patients suffering from visual fatigue, an exaggerated alternation of pupil constriction (miosis) and dilation (mydriasis), known as *hippus*, was observed in near vision, with no connection to variations in lighting or fixation.

16.3.2. *Cognitive fatigue*

Visual fatigue may take the form of ocular disturbances, but may also create other signs such as headaches [AND 02]. Studies have shown that the visual fatigue induced by viewing a stereoscopic 3D video slows the event-related potentials P100 (after 100 ms) [EMO 05] and P700 (after 700 ms) [LI 08]. These results show that visual fatigue also has an impact on the cognitive processes of the HVS. In the context of stereoscopic 3D, cognitive fatigue may take the specific form of stereopsis problems. One particular stereopsis property, the fusion range, i.e. the retinal disparity range for which left and right images can be fused, is frequently used to measure stereopsis performance. Over a longer period, after viewing a 3D stimulus for 60 min, visual fatigue manifested through a reduction in the fusion range [EMO 04, EMO 05, NOJ 04b]; its extent returned to the initial value after a rest period of 5–10 min. However, in [LAM 09], the short-term visual fatigue induced by reading a 3D text (the Wilkins text) was shown to be positively correlated with an increase in the fusion range.

The fusion range may also be increased for a longer period by training [EMO 05]: repeated visualization of 3D content may thus contribute to a limitation of the signs and symptoms of fatigue. The fusion interval may be measured in two ways. In the first method, starting from a position where fusion is possible, we may increase retinal disparity until diplopia (double vision) is reached. In the second method, we begin from a position of diplopia and reduce the retinal disparity until binocular fusion occurs. These two methods do not produce the same values and lead to fusion hysteresis, which may also be used to detect visual fatigue [EMO 04]. In [EMO 05], instances of diplopia were used to detect fatigue. Other characteristics of stereopsis, such as stereoacuity [HOF 08, SCH 87] and binocular fusion speed [HOF 08], have also been linked to visual fatigue.

16.3.3. *Signs and symptoms linked to discomfort*

As discomfort is subjective, almost all studies evaluating discomfort use questionnaires. Certain studies [HOW 97, KEN 93] have aimed at detecting the presence of a number of characteristic symptoms, while others concentrate on evaluating comfort levels using a scale. Some symptoms have shown strong correlation, notably ocular pain or irritation [KIM 11, KUZ 08, LAM 09], double vision [LAM 09, YAN 11], a loss of sharpness [LAM 09, YAN 11], focusing difficulties [KIM 11, KUZ 08], nausea [KUZ 08] and headaches [KUZ 08]. In [KIM 11], the level of cerebral activity of the frontal eye field, which plays an important role in controlling eye movements, was correlated with perceived levels of discomfort. The authors of [LI 13] showed that blinking frequency is proportional to comfort when the visual stimulus is static and inversely proportional to comfort when the stimulus moves in a plane parallel to the screen; to our knowledge, these are the only studies to have established links between an objective sign and the existence of discomfort.

16.4. Sources of visual fatigue and discomfort

16.4.1. *Ocular constraints*

16.4.1.1. *Accommodation/vergence conflicts*

Using (auto)stereoscopic screens, accommodation and vergence demands are desynchronized: accommodation is fixed on the screen, whereas vergence varies with the disparity presented on the screen (see Figure 16.2(b)). These demands enter into conflict with the oculomotor coupling of accommodation and vergence. A certain number of studies have shown that accommodation/vergence conflict introduces instabilities into the behavior of the oculomotor system, which constantly oscillates between accommodation and vergence demands. However, this hypothesis is difficult to verify due to accommodation lead and lag. In [UKA 02], convergence difficulties and oculomotor instabilities were observed when a high degree of disparity was presented on screen (up to 2.6°). In [OKA 06], the authors showed that the role of convergence accommodation increases with the degree of conflict, whereas the role of focus accommodation decreases with an increase in retinal blurring. In [FUK 09], the authors suggest that the accommodation response follows convergence accommodation in a transient manner and is then corrected in relation to retinal blurring. This conflict is generally considered to be an important source of visual fatigue. A number of observations support this idea:

– the fusion time of a binocular stimulus increases monotonously with the accommodation/vergence conflict [HOF 08];

– stereoacuity is higher in the absence of conflict [HOF 08, WAT 04];

– visual fatigue increases with the accommodation/vergence conflict [HOF 08].

Finally, the visualization of stereoscopic images induces visual discomfort when the disparity is over 60 arc-minutes [NOJ 04b].

16.4.1.2. *Depth and plane movements*

Certain studies have shown that variations of the accommodation/vergence conflict are problematic, but not the conflict itself. Tracking depthwise movements, and the visualization of 3D scenes where the foreground and background are separated by a large depth interval, leads to variations in conflicting accommodation and vergence demands in the course of visualization. In [NOJ 04b], the temporal variations in the disparity of stereoscopic images, when they are discontinuous and larger than 60 arc-minutes, are shown to induce visual discomfort. In [EMO 05], visualization of a stereoscopic 3D sequence through a system of prisms induced visual fatigue when the power of the prisms varied repeatedly. In [YAN 02, YAN 04], visual discomfort and fatigue were induced when visualized stereoscopic sequences included strong depthwise movement, even when the accommodation/vergence conflict was minimized [YAN 04]. In [SPE 06], the speed of depthwise movement was shown to be a determinant for visual comfort. Visual comfort decreased as the speed of disparity variations increased.

Planar movements, with no depthwise changes, have also been considered. In [YAN 02, YAN 04], these movements were found to have no influence on fatigue or visual comfort when viewing stereoscopic 3D sequences. In [LI 11], however, visual discomfort was seen to increase with the angular speed of a stereoscopic stimulus moving in a plane parallel to the screen, and also with the depth interval separating the foreground from the background, with no connection to the degree of accommodation/vergence conflict.

16.4.1.3. *Limited depth of field*

The depth of field is the distance interval which is perceived in a sharp, non-blurred manner. It is localized around the accommodation distance. Its projection into optical space (eye, camera, etc.) is known as the depth of focus, and is generally evaluated at ± 0.2 δ (diopters) [YAN 04]. For stereoscopic screens, the depth of field refers to the depth interval behind and in front of the screen in which objects are perceived sharply[CHE 10]. Thus, the disparity of the screen should be limited in order to reproduce 3D content within the limits of the depth of field [YAN 04].

The absolute distance range of the depth of field varies with optical parameters, notably the visualization distance, the aperture and optical aberrations. It increases monotonously with the visualization distance [CHE 10] so that the role of accommodation diminishes as the visualization distance increases. The eye aperture is given by the diameter of the pupil: the narrower the aperture, the greater the depth of field [CHA 77]. The pupil diameter itself is influenced by lighting: the more the light, the smaller the aperture and the greater the depth of field. With active (*shutter*) stereoscopic screens in particular, it is important to verify that the illumination level is sufficiently high to preserve the depth of field: active eyewear may reduce lighting levels by up to 80%. In [LI 05], the authors show that the pupillary reaction is also activated by the visualization of moving autostereograms, thus only by the disparity of the observed stimulus. Finally, optical aberrations in the eye also influence the depth of field. Refraction errors, for example, increase the likelihood of visual fatigue [BLE 05].

16.4.2. *Cognitive constraints*

16.4.2.1. *Detecting disparity: cortical anomalies*

Richards [RIC 71] suggests that each type of disparity – crossed, uncrossed or (quasi) null – stimulates a specific set of neurons (Figure 16.3). The final disparity is given by the disparity of the set (or channel) with the maximum response.

Unequal assignment of cortical neurons to the different channels may explain the observed variations in sensitivity to disparity between individuals. In degraded visualization conditions, for example, 20–30% of the population perceive crossed disparities as uncrossed disparities, and vice versa [RIC 71]. About 6–8% of the population are unable to perceive disparity; one hypothesis is that these individuals lack disparity channels [RIC 71]. Although this type of anomaly does not necessarily prevent relief perception, this can generate sporadic incoherence, inducing fatigue or discomfort.

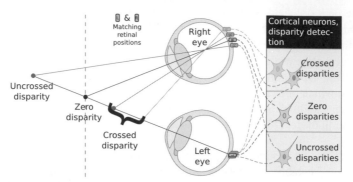

Figure 16.3. *Disparity channels: neurone based disparity calculation*

16.4.2.2. *Limited binocular fusion zones*

Binocular fusion, or stereopsis, constructs depth information in relation to a reference surface known as the horopter: the set of points in space which stimulate corresponding points on the left and right retinas. In the 19th Century, G. Vieth, followed by J. Müller, constructed a theoretical model of a surface of zero retinal disparity, i.e. the horopter, a model which later came to be known as the *Vieth–Müller circle* [SCH 06]. The set of points in space located in front of and behind the horopter for which fusion is possible forms a zone known as *Panum's area*. Outside the Panum's area, fusion is impossible, creating diplopia or binocular rivalry, and only one of the two views is seen at a time, either in alternation (binocular rivalry) or by suppressing one of the two views.

The size of Panum's area is influenced by a variety of factors. The first studies on the subject gave variable measurements, from 14 [MIT 66] to 120 [FEN 67] arc minutes. The extent of the area, initially described as an elliptical surface [SCH 81], increases in line with the eccentricity of the visual stimulus [HAM 83]. The fusion interval, therefore, increases proportionally with the viewing angle [NAG 96]. The dimensions of Panum's area increase with the size of the visual stimulus, illumination, viewing duration [WOO 74] and training [JON 89]. However, these dimensions decrease as the spatial frequency of the stimulus increases, and when disparity undergoes temporal modulations [SCH 81]. The visualization of 3D content with relief objects which are small, poorly lit, highly detailed or make frequent depthwise movements may, therefore, prove problematic, generating diplopia episodes and consequently visual fatigue [EMO 05]. It is, therefore, important to limit the disparity of 3D content in order to ensure its visualization within Panum's area. This requires us to convert screen disparity into perceived disparity: while the first value is simply generated by a horizontal shift on the screen between the two views of the same stimulus, the second value depends on the shape of the retina and the form of the horopter, and therefore varies between individuals and stimulus properties.

The horopter is a surface which curves toward the exterior (see Figure 16.4(b)), of which the backwardly inclined profile [COO 11] (see Figure 16.4(a)) seems adapted to real-world scenarios in which the closest objects are situated in the lower part of the field of vision, and the most distant objects toward the top (clouds, for example). In [NOJ 04a], the visualization of stereoscopic 3D sequences respecting this property (crossed disparities at the bottom of the screen, uncrossed disparities at the top of the screen) was shown to reduce the visual discomfort. In [COO 11], the authors measured the form of the horopter before and after a period of five days during which subjects wore distorting glasses; the results showed that the horopter does not adapt to a presented distortion.

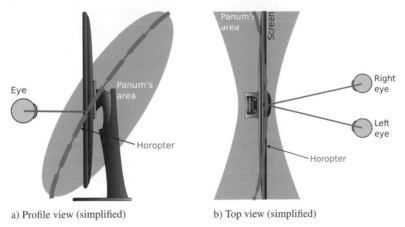

a) Profile view (simplified)   b) Top view (simplified)

Figure 16.4. *Shapes and orientations of Panum's area and the horopter*

16.4.2.3. *Duality of binocular perception*

The central and peripheral zones of the field of vision are treated in different ways by the visual cortex, and generate information of different natures. The parvocellular pathway transmits information from the center of the retina, the fovea, to the ventral stream of the visual cortex, where neuron responses are slow and sustained (perception of details and high spatial frequencies). The magnocellular pathway connects the periphery of the retina to the dorsal stream of the visual cortex, where neuron responses are rapid and transient (rough estimation of the optical flow, perception of movement). This duality is also encountered in stereopsis. In [SCH 81], a small fusion interval and high stereoacuity were observed during visualization of fixed or slow-moving stimuli presenting spatial details. Inversely, a large fusion interval and reduced stereoacuity were observed in cases involving stimuli in rapid movement presenting low detail levels. The first case seems to correspond to the use of the parvocellular pathway and the second case to the magnocellular pathway. Notions of patent and qualitative stereopsis are also used. In the first case, fusion is easy and the perceived depth increases monotonously with retinal disparity. In the second case, fusion is more difficult and only generates relative depth information (e.g. "behind" or "in front of" x), but operates in a disparity interval which largely exceeds the fusion interval. The authors of [STR 10] suggest that this dichotomy is the result of two different neuronal mechanisms. These observations may be correlated with the form of Panum's area, the extent of which increases with the retinal eccentricity of the stimulus [HAM 83]. In terms of 3D content, it is thus highly possible that the perception of depthwise movement and major disparities would be facilitated if these aspects were perceived in peripheral vision. When high stereoacuity is required for visualization, however, foveal vision should be used.

16.4.2.4. *Depth cues and cognitive conflicts*

Depth perception is based on the estimation of retinal disparity and on a set of cognitive and physiological, monocular or binocular cues [CUT 95]. The HVS involves a set of mechanisms allowing these cues to be integrated in order to produce a final depth estimation. In [HOF 08, HEL 10], the authors suggest that the quantity of depth is something that best correlates with each cue taken individually (statistical inference). Certain authors, however, suggest that a deterministic process links a depth quantity to the value of individual cues [DOM 06]. The model proposed in [RIC 09] is mostly based on retinal disparity; other cues are only used to convert depth-related information (quantified by disparity) into an absolute depth percept. Certain studies propose that the visual system constructs a mental representation of the field of vision over time [UKA 08]. This representation would be used for prediction purposes, considerably simplifying the depth cue integration process.

Stereoscopic screens reproduce the depth effect by providing artificial binocular cues. While little work has been published on this subject, certain researchers have suggested that these artificial cues and their incoherences may pose problems in terms of integration [PAT 09], and may induce cognitive fatigue or discomfort.

Studies have been conducted to evaluate the limitations of current display technology. The inversion of the two stereoscopic views creates a conflict between binocular information and monocular cues, generating discomfort [KIM 12]. The violation of the stereoscopic window also induces visual discomfort [MEN 09].

Moreover, the scene as a whole is generally reproduced in a clear, sharp manner, whereas in the real world, zones outside the field depth are perceived in a blurred manner and thus do not require binocular fusion. In the absence of precise results, care should be taken to ensure that generated depth cues are as coherent as possible [HOF 08].

16.5. Application to 3D content and technologies

16.5.1. *The comfort zone*

To limit accommodation/vergence conflict and offer a suitable depth of field, a comfort zone should generally be respected. This is the depth interval behind and in front of the screen (see Figure 16.5(a)) in which content is perceived within the fusion interval (otherwise diplopia, binocular rivalry or view suppression occurs) and within the limits of the depth of field (any object located outside these limits is perceived blurred). The comfort zone is specified in different manners: ±0.2 diopters in terms of depth of field [CHE 10, YAN 02], $\pm1°$ of screen angle disparity [SPE 06], respectively, 1% and 2% of screen size in terms of crossed and uncrossed disparities [MEN 09]. All these values are similar as shown in Figure 16.5(b).

16.5.2. *Reproduction artifacts*

Camera, postprocessing or imaging flaws may produce mapping faults between two stereoscopic views. The use of convergent (*toed-in*) cameras, for example, generates erroneous vertical disparities. In [SPE 02], the global introduction of vertical disparity was shown to induce visual discomfort, which increased with the size of the vertical shift and the viewing duration.

Trapezoid distortions (*keystoning*), notably introduced by the use of toed-in cameras, also generate discomfort when they are higher than 1 prismatic diopter [KOO 04]. The same applies to rotations over 1° and changes in scale of over 2.5%.

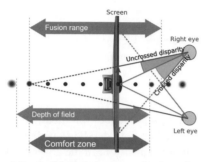 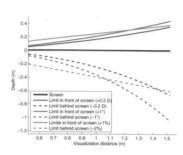

a) *Depth of field, fusion interval and comfort zone*

b) *Limits of the comfort zone for a 42" screen with a 16/9 display ratio*

Figure 16.5. *The comfort zone a) and its definitions b)*

Transmission errors, particularly packet loss, degrade left and right views in different ways and may induce binocular rivalry. In [BAR 10], the authors showed that the use of 2D error resilience mechanisms does not prevent these errors from causing visual discomfort. The resilience strategy which generates the least discomfort has been shown to be a temporary switch to 2D display. Finally, coding artifacts, particularly quantization effects, generate discomfort: in [BAR 10], the percentage of observers experiencing visual discomfort increased in line with the quantization step of the H.264 codec.

Interference between active (shutter) eyewear and certain light sources (e.g. neon lighting), which flicker with the frequency of the electrical network (50 Hz), or insufficient display refresh frequencies, generates flickering effects; these are particularly problematic when perceived in the periphery of the visual field.

When a depth map is used to synthesize stereoscopic views or when 2D to 3D conversion techniques are used, care is required to ensure temporal coherence and avoid the appearance of noise or depth oscillation in contours [CHO 12, LAM 07].

Crosstalk is a common effect encountered when using passive (polarized) and active displays when information destined for one eye is also perceived by the other eye. This generates blurring effects and may produce phantom images, leading to visual discomfort. In [KOO 04], the introduction of crosstalk of 5% produced slight discomfort, and at 25%, the discomfort was very significant. In [SIE 01], however, the author showed that controlled crosstalk can contribute to the reduction of picket fence effects when the observer changes viewpoint in front of an autostereoscopic screen.

16.5.3. *Focus and blurring effects*

Okada *et al.* [OKA 06] show that in the presence of accommodation/vergence conflicts, the role of accommodation (through focusing) reduces as the level of blurring increases. Inversely, when the visual stimulus is clear and contains high frequencies, precise accommodation is needed, requiring higher levels of effort. In [KOO 04], the authors showed that the introduction of slight blurring is a major source of discomfort. Blurring was long considered to be a marginal depth cue due to the ambiguity of its symmetrical behavior in relation to the focusing distance. In [HEL 10], the authors showed that blurring is proportional to the absolute value of disparity, and its introduction may be used to alter depth perception. The authors of [WAN 11] showed that the introduction of blurring allows an increase in perceived depth in relation to the distance between the foreground and background of a visualized scene. Thus, the introduction of blurring, for an identical perceived depth, allows a reduction in screen disparity. While further work is required on the subject, these results suggest that visual discomfort may be reduced without modifying perceived depth by reducing disparity and reducing the role of focus accommodation (and thus the accommodation/vergence conflict).

16.5.4. *Visual attention*

A number of perceptual mechanisms involved in visualizing 3D content are influenced by visual attention [HUY 11]. Visual attention carries out selection on elements presented in the field of vision and, thus, modifies perceived stimuli. The extent of Panum's area, for example, is reduced in foveal vision but it is larger in peripheral vision. Zhang *et al.* [ZHA 10] propose a 3D visual attention model based on depth, brightness, color and movement contrast. In [SOH 11], the authors propose a visual comfort model based on a visual attention model developed for stereoscopic content.

16.5.5. *Flaws or absence of the motion parallax*

In the real world, observer movements generate a motion parallax. Stereoscopic 3D screens do not reproduce this effect, and the perceived flatness may enter into conflict with the disparity presented on screen [HOW 11]. Multiview display systems, on the other hand, may reproduce motion parallax cues when the user moves between viewpoints.

16.5.6. *Exposure duration and training*

As the duration of visualization increases, visual fatigue can accumulate. In [SUZ 04], accommodation speed remained unchanged after 15 min of 3D visualization; after 30 min, however, it was reduced. In the second case, a rest period of 90–120 min was required for accommodation to return to its initial speed. In [LAM 09], some measured visual properties showed short-term effects but rapidly returned to their initial state. In [LI 08], the cerebral activity of the beta band increased with the duration of visualization of a stereoscopic sequence.

However, the performances of the visual system may also increase with the duration of visualization. In [WOO 74], the authors showed that the extent of Panum's area increased in line with visualization time. In [LAM 09], the fusion interval was shown to increase after short-term effort. Certain studies have also shown that training may contribute to a reduction in the signs and symptoms of visual fatigue. In [EMO 05], the fusion interval increased with repeated experimental sessions over the course of one month. In [JON 89], Panum's area was shown to increase with training.

16.6. Predicting visual fatigue and discomfort: first models

Another subject of research activity in this field is the automatic prediction of visual fatigue and discomfort. Recent publications have provided objective models of discomfort [CHO 12, LI 13] and perceived fatigue [CHO 10], which allow us to predict perceived effects in certain cases.

16.7. Conclusion

In this chapter, we have considered the principal mechanisms involved in binocular vision from the perspective of visual fatigue and discomfort commonly experienced when viewing 3D content. We began by discussing the terminology, which is sometimes ambiguous, with a particular focus on the terms "visual fatigue" and "visual discomfort". We then described the physical and psychological manifestations of fatigue, before studying the way in which the adaptations and limitations of the HVS are linked to fatigue and discomfort when faced with unusual stimuli. Finally, we considered the failings in 3D content and technologies which generate these unusual stimuli. Clearly, the whole of the 3D chain, from capture to viewing, via data representation, compression, transmission and display, requires careful mastery in order to limit stereoscopic faults and reproduce content in a suitable manner (within the zone of comfort, limiting the depthwise movement of objects). In this, the human factor is decisive due to the multitude of perceptual mechanisms involved. We, therefore, need to obtain the best compromise between the quantity or intensity of 3D effects and the level of visual fatigue and discomfort experienced by the majority of people.

16.8. Bibliography

[AND 02] ANDO T., TANAKA A., FUKASAKU S., et al., "Pupillary and cardiovascular responses to a video movie in senior human subjects", *Autonomic Neuroscience: Basic & Clinical*, vol. 97, no. 2, pp. 129–35, 2002.

[BAR 10] BARKOWSKY M., WANG K., COUSSEAU R., et al., *18th International Packet Video Workshop, 2010*, IEEE, Hong Kong, 2010.

[BLE 05] BLEHM C., VISHNU S., KHATTAK A., et al., "Computer vision syndrome: a review", *Survey of Ophthalmology*, vol. 50, no. 3, pp. 253–262, 2005.

[CAI 92] CAIL F., SALSI S., La fatigue visuelle, Report no. 1–3, INRS, 1992.

[CHA 77] CHARMAN W.N., WHITEFOOT H., "Pupil diameter and the depth-of-field of the human eye as measured by laser speckle", *Journal of Modern Optics*, vol. 24, no. 12, pp. 1211–1216, 1977.

[CHE 10] CHEN W., FOURNIER J., BARKOWSKY M., et al., "New requirements of subjective video quality assessment methodologies for 3DTV", *Proceedings of 5th International Workshop on Video Proceedings and Quality Metrics*, Scottsdale, AZ, vol. 2010, 2010.

[CHO 10] CHOI J., KIM D., HAM B., et al., "Visual fatigue evaluation and enhancement for 2D-plus-depth video", *17th IEEE International Conference on Image Processing*, IEEE, pp. 2981–2984, 2010.

[CHO 12] CHOI J.-H., YUN M.-J., KIM A.-R., et al., "Visual comfort measurement for 2D/3D converted stereo video sequence", *3DTV-Conference*, IEEE, no. 1, pp. 1–4, 2012.

[COO 11] COOPER E.A., BURGE J., "The vertical horopter is not adaptable, but it may be adaptive", *Journal of Vision*, vol. 11, no. 3, pp. 1–19, 2011.

[CUT 95] CUTTING J.E., VISHTON P.M., "Perceiving layout and knowing distances: the integration, relative potency, and contextual use of different information about depth", in EPSTEIN W., ROGERS S.J. (eds), *Perception of Space and Motion. Handbook of Perception and Cognition, 2nd ed.*, San Diego, CA, US: Academic Press, vol. 5, pp. 69–117, 1995.

[DOM 06] DOMINI F., CAUDEK C., TASSINARI H., "Stereo and motion information are not independently processed by the visual system", *Vision Research*, vol. 46, no. 11, pp. 1707–1723, 2006.

[EMO 04] EMOTO M., NOJIRI Y., OKANO F., "Changes in fusional vergence limit and its hysteresis after viewing stereoscopic TV", *Displays*, vol. 25, no. 2–3, pp. 67–76, 2004.

[EMO 05] EMOTO M., NIIDA T., OKANO F., "Repeated vergence adaptation causes the decline of visual functions in watching stereoscopic television", *Journal of Display Technology*, vol. 1, no. 2, pp. 328–340, 2005.

[FEN 67] FENDER D., JULESZ B., "Extension of Panum's fusional area in binocularly stabilized vision", *Journal of the Optical Society of America*, vol. 57, no. 6, pp. 819–830, 1967.

[FUK 09] FUKUSHIMA T., WOLFFSOHN J.S., GILMARTIN B., "The relationship between CA/C ratio and individual differences in dynamic accommodative responses while viewing stereoscopic images", *Journal of Vision*, vol. 9, no. 13, pp. 1–13, 2009.

[HAM 83] HAMPTON D.R., KERTESZ A.E., "The extent of Panum's area and the human cortical magnification factor", *Perception*, vol. 12, no. 2, pp. 161–165, 1983.

[HAS 01] HASEBE S., GRAF E.W., SCHOR C.M., "Fatigue reduces tonic accommodation", *Ophthalmic and Physiological Optics*, vol. 21, no. 2, pp. 151–160, 2001.

[HEL 10] HELD R.T., COOPER E.A., O'BRIEN J.F., *et al.*, "Using blur to affect perceived distance and size", *ACM Transactions on Graphics*, vol. 29, no. 2, pp. 1–16, 2010.

[HOF 08] HOFFMAN D.M., GIRSHICK A.R., BANKS M.S., "Vergence - accommodation conflicts hinder visual performance and cause visual fatigue", *Journal of Vision*, vol. 8, no. 3, pp. 1–30, 2008.

[HOW 97] HOWARTH P., COSTELLO P., "The occurrence of virtual simulation sickness symptoms when an HMD was used as a personal viewing system", *Displays*, vol. 18, no. 2, pp. 107–116, 1997.

[HOW 11] HOWARTH P.A., "Potential hazards of viewing 3-D stereoscopic television, cinema and computer games: a review", *Ophthalmic & Physiological Optics: The Journal of the British College of Ophthalmic Opticians (Optometrists)*, vol. 31, no. 2, pp. 111–122, 2011.

[HUN 01] HUNG G.K., *Models of Oculomotor Control*, vol. 55, World Scientific Publishing Co. Pte. Ltd., 2001.

[HUY 11] HUYNH-THU Q., BARKOWSKY M., CALLET P.L., "The importance of visual attention in improving the 3D-TV viewing experience: overview and new perspectives", *IEEE Transactions on Broadcasting*, vol. 57, no. 2, pp. 432–444, 2011.

[ITU 04] ITU-R-BT.500-11, Methodology for the subjective assessment of the quality of television pictures, Technical report, International Telecommunication Union, 2004.

[JAS 91] JASCHINSKI-KRUZA W., "Eyestrain in VDU users: viewing distance and the resting position of ocular muscles", *Human Factors*, vol. 33, no. 1, pp. 69–83, 1991.

[JAS 98] JASCHINSKI-KRUZA W., "Visual strain during VDU work: the effect of viewing distance and dark focus", *Ergonomics*, vol. 31, pp. 1449–1465, 1998.

[JON 89] JONES R., STEPHENS G.L., "Horizontal fusional amplitudes. Evidence for disparity tuning", *Investigative Ophthalmology & Visual Science*, vol. 30, no. 7, pp. 1638–1642, 1989.

[KEN 93] KENNEDY R., LANE N., BERBAUM K., *et al.*, "Simulator sickness questionnaire: an enhanced method for quantifying simulator Sickness", *The International Journal of Aviation Psychology*, vol. 3, no. 3, pp. 203–220, 1993.

[KIM 11] KIM D., JUNG Y., KIM E., *et al.*, "Human brain response to visual fatigue caused by stereoscopic depth perception", *17th International Conference on Digital Signal Processing (DSP), 2011*, IEEE, pp. 1–5, 2011.

[KIM 12] KIM W.-J., "Feature-based detection of inverted-stereo for stereoscopic 3D viewing comfort", *IEEE Transactions on Broadcasting*, vol. 58, no. 2, pp. 296–300, 2012.

[KOO 04] KOOI F., TOET A., "Visual comfort of binocular and 3D displays", *Displays*, vol. 25, no. 2–3, pp. 99–108, 2004.

[KUZ 08] KUZE J., UKAI K., "Subjective evaluation of visual fatigue caused by motion images", *Displays*, vol. 29, no. 2, pp. 159–166, 2008.

[LAM 07] LAMBOOIJ M.T.M., IJSSELSTEIJN W.A., HEYNDERICKX I., "Visual discomfort in stereoscopic displays: a review", *Proceedings of SPIE*, vol. 6490, SPIE, 2007.

[LAM 09] LAMBOOIJ M.T.M., FORTUIN M.F., IJSSELSTEIJN W.A., *et al.*, "Measuring visual discomfort associated with 3D displays", *Proceedings of SPIE*, vol. 7237, SPIE, 2009.

[LAM 11] LAMBOOIJ M.T.M., IJSSELSTEIJN W.A., BOUWHUIS D.G., *et al.*, "Evaluation of stereoscopic images: beyond 2D quality", *IEEE Transactions on Broadcasting*, vol. 57, no. 2, pp. 432–444, 2011.

[LI 05] LI Z., SUN F., "Pupillary response induced by stereoscopic stimuli", *Experimental Brain Research*, vol. 160, no. 3, pp. 394–397, 2005.

[LI 08] LI H., SEO J., KHAM K., *et al.*, "Measurement of 3D visual fatigue using event-related potential (ERP): 3D oddball paradigm", *3DTV Conference: The True Vision-Capture, Transmission and Display of 3D Video, 2008*, IEEE, pp. 213–216, 2008.

[LI 11] LI J., BARKOWSKY M., WANG J., *et al.*, "Study on visual discomfort induced by stimulus movement at fixed depth on stereoscopic displays using shutter glasses", *Proceedings of 17th International Conference on Digital Signal Processing*, pp. 1–8, 2011.

[LI 13] LI J., BARKOWSKY M., LE CALLET P., "Visual discomfort is not always proportional to eye blinking rate: exploring some effects of planar and in-depth motion on 3DTV QoE", *7th International Workshop on Video Processing and Quality Metrics*, pp. 1–6, 2013.

[MEN 09] MENDIBURU B., *3D Movie Making: Stereoscopic Digital Cinema from Script to Screen*, Focal Press, 2009.

[MIT 66] MITCHELL D.E., "A review of the concept of 'Panum's fusional area'", *American Journal of Optometry*, vol. 43, pp. 387–401, 1966.

[MUR 01] MURATA A., UETAKE A., OTSUKA M., *et al.*, "Proposal of an index to evaluate visual fatigue induced during visual display terminal tasks", *International Journal of Human-Computer Interaction*, vol. 13, no. 3, pp. 305–321, 2001.

[NAG 96] NAGATA S., "The binocular fusion of human vision on stereoscopic displays - field of view and environment effects", *Ergonomics*, vol. 39, no. 11, pp. 1273–1284, 1996.

[NAG 08] NAGATANI H., HIRAYAMA Y., "Evaluation of the influence on the human body of the autostereoscopic display based on the integral imaging method", *Proceedings of SPIE*, vol. 6803, SPIE, pp. 1–8, 2008.

[NOJ 04a] NOJIRI Y., YAMANOUE H., HANAZATO A., *et al.*, "Parallax distribution and visual comfort on stereoscopic HDTV", *Proceedings of SPIE*, vol. 25, SPIE, pp. 303–313, 2004.

[NOJ 04b] NOJIRI Y., YAMANOUE H., HANAZATO A., *et al.*, "Visual comfort/discomfort and visual fatigue caused by stereoscopic HDTV viewing", *Proceedings of SPIE*, vol. 5291, SPIE, pp. 303–313, 2004.

[NOR 81] NORTH R., HENSON D.B., "Adaptation to prism-induced heterophoria in subjects with abnormal binocular vision or asthenopia", *American Journal of Optometry and Physiological Optics*, vol. 58, no. 9, pp. 746–752, 1981.

[OKA 06] OKADA Y., UKAI K., WOLFFSOHN J.S., *et al.*, "Target spatial frequency determines the response to conflicting defocus- and convergence-driven accommodative stimuli", *Vision Research*, vol. 46, no. 4, pp. 475–484, 2006.

[PAT 09] PATTERSON R., "Review paper: human factors of stereo displays: an update", *Journal of the Society for Information Display*, vol. 17, no. 12, pp. 987–996, 2009.

[QUA 12] QUALINET, Qualinet white paper on definitions of quality of experience, Technical report, Dagstuhl Seminar, 2012.

[RIC 71] RICHARDS W., "Anomalous stereoscopic depth perception", *Journal of the Optical Society of America*, vol. 61, no. 3, pp. 410–413, 1971.

[RIC 09] RICHARDS W., "Configuration stereopsis: a new look at the depth-disparity relation", *Spatial Vision*, vol. 22, no. 1, pp. 91–103, 2009.

[ROO 00] ROOKES P., WILLSON J., *Perception: Theory, Development and Organisation*, Routledge Modular Psychology Series, 1st ed., London, 2000.

[SCH 81] SCHOR C.M., TYLER C.W., "Spatio-temporal properties of Panum's fusional area", *Vision Research*, vol. 21, no. 5, pp. 683–692, 1981.

[SCH 87] SCHOR C.M., TSUETAKI T.K., "Fatigue of accommodation and vergence modifies their mutual interactions", *Investigative Ophthalmology & Visual Science*, vol. 28, no. 8, pp. 1250–1259, 1987.

[SCH 06] SCHREIBER K.M., TWEED D.B., SCHOR C.M., "The extended horopter: quantifying retinal correspondence across changes of 3D eye position", *Journal of Vision*, vol. 6, no. 1, pp. 64–74, 2006.

[SIE 01] SIEGEL M., "Perceptions of crosstalk and the possibility of a zoneless autostereoscopic display", *Proceedings of SPIE*, SPIE, pp. 34–41, 2001.

[SOH 11] SOHN H., JUNG Y.J., LEE S.-I., *et al.*, "Attention model-based visual comfort assessment for stereoscopic depth perception", *2011 17th International Conference on Digital Signal Processing (DSP)*, 2011, IEEE, pp. 1–6, 2011.

[SPE 02] SPERANZA F., WILCOX L.M., "Viewing stereoscopic images comfortably: the effects of whole-field vertical disparity", *Proceedings of SPIE Stereoscopic Displays and Virtual Reality Systems*, vol. 4660, SPIE, pp. 18–25, 2002.

[SPE 06] SPERANZA F., TAM W.J., RENAUD R., *et al.*, "Effect of disparity and motion on visual comfort of stereoscopic images", *Proceedings of SPIE*, vol. 6055, SPIE, pp. 94–103, 2006.

[STR 10] STRANSKY D., WILCOX L., "Depth magnitude and binocular disparity: a closer look at patent vs. qualitative stereopsis", *Journal of Vision*, vol. 10, no. 7, p. 330, 2010.

[SUZ 04] SUZUKI Y., ONDA Y., KATADA S., *et al.*, "Effects of an eyeglass-free 3-D display on the human visual system", *Japanese Journal of Ophthalmology*, vol. 48, no. 1, pp. 1–6, 2004.

[UKA 97] UKAI K., TSUCHIYA K., ISHIKAWA S., "Induced pupillary hippus following near vision: increased occurrence in visual display unit workers", *Ergonomics*, vol. 40, no. 11, pp. 1201–1211, 1997.

[UKA 02] Ukai K., Kato Y., "The use of video refraction to measure the dynamic properties of the near triad in observers of a 3-D display", *Journal of the British College of Ophthalmic Opticians (Optometrists)*, vol. 22, no. 5, pp. 385–388, 2002.

[UKA 08] Ukai K., Howarth P.A., "Visual fatigue caused by viewing stereoscopic motion images: background, theories, and observations", *Displays*, vol. 29, no. 2, pp. 106-116, 2008.

[VIN 07] Vink P., De Looze M.P., "Crucial elements of designing for comfort", in Schifferstein H. N.J., Hekkert P., (eds), *Product Experience*, Elsevier Science, pp. 441–460, 2007.

[WAN 02] Wann J.P., Mon-Williams M., "Measurement of visual aftereffects following virtual environment exposure", Stanney K.M., (ed.), *Handbook of Virtual Environments: Design, Implementation, and Applications*, Chapter 37, Lawrence Erlbaum Associates, Mahwah, NJ, USA, pp. 731–749, 2002.

[WAN 11] Wang J., Barkowsky M., Ricordel V., "Quantifying how the combination of blur and disparity affects the perceived depth", *SPIE Electronic Imaging*, vol. 7865, pp. 1–10, 2011.

[WAT 04] Watt S.J., Akeley K., Girshick A.R., *et al.*, "Achieving near-correct focus cues in a 3-D display using multiple image planes", *ACM Transactions on Graphics - Proceedings of ACM SIGGRAPH 2004*, vol. 23, no. 3, 2004.

[WOO 74] Woo G., "The effect of exposure time on the foveal size of Panum's area", *Vision Research*, vol. 14, no. 7, pp. 473–480, 1974.

[YAN 02] Yano S., Ide S., Mitsuhashi T., *et al.*, "A study of visual fatigue and visual comfort for 3D HDTV/HDTV images", *Displays*, vol. 23, no. 4, pp. 191–201, 2002.

[YAN 04] Yano S., Emoto M., Mitsuhashi T., "Two factors in visual fatigue caused by stereoscopic HDTV images", *Displays*, vol. 25, no. 4, pp. 141–150, 2004.

[YAN 11] Yang S., Sheedy J.E., "Effects of vergence and accommodative responses on viewer's comfort in viewing 3D stimuli", *Proceedings of SPIE Stereoscopic Displays and Applications XXII*, vol. 7863, SPIE, pp. 78630Q1–13, 2011.

[ZHA 10] Zhang Y., Jiang G., Yu M., *et al.*, "Stereoscopic visual attention model for 3d video", *16th International Conference on Advances in Multimedia Modeling*, Chongqing, China, pp. 314–324, 6–8 January 2010.

Chapter 17

2D–3D Conversion

17.1. Introduction

Native stereoscopic filming may be considered too costly in terms of time or resources. At times, filming may already have taken place, and the scene may no longer exist. In other cases, *a posteriori* 3D conversion is the only way of producing 3D content. The recent release of 3D versions of "older" blockbusters, such as *Titanic*, *Top Gun* or *The Lion King*, has made use of this technique. The conversion of "flat" 2D content into 3D content involves creating missing information. The process involves an automatic aspect, where parallax is created from other depth cues present in the scene, and an aspect carried out by human operators, adding a creative dimension to the procedure.

Methods developed for 2D–3D conversion may also be used for parallax correction in existing, but unsatisfactory, stereoscopic content. The 2D–3D conversion industry is currently expanding, with an estimated annual turnover of billions of dollars. Companies, such as Prime Focus, employ 1,200 workers in India for the conversion process alone. A number of directors have chosen to use this technique, including George Lucas for the *Star Wars* films and James Cameron for *Titanic*.

Although it presents certain advantages, 2D–3D conversion is not a simple process; situated on the border between technology and art, it requires continuous human intervention. The cost of high-quality conversion currently

Chapter written by David GROGNA, Antoine LEJEUNE and Benoît MICHEL.

ranges from $50,000 to $150,000 per minute. Pioneers in the domain include In-Three Inc., Legend3D, Sony Pictures Imageworks and Passmore Lab.

2D–3D conversion is carried out almost exclusively using a depth map. Various methods of calculating a depth map from a stereoscopic pair have been put forward, including work by Lucas and Kanade [LUC 81], Horn and Schunck [HOR 81], Periaswamy and Farid [PER 03], Wu *et al.* [WU 00], Alvarez *et al.* [ALV 00], Schmidt [SCH 02] and Ran and Sochen [RAN 00]. While the domain has been explored in detail, the generation of a depth map from a single image is a problem, which presents an infinite number of solutions, and the proposed methods cannot, therefore, claim to offer universally acceptable solutions.

17.2. The 2D–3D conversion workflow

2D–3D conversion involves creating the second image of a stereoscopic pair from the first image. The process may be divided into four stages [MIC 11]:

– Depth map generation: the depth map is a bitmap image of the same size as, and in exact correspondence with, the starting image. Each pixel is defined as the distance between an object visible in the scene and the camera, rather than a color. This map may easily be shown in gray scale and manipulated visually in the same way as any monochrome image.

– Segmentation: in this crucial stage, we define the contours of the different objects that make up a scene and must appear in different depth planes in the image. Thus, we separate the main objects or characters, often situated in the medium plane, from the background. While this operation is not particularly difficult for rigid objects [XU 11], it is more complex for blurred or semi-transparent images, such as smoke clouds, and for all objects, which do not cover the whole of the pixels in which they appear, such as hair, fur and netting. The segmentation operation must precisely define discontinuity lines in the depth map.

– Missing image generation: a new image is generated by lateral shifting of the pixels in the original image, over a distance defined by the value of the depth map at each specific point. The depth level corresponding to a null parallax, which will show the pixel in the screen plane, is defined in the *depth script*. This decision is essentially of an artistic nature. The image generation stage is automatic. The values of the maximal positive parallax respect the limit imposed for the background in order to avoid spectator view divergences. The image generation procedure using a depth map is often known as *depth image based rendering* (DIBR). Various open-source implementations are available, notably using MATLAB [DA 10]. As the horizontal shifts, used to generate new pixel, are generally fractional, the source and destination images are usually oversampled before DIBR. Upsampling by a factor 5 is often used, as the human visual system is able to detect shifts of the order of one-fourth of a pixel (for high-definition images with a width of 2 K pixels) [GRA 12]. Note that it is also possible to generate two symmetrical left and right images from a "central" view, creating lateral views by shifts with opposite signs and a distance reduced by half. This solution offers advantages in that artifacts are distributed equally across the two views, and their size is reduced.

– Artifact removal: the pixel shifts involved in the previous stage remove certain pixels, generating holes in the image. The missing pixels must be recreated by a *disocclusion filling* procedure using a variety of possible sources, including the previous or following images in the temporal sequence, or adjacent pixels in the current image. Pixels are recreated by duplication, interpolation or extrapolation using these sources.

17.3. Preparing content for conversion

17.3.1. *Depth script*

Before converting a whole film or a series of sequences for insertion into an existing production, a *depth script* is created, where sequences with depth effects are clearly identified, alongside maximum values for the *depth budget* that will be applied to the sequences for conversion. Note that decisions made at this level are essentially artistic, but they must also take account of technical limits imposed on the production, such as the maximum positive parallax, which depends on the size of the screen used for projection. This *depth grading* procedure must also impose values for each sequence, or each character or object present in a sequence, which will remain comfortable for the spectator [SEI 12]. Thus, positive parallaxes will be lower in productions for a juvenile audience, as children have a smaller interocular distance than adults [DOD 04]. Decisions affecting the depth budget also take into account the temporal succession of sequences in order to avoid sudden changes in the distance of the zone of interest during scene changes.

17.3.2. *The advantage of video over fixed images*

Image sequences are clearly more practical than separate fixed images when carrying out conversion. In a video or cinema sequence, a considerable amount of redundancies exist between consecutive images, providing an important information source for the artifact removal stage described above. This is particularly true in cases where the camera is subject to lateral horizontal movement during the sequence. From a conversion perspective, the most useful sequences are those which contain a predictable horizontal motion parallax; the most typical cases of this type include a camera mounted onto a traveling chariot, moving perpendicular to the view axis. A number of other cases offer usable motion parallaxes, such as objects, vehicles or characters, in relative movement with each other or with the background. The usefulness of these movements is increased by regularity; they include translations (vehicles on a road) and rotations (fairground rides, fan blades). Motion parallaxes are easier to use for objects with clear borders, which are easy to segment precisely, such as vehicles, buildings or manufactured objects.

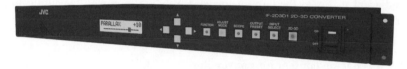

Figure 17.1. *The professional JVC IF-2D3D1 package performs automatic 2D–3D conversion in addition to other stereoscopic processes*

17.3.3. *The automatic conversion trap*

2D–3D conversion combines complex algorithms with knowledge and abilities of experienced human operators. To date, no 100% automatic procedure has been able to provide wholly satisfactory results in all cases.

A certain number of professional 2D–3D converters are available on the market, but none of these converters can guarantee perfect results for video sequences of unknown origin. However, some converters offer acceptable results in highly constrained environments, such as a football stadium with fixed cameras, controlled lighting and a scene geometry containing few different depth planes. The use of this type of equipment for live broadcast, however, remains risky.

Certain specific error types are generally encountered with 2D–3D converters. Objects in the foreground of an image may appear in the background and vice versa. Transparency, and fine structures, which are difficult to segment, such as rain, hair or the holes in a tennis racket, are generally badly interpreted or completely ignored by converters. The segmentation process may consider fine structures as noise and integrate them into the background, or over-segment them, integrating parts of the background. In both the cases, the results are unconvincing in terms of depths.

A variety of simple and inefficient converters are available to the public, both as software [GMB 10] and specific devices [IPP 11]. The simplest of these methods consists of taking the stereoscopic pair of the following image or the previous image of a video sequence as the second image. If the camera captures a regular panoramic lateral view, the disparity between two consecutive images produces a 3D effect; the faster the camera moves, the stronger the effect is. This approach rarely works and, if the camera moves in a vertical plane, the results are very uncomfortable. Moreover, if the camera changes direction, the depth effect is inverted and the background switches to the foreground. A more sophisticated variation of this method only conserves the horizontal image shift component, reducing undesirable effects, but this does not prevent the 3D effect from disappearing when the camera ceases to move.

17.3.4. *Specific cases of automatic conversion*

A simple mechanism whereby a left image is created using the previous right image in chronological order can produce good results. This is the case for lateral views taken from an airplane in constant linear flight, a method used from before World War II in aerial reconnaissance. A number of refinements have been used to improve the effectiveness of this technique [KNO 07, MAT 97, MOU 05].

In the case of sequences containing considerable nonlinear movement, certain methods can produce acceptable results. These methods, based on the use of motion vectors used in MPEG compression of video streams for television, give variable, and sometimes satisfactory, results; one example of this type was developed by Ideses *et al.* [IDE 07].

Static capture in closed environments produces favorable conditions for real-time automatic conversion. This is the case for a majority of sporting events [GRA 10, SCH 11]. These real-time conversion techniques use *a priori* knowledge of the scene, including distances or reference points. After an initial automatic camera calibration phase, these algorithms generally calculate a background model in the form of an image mosaic. This background is made up of all of the static elements in a scene. The following step consists of creating a depth map, which will not (or barely) change over time and may even be reused as long as the location remains unchanged. The calculation of this depth map generally requires additional information, such as the length of pitch lines or the position of certain points. Dynamic foregrounds are obtained by differences with the background model. Different procedures are used to associate depth with these foreground images; the simplest method is to associate each silhouette with the depth of the ground element at its base.

17.3.5. *Optimal content for 2D–3D conversion*

In many cases, sequences or even whole movies are filmed in 2D, and then converted to 3D in postproduction. This reduces filming costs, although the conversion phase greatly increases the total cost of the film. In these cases, framing and scene setting are carried out as if 3D filming was used, avoiding foregrounds that cut across frame edges and depth jumps in scene changes and prioritizing long shots.

We should note, however, that even semi-automatic conversion can be extremely labor-intensive for certain types of scenes. The complexity and, *a fortiori*, the cost of conversion increases according to the number of depth layers used in a scene. If one of the essential elements in the scene has a highly complex geometric structure, such as a plant or a crowd of people, a stereoscopic camera will always give better depth rendering. Scenes filmed using a crane or steadicam may also be difficult to convert due to the presence of irregular and nonlinear movements.

In the same way, transparency and atmospheric phenomena, such as rain, snow or fog, further complicate the conversion process. First, the segmentation between the foreground and background ceases to be clearly defined: a single pixel may be part of several planes at once. This complicates the creation of the depth map and the generation of the stereoscopic image pair. Secondly, this type of effect can drastically increase the number of depth layers used in a scene.

In other cases, 2D–3D conversion may be highly efficient [DEJ 08]. In the case of concerts or other scenes using significant lighting effects, 2D–3D conversion may be used to avoid asymmetry in light reflections between left and right images. Conversion is also used when the equipment used is poorly suited to 3D filming (tele-photo lenses which compress scenes, macro etc.). Nevertheless, certain rules should be followed to increase the chances of successful conversion:

– Keep the horizon and other main planes of reference strictly horizontal.

– Plan long sequence shots. Shots of less than 3 s contain insufficient depth information for effective processing. Moreover, observers take several seconds to assimilate the various depth levels in a scene.

– Use suitable lighting for the background as well as the foreground. Dark zones lack detail and are difficult to convert.

– Ensure that all objects in the field are clearly textured and contrasted. In cases of untextured objects, it is impossible to define a parallax between homologous points in two images and thus the depth.

– Leave as many monoscopic depth cues as possible in the field: vanishing lines, geometric objects, etc.

– Use a wide-field lens rather than a tele-photo lens, which considerably reduces most depth cues.

– Use regular camera movements. Horizontal and panoramic traveling provide large quantities of perspective information.

When a movie is filmed with the intention of carrying out 3D conversion in post-production, efforts must clearly be made to simplify the conversion process. Thus, scenes including effects, made up of several shots or including CGI elements are pre-separated into superposed layers; for elements created using 3D modeling software, a depth map is provided automatically.

The Walt Disney Pictures movie *G-Force* is an example of partial conversion, carried out by In-Three, pioneers in 2D–3D in Hollywood, using a procedure called *dimensionalization*. In *G-Force*, scenes filmed in 2D were converted and then integrated into a synthesized, computer-generated world.

17.4. Conversion stages

The purpose of the industrial video conversion workflow is to create the second view of a stereoscopic pair using the first view. For reasons of efficiency, the various stages in the procedure are carried out by dedicated workers or programs. As automatic algorithms are unable to produce perfect results for each stage, visual checks by experienced staff are necessary after each of the stages mentioned below:

1) Detection of key images (in the original, left view), between which movements in different depth planes are sufficiently linear and/or predictable.

2) Segmentation of foregrounds, characters and mid-shots.

3) Evaluation of the depth Z of the center of gravity of each element.

4) Spatial propagation of Z to all of the pixels of each element in the image.

5) Temporal propagation of the segmentation and depth maps of each element to images between two key images.

6) Generation of right images for the whole sequence.

7) Correction of disocclusion artifacts produced during the previous stage.

All of these stages, with the exception of step 6 which is completely automatic, are semi-automatic, meaning that they use algorithms and computing tools manipulated by human users. One example of this is the automatic detection of contours, which are then corrected using spline/bezier tools.

17.4.1. *The segmentation stage*

The detection of key images is a manual operation which does not involve a complex algorithm. Segmentation, the crucial first stage in the workflow, is different. Each key image is split into main elements which are non-contiguous in terms of depth: sky, ground, objects and characters. The segmentation of an image into distinct elements is easier in a video sequence than in a single fixed image, as the movement of objects in relation to one another generates a movement parallax, allowing contour determination. The segmentation program also determines the relative distance between objects: the character is in front of the table, which is in front of the sky, etc. We thus establish a classification of values between the various elements of the final depth map. A variety of segmentation methods may be used based on different visual criteria such as color, shape and other characteristics of objects and characters [AHA 06, CHE 98, FRE 97, GOR 97, KIN 01, MAL 93, VIO 04]. It is easy for the human eye to follow the silhouette of a moving object in a video sequence. The most promising automatic segmentation methods are based on optical flows.

Optical flow methods calculate the trajectory of each pixel in an image according to its movement history, and thus determine where and when an element appears and disappears and when it is hidden by another object. Precise optical flow calculation, applied to all pixels of an image, was long considered to be impossible due to its considerable requirements in terms of processing power. The development of massively parallel graphics processing units (GPUs) has changed this, and calculations of this type may now be carried out in real time. By knowing the precise trajectory of each pixel, we may interpolate their positions and thus detect the contours of moving objects in front of a static or slow-moving background.

Segmentation is not easy: in many cases, the edges of objects or characters are not clear. Partially transparent elements, such as hair blown by the wind, lace veils, clouds of smoke or the strings of a tennis racket are difficult to separate from their background.

17.4.2. *Depth maps: calculation and propagation*

Once the relative distances between the elements of the scene have been obtained, each element is assigned an absolute distance in relation to the camera. This stage may, clearly, be carried out manually, but an intelligent program can prove effective and make use of a number of depth cues used by the human brain: *a priori* knowledge of objects, the relative size of several identical objects, etc. A human head in closeup, for example, will clearly be close; a human silhouette of one-fourth of the height of the screen will be approximately 10 m away, a car of a few pixels in length will be in the background, and the sky will be considered to be in the background.

An automatic algorithm begins by seeking a horizon as exterior shots are often used. It then applies a first estimation: the sky is in the background, and the ground approaches in a linear manner from the horizon to the foot of the camera. All possible depth cues are explored and used, whether automatically or manually:

– If the camera offers a limited depth of field, blurring may be used to evaluate which elements form the center of interest in the image and are thus close to the zero parallax distance.

– Perspective, receding lines and the position of receding points are very useful in locating the relative depth of buildings, roads, edges of sports pitches, rooms, etc.

– For objects with a strong incline in relation to the image plane, we do not define a depth, but a depth gradient. This is typically the case for the ground, or for walls.

– Movements in highly dynamic scenes are used to determine the distance of an object of known size, such as a vehicle, a ball or a character. If the position of the ground has already been determined, for example, moving persons will be located at the distance where their feet touch the ground.

a)

b)   c)

Figure 17.2. *a) Starting image of a depth map conversion; b) depth map with two segmented elements; c) the same depth map after depth matching between two points at the same distance from the two elements*

At this stage, we obtain a depth map made up of several planes that can be recognized by their different brightnesses. Each element in the image then needs to be refined; these elements are generally memorized in several distinct superposed layers, and the depth of each is determined by imposing a light level variation in the depth map, giving an inclination or roundness. Once again, monocular depth cues are used to add details to the result, such as shadow and light effects showing the shape of a face. *A priori* knowledge of objects and characters is also very important. This knowledge is used, for example, to round out faces or balls by directly modifying the depth map by hand or by creating computer models of these elements from which the depth map may be extracted. This operation is generally supervised and corrected by hand. Thus, the operator ensures that characters are correctly anchored to the ground by giving their feet the same depth (and thus the same shade in the depth map) as the ground below them. At the end of this stage, the key image has a perfect depth map, and a first result may be visualized on screen before proceeding to the next step.

Next, using these depth maps of key images of a video sequence, interpolation techniques are applied to create depth maps corresponding to intermediate images. Simple linear interpolation may be used if the movement is sufficiently uniform. In all cases, the aim is for the depth values between two key images to follow movement, while remaining coherent in relation to the depth specified in the key images. The quality of segmentation and of movement estimation is crucial in ensuring successful interpolation. Temporal filtering of depth maps may also be used to improve quality, increasing the coherence of depth across a whole sequence [BLE 09]. Manual validation is always helpful, as a movement that initially appears linear may not, in fact, be completely linear; segmentation contours, depth values and other parameters may then need to be adjusted.

17.4.3. *Missing image generation*

All of the original images are now associated with a depth map. At this stage, each pixel of the original image needs to be shifted horizontally over a distance proportional to its depth. In professional workflows, the values of Z are usually coded in the form of 16-bit integers. The horizontal shifts needed for depth rendering are generally less than 3% of the width of the image and are strongly nonlinear as a function of depth. Thus, a horizontal shift of one pixel may correspond to several tens of meters, and shifts of a fraction of a pixel are required for correct rendering.

Each sequence of a film must follow a fixed depth budget, using the minimum and maximum authorized parallax values for the shot. This budget is determined by a stereographer as a function of the size of the screen used for image projection and of the desired depth amount from an artistic perspective. The stereographer also determines the value corresponding to the screen plane. The face of the main actor, for example, may be set at parallax zero. The shade of gray of this point in the depth map will thus be chosen as the reference point.

As the depth budget is often only a few pixels deep, or at most tens of pixels behind the screen plane, fractional shifts are often used. It is not rare to need to situate tens of objects or individuals at different distances using only a few pixels of parallax difference.

Once the values of gray from the depth map corresponding to minimum, zero and maximum parallax values have been fixed, an automatic procedure easily applies the desired shift transformation to each pixel of each image in a sequence. On-screen 3D verification of the whole sequence is then possible.

The horizontal shift of large parts of an image, which may be substantial, is not without its problems. When two neighboring pixels shift over different distances, two situations may arise: superposition, or a hole. In the first case, as the shifting program begins with the furthest pixels, the closer pixel is layered without creating an artifact: the closest objects always hide objects that are further away. In the second case, part of the image will contain no information, as the shift shows part of the scene that was not present in the original image; this creates a hole. Automatic methods use various heuristics, such as simple duplication of a neighboring pixel to fill the hole. Other heuristics synthesize the missing pixel by interpolating or extrapolating the color of neighboring pixels. The most intelligent algorithms search for the missing information in previous and subsequent images of the same sequence using heuristics such as those described above in case of failure. Once a section of an image has been constructed, the most important point is to avoid retinal rivalry: it is essential to check that the added pixel and its homologue in the original image do not present excessive luminosity or color differences.

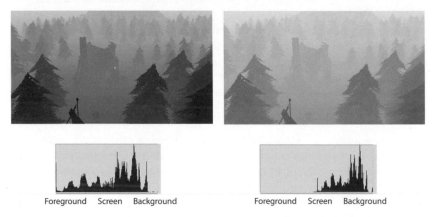

Foreground   Screen   Background   Foreground   Screen   Background

Figure 17.3. *Left: a depth map using the whole depth budget. Right: a depth map where the whole scene has been relegated behind the screen. The histogram of the depth map is shown under each image*

Methods which create multi-layer depth maps simplify the problem as each layer may be corrected independently of the others. The result is never perfect, but visual examination (on a 3D display) allows the most significant errors to be detected. As a general rule, this stage is semi-automatic using algorithmic correction followed by visual checks and, where necessary, manual correction by an experienced operator.

17.5. 3D–3D conversion

Interestingly, the demand for 2D–3D conversion has led to the emergence of a number of algorithms and methods which may be used to regenerate stereoscopic pairs using existing, but unsatisfactory, stereoscopic content. Correction is often required in cases where there are errors in scene geometry (the foreground is too close) or in geometric calibration of camera settings (interaxial distance or convergence errors). As filming costs can be very high, notably for movies, 3D sequences may be recreated in post-production using depth maps calculated precisely using stereoscopic originals, using one of the methods described above.

In these cases, the depth budget used is different to that used in the original, and applied to one of the two original images following a process similar to that used in 2D–3D conversion. The advantage of two original images is exploited in the artifact correction phase, providing an additional source of background information. Legend3D are specialists in this 3D–3D conversion workflow, with its StereoWorks division created in 2012.

17.6. Conclusion

2D–3D conversion is on the border between technical and artistic activities, and human participation remains essential. A *depth script* is established for each sequence for conversion prior to execution of a variety of partly manual and partly automated stages. The 2D–3D conversion workflow is broken down into a series of stages, notably depth map generation, segmentation, missing image generation and artifact suppression.

Certain commercial solutions offer fully automated 2D–3D conversion, but the results are generally unsatisfactory, with the exception of very specific cases where the geometry of the scene is subject to strong constraints, movements are linear and predictable and segmentation is simple. Not all content is equally suited to 2D–3D conversion. Conversion is notably facilitated when planned during the filming phase. In these cases, the costs and complexity of filming are identical to those involved in 2D filming, and the conversion stage, with the benefits of favorable framing, staging and other conditions, does not have excessive effects on the production budget.

17.7. Bibliography

[AHA 06] AHARON M., ELAD M., BRUCKSTEIN A., "K-SVD: An algorithm for designing overcomplete dictionaries for sparse representation", *IEEE Transactions on Signal Processing*, vol. 54, no. 11, pp. 4311–4322, November 2006.

[ALV 00] ALVAREZ L.J., DERICHE R.D., SÁNCHEZ J., *et al.*, Dense disparity map estimation respecting image discontinuities: a PDE and scale-space based approach, Research report no. RR-3874, INRIA, 2000.

[BLE 09] BLEYER M., GELAUTZ M., "Temporally consistent disparity maps from uncalibrated stereo videos", *Proceedings of the 6th International Symposium on Image and Signal Processing and Analysis (ISPA)*, Salzburg, pp. 383–387, 16–18 September 2009.

[CHE 98] CHEN S., DONOHO D., SAUNDERS M., "Atomic decomposition by basis pursuit", *SIAM Journal on Scientific Computing*, vol. 20, no. 1, pp. 33–61, 1998.

[DA 10] DA SILVA V., "Depth image based stereoscopic view rendering for MATLAB", available at http://www.mathworks.com/matlabcentral/fileexchange/27538-depth-image-based-stereoscopic-view-rendering, 2010.

[DEJ 08] DEJOHN M., SEIGLE D., A summary of approaches to producing 3D content using multiple methods in a single project, Report, In-Three, 2008.

[DOD 04] DODGSON N.A., "Variation and extrema of human interpupillary distance", *Proceedings of SPIE, Stereoscopic Displays and Virtual Reality Systems XI*, vol. 5291, San Jose, CA, pp. 36–46, May 2004.

[FRE 97] FREUND Y., SCHAPIRE R.E., "A decision-theoretic generalization of online learning and an application to boosting", *Journal of Computer and System Sciences*, vol. 55, no. 1, pp. 119–139, 1997.

[GMB 10] GMBH E.M., "MakeMe3D software", available at http://www.makeme3d.net/convert_2d_to_3d.php, 2010.

[GOR 97] GORODNITSKY I.F., RAO B.D., "Sparse signal reconstruction from limited data using FOCUSS: a re-weighted minimum norm algorithm", *IEEE Transactions on Signal Processing*, vol. 45, no. 3, pp. 600–616, March 1997.

[GRA 10] GRAU O., VINAYAGAMOORTHY V., "Stereoscopic 3D sports content without stereo rigs", *SMPTE Motion Imaging Journal*, vol. 119, pp. 51–55, 2010.

[GRA 12] GRAZIOSI D., TIAN D., VETRO A., "Depth map up-sampling based on edge layers", *Signal Information Processing Association Annual Summit and Conference (APSIPA ASC)*, Hollywood, CA, pp. 1–4, 3–6 December 2012.

[HOR 81] HORN B.P., SCHUNK B.G., "Determining optical flow", *Artificial Intelligence*, vol. 17, pp. 185–203, 1981.

[IDE 07] IDESES I., YAROSLAVSKY L., FISHBAIN B., "Real-time 2D to 3D video conversion", *Journal of Real-Time Image Processing*, vol. 2, pp. 3–9, 2007.

[IPP 11] IPP, "3D media converter box", available at http://ippstore.com/3D_Media_Converter_Box.html, 2011.

[KIN 01] KINGSBURY N., "Complex wavelets for shift invariant analysis and filtering of signals", *Applied and Computational Harmonic Analysis*, vol. 10, no. 3, pp. 234–253, 2001.

[KNO 07] KNORR S., SIKORA T., "An image-based rendering (IBR) approach for realistic stereo view synthesis of TV broadcast based on structure from motion", *IEEE International Conference on Image Processing, ICIP 2007*, San Antonio, TX, vol. 6, pp. VI–572–VI–575, 16 September–19 October 2007.

[LUC 81] LUCAS B., KANADE T., "An iterative image registration technique with an application to stereo vision", *Proceedings of the International Joint Conference on Artificial Intelligence (IJCAI)*, Vancouver, Canada, pp. 674–679, April 1981.

[MAL 93] MALLAT S., ZHANG Z., "Matching pursuits with time-frequency dictionaries", *IEEE Transactions on Signal Processing*, vol. 41, no. 12, pp. 3397–3415, December 1993.

[MAT 97] MATSUMOTO Y., TERASAKI H., SUGIMOTO K., *et al.*, "Conversion system of monocular image sequence to stereo using motion parallax", *Proceedings of SPIE 3012, Stereoscopic Displays and Virtual Reality Systems IV*, pp. 108–115, 15 May, 1997.

[MIC 11] MICHEL B., "La conversion 2D–3D", in *La Stéréoscopie Numérique*, Eyrolles, Chapter 5, 2011.

[MOU 05] MOUSTAKAS K., TZOVARAS D., STRINTZIS M., "Stereoscopic video generation based on efficient layered structure and motion estimation from a monoscopic image sequence", *IEEE Transactions on Circuits and Systems for Video Technology*, vol. 15, no. 8, pp. 1065–1073, August 2005.

[PER 03] PERIASWAMY S., FARID H., "Elastic registration in the presence of intensity variations", *IEEE Transactions on Medical Imaging*, vol. 22, no. 7, pp. 865–874, July 2003.

[RAN 00] RAN A., SOCHEN N.A., "Differential geometry techniques in stereo vision", *6th European Workshop on Computational Geometry*, pp. 98–103, 13–15 March 2000.

[SCH 02] SCHMIDT J., NIEMANN H., VOGT S., "Dense disparity maps in real-time with an application to augmented reality", *Proceedings of the 6th IEEE Workshop on Applications of Computer Vision (WACV 2002)*, IEEE, Orlando, FL, pp. 225–230, December 2002.

[SCH 11] SCHNYDER L., WANG O., SMOLIC A., "2D to 3D conversion of sports content using panoramas.", in MACQ B., SCHELKENS P. (eds), *ICIP*, IEEE, pp. 1961–1964, 2011.

[SEI 12] SEIGLE D., "Depth grading in 3D creation", available at http://www.docstoc.com/docs/94241109/Depth-Grading-in-3D-Creation, 2012.

[VIO 04] VIOLA P., JONES M., "Robust real-time face detection", *International Journal of Computer Vision*, vol. 57, no. 2, pp. 137–154, 2004.

[WU 00] WU Y.-T., KANADE T., LI C.-C., *et al.*, "Image registration using wavelet-based motion model", *International Journal of Computer Vision*, vol. 38, no. 2, pp. 129–152, 2000.

[XU 11] XU F., LAM K.-M., DAI Q., "Video-object segmentation and 3D-trajectory estimation for monocular video sequences", *Image and Vision Computing Journal*, vol. 29, no. 2–3, pp. 190–205, 2011.

PART 5

Implementation and Outlets

Chapter 18

3D Model Retrieval

18.1. Introduction

The use of three-dimensional (3D) models in multimedia data is expanding rapidly. The appearance of 3D scanners, the increasing power and simplicity of 3D creation tools and the arrival of 3D television (3DTV) all make it increasingly easy to create realistic and detailed 3D models. One question raised by the designers of virtual worlds concerns the best use of huge existing collections of 3D models, and efficient retrieval mechanisms for use with these collections.

The most immediately apparent solution to this problem is one already used to index images (by Google Images, for example): textual collection indexing, whereby each element in a collection is described using one or more keywords. This solution, although it appears simple, presents a number of drawbacks, including the time involved in tagging or labeling a collection and the subjective nature of keywords, which will inevitably be linked to the perception and culture of the tagging operator.

The most efficient method, and certainly the most intuitive method for final users of the retrieval system, is known as *search by example*. This method consists of characterizing 3D models using shape descriptors which are invariant to certain geometric transformations, notably translations, rotations and homothetic transformations. The user expresses a request in the

Chapter written by Jean-Philippe VANDEBORRE, Hedi TABIA and Mohamed DAOUDI.

form of an example of the object to find in a collection. A metric is then used to compare descriptors in the request and those of the models in the collection. To compare the request to these models, the metric must express a distance, which translates as a *visual distance* between two objects. The process of retrieving 3D models from a data collection using an example is known as indexing.

3D models are represented in different ways according to needs: point clouds, surface or volumetric representations, etc. Surface representations, which only take account of the external surface of the represented object, are the most widespread. These representations generally take the form of a polygon mesh, which constitutes a discrete sampling of a continuous surface. Polygon meshes are ideal when representing a 3D object for a computer graphics card. In this chapter, we will focus on the representation of 3D models as polygonal, usually triangular, meshes.

We will also limit ourselves to consider the shape of objects. The colors and/or textures associated with the illustrations presented in this chapter are purely decorative, with no relevance or impact on the methods presented.

We will begin by discussing the general principles of 3D indexing before considering global 3D shape descriptors, 2D view descriptors and local 3D shape descriptors. We will then present the principles of 3D shape retrieval in video sequences, followed by methods for evaluating 3D indexing and recognition. Finally, we will discuss certain key applications of indexing and 3D recognition.

18.2. General principles of shape retrieval

The 3D model indexing process may be divided into two stages; before retrieval can be carried out, the collection needs to be prepared.

The first stage consists of *indexing* the collection of 3D models to search. During this stage, each 3D model is associated with one or more concise geometric descriptors describing the form of the object. These geometric descriptors must be invariant to geometric transformations such as translations, rotations and scaling (homothety). As none of these transformations affects the shape of 3D objects, they should also not affect geometric descriptors. These descriptors are known as 3D model indexes. This first stage is known as the offline stage, as it precedes the search stage. It may be relatively long, as calculations must be carried out for each 3D model in the collection; however, this is not a major drawback as the stage is carried out only once.

The second stage is the search by example itself. The user expresses a request – i.e. the desired 3D object – by presenting an example in one form or another: full 3D model, photograph or drawing of the object, etc., depending on the method used by the retrieval system. The system then calculates one or more geometric descriptors of the sort used in the offline stage. Using a distance, it then compares calculated descriptors with the indexes of each 3D model in the collection. Results are presented in the form of a list of 3D models by order of relevance, from that which most resembles the request (i.e. the 3D model with the minimum distance value) to that which is least like the request (i.e. the model with the maximum distance value). This second stage is known as the *online* stage, and is carried out for each new user request.

Figure 18.1 shows the two stages of 3D model indexing.

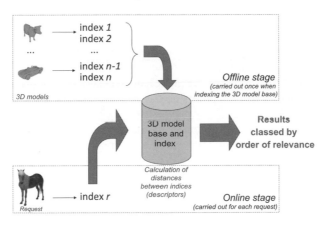

Figure 18.1. *The two stages of 3D model indexing. The offline stage is carried out once, whereas the online stage is carried out for each request*

During the *offline* stage, the main difficulty lies in finding or combining efficient geometric descriptors which are invariant to geometric transformations. For the *online* stage, the use of a suitable mathematical distance, which may be assimilated to a visual distance, between 3D objects is the most critical aspect.

18.3. Global 3D shape descriptors

Global descriptors are based on a shape function which measures the geometric properties of a 3D object as a whole, breaking down an object into several elementary subobjects or calculating the distances between points on the surface of the object, etc.

18.3.1. *Distribution of a shape descriptor*

Osada *et al.* [OSA 02] divide 3D mesh model indexing into three steps: selection of a discriminant shape function, efficient construction of the distribution of this function for each model in the collection and calculation of similarity distances between distribution pairs.

The authors tested five shape functions, chosen on the basis of simplicity of calculation: measurement of angles between three random points on the surface of the 3D model, measurement of the Euclidean distance between a fixed point in the model (the center of gravity, for example) and a random point on the surface, measurement of the Euclidean distance between two random points on the surface of the model, measurement of the area of triangles of three random points on the surface of the model and measurement of the tetrahedral volume between four random points on the surface of the model.

These descriptors are invariant to geometric transformations. To obtain invariance to scaling, the authors chose to normalize distributions. Further testing was carried out using the third function (Euclidean distance between two random points on the surface of the model), known as D2, which is the best known and most efficient of these descriptors, showing good results for the chosen approach when used for pre-classification before application of a more precise, but more costly, search method.

18.3.2. *Spherical harmonics*

Spherical harmonics constitute a 3D shape descriptor based on the frequency decomposition of a 3D signal. This descriptor emerged as the result of work by the Leipzig team, who applied a Fourier transform to sphere S2 [SAU 01], applying the spherical harmonic formulas proposed in [HEA 96]. To remove the rotation invariance problem, the Princeton team then proposed the application of the spherical harmonic decomposition of spherical functions defined by the intersection of the surface of the 3D object with a set of concentric spheres [KAZ 02].

18.4. 2D view oriented methods

The general idea used in oriented views is that if two objects are similar, their views will also be similar from all viewpoints. Based on the fact that two objects that are similar in shape terms will have similar sets of extracted views, these methods can be used to classify 3D objects based on a 2D request, or using a 3D request model from which characteristic views are extracted. Two important issues must be addressed: the choice of views to represent the 3D object and the distance used to compare two 3D objects based on these views.

Chen *et al.* [CHE 03] proposed the *light field descriptor* method, where a 3D object is characterized by 10 sets of 10 orthographic views taken from the first 10 faces of a dodecahedron centered on the object. These views are then characterized by Zernike moments and Fourier coefficients; they may then be compared using a Euclidean distance. Filali *et al.* [FIL 07] suggested determining a number of views as a function of the complexity of the 3D object by using an information criterion. Subsequently, the problem of comparing two objects based on their views is formalized using a Bayesian approach. Figure 18.2 shows an example of results obtained using this method [FIL 07]. The image in the top left of the figure shows a 2D view of a 3D model of a human being. The figure shows the first 15 results provided by the search engine.

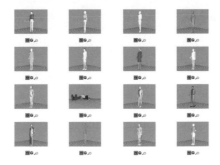

Figure 18.2. *Example of results obtained using the 2D view oriented approach developed by the MIIRE team [FIL 07]*

18.5. Local 3D shape descriptors

Unlike global descriptors, which tend to describe the shape of 3D objects in a global manner, local descriptors take account of the local geometric characteristics of the shape of these objects. In this section, we will present the main local descriptors used for indexing and similarity research of 3D objects.

18.5.1. *3D shape spectrum descriptor*

The 3D shape spectrum descriptor (3D-SSD) put forward by Zaharia and Prteux [ZAH 03] is defined as the distribution of the shape index across all facets of the mesh. The shape index is a well-known differential geometric criterion (see [KOE 90]), defined as follows:

Let p be a point on surface S, with curves k_p^1 and k_p^2 associated with each point p of the surface. The shape index I_p is defined for any point p by:

$$I_p = \frac{2}{\pi} \arctan \frac{k_p^1 + k_p^2}{k_p^1 - k_p^2} \quad \text{with } k_p^1 \geq k_p^2 \qquad [18.1]$$

where k_p^1 and k_p^2 are the principal curvatures of the point p on the surface S. The shape index I_p allows us to group different local geometric forms into families of elementary surfaces such as cylinders, spheres, saddles, etc. While initially used as a global descriptor, the shape spectrum has also been used as a local descriptor to characterize surface details in 3D objects.

18.5.2. *3D shape contexts*

Initially introduced for object recognition in 2D images, Körtgen *et al.* [KÖR 03] used shape contexts in 3D. Their idea was to draw N points, uniformly distributed across the surface of the 3D object, to which distributions characterizing the shape of the object are associated. Each point is associated with a histogram of the distances to the other $N - 1$ points. To conserve a notion of spatial coherence, the authors suggest partitioning the space into C cells according to three classic types of decomposition. Each point is localized by the cell in which it is located. Depending on the type of representation chosen (shell, sector or combined), a normalization of the object pose may or may not be required. The shape context descriptor is very large; descriptor comparison consists of reducing its size, using vector quantization techniques.

18.5.3. *Spin images*

Spin images were proposed by Johnson *et al.* [JOH 99] for 3D object recognition. Spin image representation takes account of the global shape characteristics of an object, while remaining robust to occlusion. Spin images are classified as local descriptors as they may be calculated for local characteristic points. For each characteristic point p of a mesh, a spin image is constructed by projecting the other points in the plane tangent to the object onto p. The size of the spin image is chosen arbitrarily. The similarity between two 3D objects may be calculated using the correspondence minimizing the sum of the distance between the spin images.

18.5.4. Heat kernel signature

The heat kernel signature (HKS) is a descriptor used to analyze the shape of deformable 3D objects. This descriptor belongs to the group of spectral analysis methods for 3D shapes. It may be calculated for a local or global surface of a 3D object.

The HKS was introduced by Sun *et al.* [SUN 09] in 2009 and is based on the heat kernel, a fundamental solution to the heat equation, which itself is based on the concept of heat diffusion across a surface. Given an initial heat distribution $f_0(p)$ over a surface, the heat kernel $h_t(p, q)$ describes the quantity of heat transferred between p and q after a time t. This description is invariant to isometric transformations. The HKS represents the geometric properties of a surface as a function of time t. For a low value of t, the HKS is considered to be a local descriptor; however, as t increases, the HKS increasingly characterizes the global geometric properties of the shape.

As $h_t(p, q)$ is defined for a pair of points in a temporal domain, the use of heat kernels in this form as a 3D descriptor can be complex. For this reason, Sun *et al.* [SUN 09] concentrated on the temporal domain alone, considering only $h_t(p, p)$. The HKS inherits most properties of heat kernels, under certain conditions.

The heat diffusion equation for a compact Riemannian manifold M (which may include a border) is given by:

$$\left(\Delta + \frac{\partial}{\partial t}\right) f(p, t) = 0 \qquad [18.2]$$

where Δ is the Laplace–Beltrami operator and $f(p, t)$ is the heat distribution at a point p at instant t. The solution to this equation may be expressed as follows [SUN 09]:

$$f(p, t) = \int h_t(p, q) f_0(q) dq \qquad [18.3]$$

The Eigen decomposition of the heat kernel is expressed as follows:

$$h_t(p, q) = \sum_{i=0}^{\infty} \exp(-\lambda_i t)\phi_i(p)\phi_i(q) \qquad [18.4]$$

where λ_i and ϕ_i are the ith eigenvalues and eigenfunctions of Δ.

The heat kernel entirely characterizes the surface to within an isometry. For any surjective map $T: M \to N$ between two Riemannian manifolds M and N, if $h_t(p, q) = h_t(T(p), T(q))$, then T is an isometry, and vice-versa. For the concise descriptor proposed by Sun et al. [SUN 09], the calculation of the HKS is restricted to the temporal domain:

$$h_t(p, p) = \sum_{i=0}^{\infty} \exp(-\lambda_i t)\phi_i^2(p) \qquad [18.5]$$

Authors such as El Khoury et al. [ELK 12] have also used the HKS with various modifications: the diffusion distance is used to calculate characteristic points on the 3D model, then the commute-time distance is used to calculate local descriptors in relation to characteristic points.

18.6. Similarity between 3D shapes

In this section, we will present two approaches used to calculate the similarity between 3D shapes: reeb graphs and the bag-of-words method.

18.6.1. Reeb graphs

Structural approaches consist of representing a 3D object in the form of a graph. This representation allows us to code information linked to the general structure of 3D objects. As a graph characterizes the global structure of a 3D object, it may be used as a powerful tool in searching for and comparing 3D shapes. However, structural representation loses its discriminatory strength when comparing objects to some topological transformations. In this section, we will present a number of structural approaches, particularly those based on 3D object segmentation and those using Reeb graphs.

Funkhouser et al. [FUN 04] suggest segmenting objects to enable partial research. Gal and Cohen-Or [GAL 06] used salient characteristic points to extract a signature characterizing 3D objects. Points are selected based on the variance of curvature in their neighborhood. Thus, each region is described by a local descriptor. The authors used hash-chaining techniques to accelerate the mapping of similar regions.

Based on Morse's theory, Reeb graphs [REE 46] are used to capture the topology of 3D objects. This type of graph has been used in 3D searches and indexing by a number of authors. In [HIL 01], for example, Hilaga *et al.* propose the use of multiresolution Reeb graphs to match 3D objects. The multiresolution Reeb graph represents the skeleton and topological outline of a 3D shape at different resolution levels. The similarity between shapes is calculated using a coarse-to-fine strategy, while preserving the coherence of the graph structures, thus establishing correspondences between parts of objects. This type of representation can be modified as a function of the size of database used. It may be adapted for use with large databases by limitation to low-resolution levels; it may also be extended to higher levels of resolution depending on the desired processing time and the level of details.

DEFINITION 18.1 (REEB GRAPH).– Take $f:M \to \mathbb{R}$ a Morse function defined for a compact manifold M. The Reeb graph $R(f)$ is the quotient space defined by $M \times \mathbb{R}$ for the equivalence relationship $(p_1, f(p_1)) \sim (p_2, f(p_2))$ if and only if $f(p_1) = f(p_2)$ and p_1, p_2 belong to the same connected component $f^{-1}(f(p_1))$.

Figure 18.3 shows an example of a Reeb graph obtained for the height function f. Biasotti *et al.* [BIA 03] propose a graph mapping method based on the propagation of mapped subgraphs. The method presents quadratic complexity and uses extended Reeb graphs or ERGs: Reeb graphs are oriented and possess edge information.

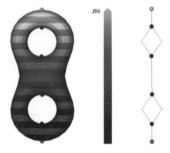

Figure 18.3. *Example of a Reeb graph obtained by the Morse height function*

Tierny *et al.* [TIE 09] propose the use of Reeb graphs to extract signatures of 3D objects. A Reeb graph of the input surface is first calculated in order to segment it into controlled topology maps, known as Reeb charts, which take the form of either a disk or a ring. For each Reeb chart, an application toward the canonical planar domain is defined. The stretch signature of each application is calculated, based on an estimation of area distortion. Finally, the input surface is represented by the set of stretch signatures. An application to evaluate similarities insensitive to pose and to partial transformations by comparing the signatures of different Reeb charts has also been proposed.

18.6.2. *Bag-of-words*

Bag-of-words representation is a widely used description in information searches. This representation method was first used to index text documents, then for the representation of 2D images and, more recently, for 3D objects. In bag-of-words–based approaches, a document (text, image or 3D object) is described using a dictionary. In the case of 3D objects, we generally talk about visual words or elementary shapes. These methods require several stages: first the local description of the 3D model and then the calculation of invariant local descriptors. A quantization method is generally used to construct classes, each of which has representatives, known as the dictionary of 3D visual words. A histogram of these words is then constructed. Any specific 3D object is represented by the histogram of occurrences of the visual words, which it includes. Thus, we have a vector of the same size as the dictionary of which component i indicates the number of occurrences of the ith word in the dictionary in the object. The constitution of the dictionary is, therefore, a critical stage for determining the performance of systems using these representations.

Several authors [BRO 11, LAV 12, TAB 12] have used this approach for indexing and establishing correspondences between 3D objects. Bag-of-words–based representation is well suited to searching large databases. Figure 18.4 summarizes the different stages involved in the bag-of-words approach. In addition, Tabia *et al.* [TAB 11] introduced the notion of signification of 3D parts to map. The authors initially consider each part as a source of information on a 3D object. Next, the information source is weighted by significance. Finally, the information sources relating to all parts of the 3D object are combined in a belief function framework to produce a final decision concerning the shape of the object. Bronstein *et al.* [BRO 09] used distributions of geodesic distances on the object surface.

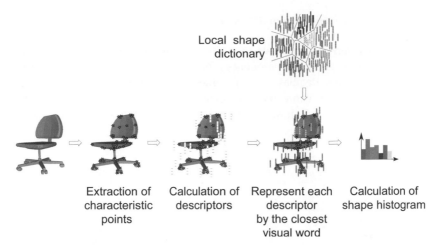

Figure 18.4. *Bag-of-words approach. A set of characteristic points is first extracted from the 3D object. Next, a descriptor is calculated for each characterized point; each descriptor is then represented by the closest visual word in the shape dictionary. Finally, a histogram is calculated showing the number of occurrences of each visual word.*

18.7. Shape recognition in 3D video

In recent years, new activities have emerged on the basis of the retrieval of actions and movements and the recognition of facial expressions in 3D video sequences. These research activities have been facilitated by the arrival of sensors that are able to produce very high-resolution 3D sequences. The availability of dynamic model databases, as shown in Table 18.1, has greatly contributed to the development of research activities in this domain.

Bases	Context of use
IXMAS [WEI 06]	3D Actions
i3DPost [GKA 09]	3D Action and dynamic facial expressions
BU-4DFE [YIN 08]	Dynamic facial expressions
Hi4D-ADSIP [MAT 12]	Dynamic facial expressions

Table 18.1. *Dynamic 3D bases (3D+time)*

18.7.1. *Action recognition in 3D videos*

Using a video sequence, we should be able to find 3D models of similar poses or, more generally, sequences or subsequences containing similar actions in a video database. A 3D video is a sequence of frames, each consisting of a mesh (geometry and connectivity), in this case of a human in a given pose. Figure 18.5 shows examples of frames (meshes) taken from a 3D video sequence.

Figure 18.5. *Examples of 3D meshes extracted from a 3D video sequence*

Huang *et al.* [HUA 10] consider searches for models with similar shapes and poses in temporal sequences of moving human beings. The aim of the study was to discriminate between different poses of the same object, rather than to distinguish different classes of objects. Static shape descriptors, such as shape distributions, spin images and spherical harmonics, were evaluated and compared using synthetic and real databases. The results showed that the best performances were obtained using shape distributions. These same descriptors were then extended into temporal descriptors by convolution, using a time filter which allows us to take into account the frame and its vicinity within a fixed window. The authors showed that the use of this filter improves results when identifying similar frames in 3D videos.

Tung *et al.* [TUN 12] used Reeb graphs to analyze 3D videos. The video was made up of meshes of moving people, and each frame was represented by a Reeb graph of its topology. The similarity of two frames was then estimated by mapping the nodes of these diagrams. The authors proposed a topological dictionary for use in coding and describing the content of a 3D video. Their model is based on a description and topological classification using multiresolution Reeb graphs, and a moving Markov graph to model states relating to changes in topology.

Slama *et al.* [SLA 13] proposed a new descriptor, known as the *extremal human curves* (EHC) descriptor, extracted from the mesh of a 3D model. First, the extremities of the body are detected, then the shortest pathways along the mesh connecting two extremities are identified, exploiting their properties of invariance in relation to changes in pose and topology. Finally, we obtain a collection of curves representing the pose of the body. To compare two poses, the authors compared the corresponding curves of two models through a framework based on Riemannian geometry. The curve comparison estimates the deformation and produces a similarity score for models in a video sequence, or a database of human body images with pose variations.

18.7.2. *Facial expression recognition in 3D videos*

The analysis of 3D facial expressions has applications in a number of domains, such as user–machine interfaces, facial animation and, more generally, in "ambient intelligence". The problem consists of analyzing frame (mesh) deformations in a sequence of faces. A classifier is used to assign this deformation vector to one of six classes (six facial expressions). In [DRI 12], the authors proposed a vector known as the *deformation vector field* (DVF) and the *Random Forest* algorithm for classifying facial expressions.

18.8. Evaluation of the performance of indexing methods

The evaluation of the performance of indexing methods is an important element to take into account in choosing an approach.

3D model indexing and retrieval represent two distinct steps, as discussed in section 18.2. Indexing constitutes the offline stage. This step is carried out only once, and its performance is based solely on the choice of the descriptor(s). These choices have an effect on retrieval performances, i.e. when a user expresses a request and the online stage begins.

Our attention, therefore, must be focused on the online stage. In this case, the term "performance" refers to two aspects of the performance of any search engine: the request processing time and the speed at which responses are provided, and – especially – the relevance of the results produced in response to a request. To judge the relevance of results during test phases, a collection of ground truth data is required. 3D models from the collection are grouped into several classes by experts. Each 3D model in the collection is then used as a request submitted to the search engine, then we evaluate the relevance of results: a model response that falls into the same class as the request is judged to be relevant, while responses outside this class are considered irrelevant.

18.8.1. *Statistical evaluation tools*

Statistical evaluation tools are available for use in evaluating any search engine. Examples include:

– *Nearest Neighbor* (NN): this is an evaluation of the closest neighbor produced by the search engine and is simply the percentage of closest matches that belong to the same class as the request.

– *First Tier* (FT) and *Second Tier* (ST): percentage of models in the same class as the request appearing in the first K results. K depends on the size of the class in question.

– *Recall* (R) and *Precision* (P): recall and precision are defined as follows: $R = N/Q$ and $P = N/A$. N is the number of relevant results in the first A results. Q is the number of relevant models in the whole collection, i.e. the number of models in the class to which the request belongs.

– *Recall versus Precision plot*: recall is often expressed as a function of precision. The recall versus precision plot is an interesting visual tool, offering a global or class-specific idea of the performance of a retrieval method.

Other tools are also available, such as ROC curves, *E-Measure* and *Discounted Cumulative Gain*. More details are given in [SHI 04], for example.

18.8.2. *Benchmarks*

Benchmarks include work from the University of Princeton[1], presented in the article by Shilane *et al.* [SHI 04]. This benchmark is made up of 1,814 meshed 3D models from a variety of sources, essentially found on the Internet. The models are grouped into 92 classes, from airplanes to automobiles, including furniture, plants, humanoids, etc. It also provides standard statistical tools for evaluating and comparing the performance of research methods. Established in 2004–2005, its use is waning; annual international competitions, the *Shape Retrieval Contests* (SHREC), are now preferred.

The SHREC competition was established in 2006 in the context of a European excellency network, AIM@SHAPE[2], and shows the growing

---

1 http://shape.cs.princeton.edu/benchmark/.

2 http://www.aimatshape.net/.

interest of the scientific community for retrieval and recognition of 3D models. Calls for proposals are sent out each year. Researchers then propose *tracks* in different domains linked to 3D shape recognition. These have included protein models (2007, 2010), CAD models (2007, 2008), 3D face models (2007, 2008, 2010), partial matching (from 2007 to 2010), 3D retrieval using machine learning (2009, 2010), non-rigid shapes (2010, 2011), architectural models (2010), 3D mesh segmentation (2012), sketch-based 3D shape retrieval (2012), generic 3D model retrieval (every year), etc.

Data collections, performance measurement tools and the results of these contests are increasingly used in articles to present the strengths and weaknesses of new methods. Over the years, there has been growing interest in methods capable of accounting for the partial aspects of 3D models, both in indexed collections and in requests.

18.9. Applications

3D model indexing and recognition techniques have numerous applications, both in purely multimedia and 3D domains and elsewhere. Some of these applications will be explored in this section.

18.9.1. *Browsing a collection of 3D models*

As we highlighted in section 18.1, the use of simple keywords is not the best way for characterizing a request. 3D model indexing and recognition help users to navigate a large collection of data. By specifying the search subject using a drawing, a photo or even a full 3D model, the retrieval system is able to provide the 3D models from the collection that best respond to the request.

Search engines are currently available online, including those developed by the University of Princeton[3] or the MIIRE team at the LIFL[4], illustrated in Figure 18.6.

18.9.2. *Modeling by example*

The most frequent question asked by 3D designers concerns the best way of exploiting the vast collections of existing 3D models. The creation of a new

---

3 http://shape.cs.princeton.edu/search.html.
4 http://www-rech.telecom-lille1.eu/3dretrieval/.

3D model is a long and costly process. 3D model indexing and recognition can constitute a design assistance tool. A search engine may be used to sort through existing models in their collection, and the results produced may act as a starting point for new creations.

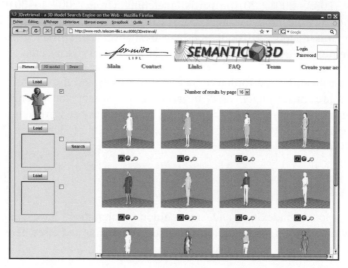

Figure 18.6. *Web interface of the MIIRE team's 3D search engine [FIL 07]*

This simple inspiration suggestion may be extended to a more active reuse of the results turned up by a search engine. Modeling, for example, first proposed by Funkhouser *et al.* [FUN 04], is an approach where the creation of new 3D models is assisted by existing model retrieval. It consists of providing designers with tools to assemble pieces of 3D models in order to create new objects. The indexing used in this application must evidently take account of the partial nature of 3D models in requests and results. To this end, Tierny *et al.* [TIE 09] propose an approach whereby 3D models are broken down into different parts using Reeb graphs. Figure 18.7 shows an example of a new 3D model created by joining parts of other retrieved models.

18.9.3. *Decision-making tools*

3D model recognition may also be used in other types of applications.

The automobile industry, for example, produces large numbers of 3D objects, used to precisely model vehicles before production or even full-scale tests. Once the car is on the market, these models are then used to produce illustrated user manuals or technical repair sheets used by technicians in maintaining and repairing the vehicles. 3D model searches using a photo or 3D scan of a mechanical part can assist the technician in finding similar pieces or suitable repair notices for the component. The RNRT SEMANTIC-3D[5] project considered the automatic search approach for automobile parts, alongside other issues, in collaboration with Renault.

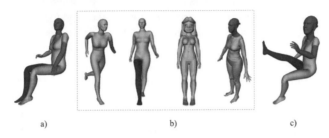

a)   b)   c)

Figure 18.7. *Example modeling. a) Initial request; b) results produced by the search engine; and c) use of parts of results to create a new 3D object*

18.9.4. *3D facial recognition*

The automatic recognition of human faces based on the use of 2D images has undergone major developments over the recent years, and several techniques have been proposed. In spite of the results obtained in the domain, robust facial recognition remains very difficult. Current methods are effective when the conditions in which test images are obtained are similar to those used for the images in the learning stage. However, significant variation created by changes in lighting and capture conditions causes serious problems for a large number of current recognition systems. One solution consists of using 3D facial information, which gives better information on the characteristics of a face in 3D space. This information produces invariance in relation to lighting and view-capture conditions. Moreover, recent advances in 3D imaging (acquisition tools, modeling software, graphics maps, etc.) have enabled the creation and storage of 3D facial images. Nevertheless, faces cannot be treated as rigid images, as they are subject to deformations resulting from facial expressions. Several approaches have been proposed recently to develop algorithms that are robust to non-rigid transformations [DRI 13, PEI 13].

---

5 http://liris.cnrs.fr/semantic-3d/.

18.10. Conclusion

In this chapter, we have shown the growing interest in analyzing the shape of 3D objects in recent years, particularly in terms of retrieving objects from large 3D model databases. This is essentially due to the spectacular progress made by 3D sensors, which are now able to produce high-quality static and dynamic 3D models. These models are now used in a variety of applications, such as CAD, special effects for cinema, video games, archeology, bioinformatics, virtual reality and simulation. 3D model retrieval methods use various mathematical tools, such as spectral methods and differential/Riemannian geometry, machine-learning mechanisms and numerous computing aspects (multidimensional data structures and Human-Machine Interfaces (HMIs)). A number of issues remain open, such as pose and isometric transformation invariant 3D object recognition, and the recognition of partially represented or hidden objects. Red-Green-Blue and Depth (RGB-D) captors, such as the Microsoft Kinect, open interesting perspectives and new lines of research.

18.11. Bibliography

[BIA 03] BIASOTTI S. *et al.*, "3D shape matching through topological structures", *Discrete Geometry for Computer Imagery*, vol. LNCS 2886, Springer-Verlag, pp. 194–203, 2003.

[BRO 09] BRONSTEIN A.M., BRONSTEIN M.M., BRUCKSTEIN A., *et al.*, "Partial similarity of objects, or how to compare a centaur to a horse", *International Journal of Computer Vision*, vol. 84, no. 2, pp. 163–183, August 2009.

[BRO 11] BRONSTEIN A.M., BRONSTEIN M.M., GUIBAS L.J., *et al.*, "Shape google: geometric words and expressions for invariant shape retrieval", *ACM Transactions on Graphics*, vol. 30, pp. 1–20, 2011.

[CHE 03] CHEN D.-Y., TIAN X.-P., SHEN Y.-T., *et al.*, "On visual similarity based 3D model retrieval", *Computer Graphics Forum*, vol. 22, no. 3, pp. 223–232, September 2003.

[DRI 12] DRIRA H., BEN AMOR B., DAOUDI M., *et al.*, "3D dynamic expression recognition based on a novel deformation vector field and random forest", *21st International Conference on Pattern Recognition*, Tsukuba Science City, Japan, November 2012.

[DRI 13] DRIRA H., BEN AMOR B., SRIVASTAVA A., *et al.*, "3D face recognition under expressions, occlusions and pose variations", *IEEE Transactions on Pattern Analysis and Machine Intelligence*, vol. 35, no. 9, pp. 2270–2283, September 2013.

[ELK 12] EL KHOURY R., VANDEBORRE J.-P., DAOUDI M., "Indexed heat curves for 3D-model retrieval", *21st International Conference on Pattern Recognition (ICPR)*, Tsukuba Science City, Japan, 11–15 November 2012.

[FIL 07] FILALI ANSARY T., DAOUDI M., VANDEBORRE J.-P., "A Bayesian 3D search engine using adaptive views clustering", *IEEE Transactions on Multimedia*, vol. 9, no. 1, pp. 78–88, January 2007.

[FUN 04] FUNKHOUSER T., KAZHDAN M., SHILANE P., *et al.*, "Modeling by example", *ACM SIGGRAPH 2004 Papers*, pp. 652–663, 2004.

[GAL 06] GAL R., COHEN-OR D., "Salient geometric features for partial shape matching and similarity", *ACM Transactions on Graphics*, vol. 25, no. 1, pp. 130–150, January 2006.

[GKA 09] GKALELIS N., KIM H., HILTON A., *et al.*, "The i3DPost multi-view and 3D human action/interaction database", *Proceedings of the 2009 Conference for Visual Media Production (CVMP '09)*, IEEE Computer Society, pp. 159–168, 2009.

[HEA 96] HEALY D., JR., ROCKMORE D., KOSTELEC P.J., *et al.*, "FFTs for the 2-sphere – improvements and variations", *The Journal of Fourier Analysis and Applications*, vol. 9, pp. 341–385, 1996.

[HIL 01] HILAGA M., SHINAGAWA Y., KOHMURA T., *et al.*, "Topology matching for fully automatic similarity estimation of 3D shapes", *28th Conference on Computer Graphics and Interactive Techniques (SIGGRAPH '01)*, pp. 203–212, 2001.

[HUA 10] HUANG P., HILTON A., STARCK J., "Shape similarity for 3D video sequences of people", *International Journal of Computer Vision*, vol. 89, no. 2–3, pp. 362–381, September 2010.

[JOH 99] JOHNSON A., HEBERT M., "Using spin images for efficient object recognition in cluttered 3D scenes", *IEEE Transactions on Pattern Analysis and Machine Intelligence*, vol. 21, no. 5, pp. 433 –449, May 1999.

[KAZ 02] KAZHDAN M., FUNKHOUSER T., "Harmonic 3D shape matching", *ACM SIGGRAPH 2002 Conference Abstracts and Applications*, 2002.

[KOE 90] KOENDERINK J., *Solid Shape, Artificial Intelligence*, MIT Press, 1990.

[KÖR 03] KÖRTGEN M., NOVOTNI M., KLEIN R., "3D shape matching with 3D shape contexts", *7th Central European Seminar on Computer Graphics*, 2003.

[LAV 12] LAVOUÉ G., "Combination of bag-of-words descriptors for robust partial shape retrieval", *The Visual Computer*, vol. 28, no. 9, pp. 931–942, September 2012.

[MAT 12] MATUSZEWSKI B.J., QUAN W., SHARK L.-K., *et al.*, "Hi4D-ADSIP 3-D dynamic facial articulation database", *Image and Vision Computing*, vol. 30, no. 10, October 2012.

[OSA 02] OSADA R., FUNKHOUSER T., CHAZELLE B., *et al.*, "Shape distributions", *ACM Transactions on Graphics*, vol. 21, no. 4, pp. 807–832, October 2002.

[PEI 13] PEIJIANG L., YUNHONG W., DI H., *et al.*, "Learning the spherical harmonic features for 3-D face recognition", *IEEE Transactions on Image Processing*, vol. 22, no. 3, pp. 914–925, 2013.

[REE 46] REEB G., "Sur les points singuliers d'une forme de Pfaff complètement intégrable ou d'une fonction numérique",   *Comptes Rendus Acad. Sciences*, vol. 222, pp. 847–849, 1946.

[SAU 01] SAUPE D., VRANIC D.V., "3D model retrieval with spherical harmonics and moments", *Pattern Recognition, 23rd DAGM-Symposium*, Munich, Germany, 12–14 September, 2001, *Proceedings*, Lecture Notes in Computer Science, vol. 2191, Springer, pp. 392–397, 2001.

[SHI 04] SHILANE P., MIN P., KAZHDAN M., *et al.*, "The Princeton shape benchmark", *Shape Modeling International*, Genova, Italy, June 2004.

[SLA 13] SLAMA R., WANNOUS H., DAOUDI M., "Extremal human curves: a new human body shape and pose descriptor", *10th IEEE Conference on Automatic Face and Gesture Recognition*, 2013.

[SUN 09] SUN J., OVSJANIKOV M., GUIBAS L., "A concise and provably informative multi-scale signature based on heat diffusion",   *Proceedings of the Symposium on Geometry Processing, SGP '09*, 2009.

[TAB 11] TABIA H., DAOUDI M., VANDEBORRE J.-P., *et al.*, "A new 3D-matching method of nonrigid and partially similar models using curve analysis",   *IEEE Transaction Pattern Analysis and Machine Intelligence*, vol. 33, no. 4, pp. 852–858, April 2011.

[TAB 12] TABIA H., DAOUDI M., COLOT O., *et al.*, "Three-dimensional object retrieval based on vector quantization of invariant descriptors", *SPIE Journal of Electronic Imaging*, vol. 21, no. 2, April–June 2012.

[TIE 09] TIERNY J., VANDEBORRE J.-P., DAOUDI M., "Partial 3D shape retrieval by Reeb pattern unfolding", *Computer Graphics Forum - Eurographics Association*, vol. 28, no. 1, pp. 41–55, March 2009.

[TUN 12] TUNG T., MATSUYAMA T., "Topology dictionary for 3D video understanding", *IEEE Transactions on Pattern Analysis and Machine Intelligence*, vol. 34, no. 8, pp. 1645–1657, August 2012.

[WEI 06] WEINLAND D., RONFARD R., BOYER E., "Free viewpoint action recognition using motion history volumes", *Computer Vision and Image Understanding*, vol. 104, no. 2, pp. 249–257, November 2006.

[YIN 08] YIN L., CHEN X., SUN Y., *et al.*, "A high-resolution 3D dynamic facial expression database", *Automatic Face and Gesture Recognition*, Amsterdam, The Netherlands, September 2008.

[ZAH 03] ZAHARIA T.B., PRÊTEUX F.J., "Descripteurs de forme pour l'indexation de maillages 3D", *Technique et Science Informatiques*, vol. 22, no. 9, pp. 1077–1105, 2003.

Chapter 19

3D HDR Images and Videos: Acquisition and Restitution

19.1. Introduction

The human eye is able to perceive up to 10 orders of magnitude of light intensity (10^{10} cd m^{-2}), but only 5 simultaneously (see [FER 01] and Chapter 2). This order of magnitude is reduced to 2 when displaying images on standard screens. Images acquired up to now, known as *low dynamic range* (LDR) images, contain a limited range of light intensities. This restriction is highlighted in scenes involving back lighting, for example. For this reason, the development of images with high color dynamics, or *high dynamic range* (HDR), is increasingly important.

This type of image has recently been the subject of considerable research effort, focusing on acquisition, storage, display and use. Specific HDR cameras already exist, but are either still at the experimental stage or too costly. Methodologies have been established to compensate for the absence of specific material. An introduction to HDR images and classic acquisition methods is presented in Chapter 2, in which we see that static image capture has been the subject of particular attention. Current sensors allow us to acquire and directly store a wider dynamic range of colors (up to 16 bits for still cameras). HDR video has recently attracted much attention, but video

Chapter written by Jennifer BONNARD, Gilles VALETTE, Céline LOSCOS and Jean-Michel NOURRIT.

sensors remain limited in terms of color intensity ranges (mainly represented in 12 bits). For now, it is difficult to transmit and store HDR video data in the absence of effective formats. The domain of HDR video is also relatively confidential. The number of known solutions for acquisition is very limited, as we will see in this chapter. For our purposes, the phrase "3D video" will refer to multiscopic video content (see Chapter 4).

This chapter is divided into two main sections, concerning acquisition and rendering, respectively. In section 19.2, we provide a classification of acquisition methods based on the domain in question, organized according to criteria: number of views in the scene, simultaneous or spread acquisition, acquisition of a static scene or a scene with variable representations over time. As no display technology currently permits HDR rendering, in section 19.3, we consider the possibilities of adapting existing technologies.

19.2. HDR and 3D acquisition

As we saw in Chapter 2, multiplying viewpoints during acquisition gives us the immediate ability to generate depth perception. Consequently, if we have hardware capable of native HDR data acquisition for a scene and repeat acquisition from several view points, it becomes possible to directly operate 3D HDR capture. This repetition might be obtained by moving or duplicating hardware, allowing simultaneous capture of different points of view and thus enabling HDR 3D video capture.

Unfortunately, little HDR-enabled hardware is currently available, and existing hardware is not suited to HDR video capture. Spheron[1] has developed panoramic HDR view capture equipment and an HDR video camera, although the latter is still at prototype stage and considerable quantities of data are involved. Weiss[2] offers a fully automated device, the Civetta, allowing acquisition of spherical HDR images over 360° with a resolution of 100 megapixels.

The current impossibility of obtaining native HDR data in the context of multi-viewpoint acquisition means we must use an HDR value estimation method to produce 3D HDR images. Methods for obtaining HDR images using LDR capture materials consist of combining several exposures of the same scene in order to conserve [AGG 04, MER 07] or

---

1 www.spheron.com.
2 www.weiss-ag.com.

estimate [DEB 97, MAN 95] the best brightness values for each zone in the image. The simultaneous manipulation of several images requires data to be calibrated both geometrically and colorimetrically. The level of precision used in this calibration phase varies between methods. In certain cases, we must estimate and apply the inverse response curve of the camera (see Chapter 2).

These methods, using multiple exposures, present strong analogies with the use of multiple viewpoints of the same scene when acquiring depth, or the acquisition of several instants of a dynamic scene to produce video. These analogies are shown in Figure 19.1(a), where 3D HDR video methods are divided using three axes: one corresponding to different exposures, a second to different viewpoints and a third to different instants. Note that the origin of these axes is not set at 0, but at 1: one exposition, one viewpoint and one instant. Each of the axes defines a specific type of acquisition: HDR images, 3D images and video. By choosing two axes, we create a plane showing other specific types of acquisition: HDR video, 3D video or 3D HDR images. Finally, the whole space (three axes) corresponds to 3D HDR video.

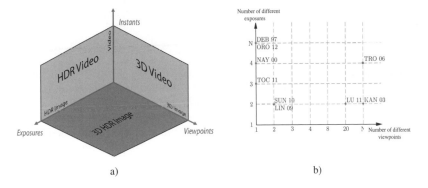

a)   b)

Figure 19.1. *a) Spatial division of 3D HDR video methods. The origin of the axes does not correspond to a value of 0, but to 1; b) plane methods corresponding to 3D HDR images, according to the number of exposures and viewpoints used. The methods on the vertical axis are purely 2D HDR, and those on the horizontal axis are purely 3D*

We will use this division into 1D and 2D subspaces to present different models described in literature on the subject. We will consider only one 1D subspace, HDR images, as the others are well known (videos) or covered elsewhere in this book (3D images). The same applies to the 2D subspace covering 3D video. We will, however, specify methods for the other 2D subspaces: 3D HDR imaging and HDR video. We will finish by discussing the possibility of extending some of these methods to the whole space, adding one or two dimensions in order to obtain 3D HDR video.

The methods presented in this chapter are classified in Figure 19.1(b) according to the number of viewpoints and the number of different exposures used during acquisition to construct HDR data. Section 19.2.1 presents methods that aim to acquire images with a single camera. Section 19.2.2 discusses a method that allows the acquisition of HDR video. In section 19.2.3, we will consider methods involving 3D HDR content.

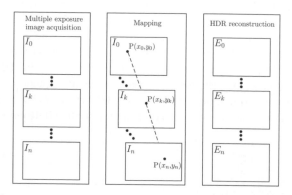

Figure 19.2. *General overview of HDR reconstruction methods based on the acquisition of multiple exposure images. Three stages are involved: (1) acquisition of n LDR images I_0, ..., I_n with different exposures from one or more viewpoints; (2) pixel mapping on these images by aligning images acquired from the same viewpoint, recalibrating data if the content changes, or correspondence mapping if the content is the same but the viewpoint differs; (3) reconstruction of one or more HDR images E_k using recalibrated LDR data. HDR image E_k corresponds to the viewpoint of LDR image I_k*

For any space, the HDR reconstruction methods considered in this chapter mostly follow the acquisition pattern illustrated in Figure 19.2, divided into three stages. In the first stage, a series of LDR images are obtained with different exposures. Stage 2 consists of pixel mapping, followed by stage 3, which uses the HDR value reconstruction algorithm. The number n of LDR input images and the number of HDR output images vary depending on the chosen method. Typically, in the 1D subspace, the viewpoint will be the same for all images I_k and a single image, E, will be generated. In the HDR video 2D subspace, there will be as many generated images E_k as there are images in the final video sequence. Images I_k will vary in terms of viewpoint and exposure, and their number will not necessarily be the same as the number of generated images E_k. In the 3D HDR image subspace, the images I_k will represent the same content, but from different exposures and points of view. Generally, the number of generated HDR images E_k will be the same as the number of input images I_k. Similarly, the mapping process varies based on the input data I_k and the HDR reconstruction objectives. The mapping process consists of aligning images if the viewpoint and content are the same I_k, data recalibration if the viewpoint is the same or similar but the content is different and correspondence mapping if the content is the same but the viewpoint differs.

19.2.1. *1D subspace: HDR images*

Numerous studies have considered the reconstruction of HDR values based on the acquisition of several images with different exposures from the same viewpoint [DEB 97, MAN 95, MIT 99]. Other approaches are mentioned in [LOS 10] and [REI 10]. Certain photographic cameras have an autobracketing function, which allows users to acquire images with different exposures using an automatic procedure, e.g. underexposed, normally exposed and overexposed views as shown in Figure 19.3. Depending on the camera, up to nine differently exposed images may be acquired using this method (see Chapter 2). In cases where this function is not available, the exposure time may be adjusted manually in order to acquire the required number of images of a scene. Whatever method is chosen, use of a tripod and a timer (or remote control) is recommended in order to stabilize the device and minimize the risk of shifts between images, leading to better results.

a) Underexposed   b) Intermediate   c) Overexposed

Figure 19.3. *Images of different exposures acquired using the autobracketing function on a photographic camera*

As we have already seen, in the absence of native HDR acquisition methods, we need to use an HDR value estimation method. We will presume that we have access to a series of images taken from the same viewpoint, but with different exposures. These images are perfectly aligned, and a point of the scene is projected at the same pixel coordinates (i, j) for all images. We have an additional set of information concerning the amount of light, recorded by the camera, coming from this point. The estimation of the HDR value for this point consists of combining these sets of information. A common method used for this operation was developed by Debevec and Malik [DEB 97], and consists of calculating a weighted average $E(i, j)$ (see equation [19.1]) of luminance values (HDR values) for the three color components for corresponding pixels in each image, with a weighting function w based on the pixel saturation level:

$$E(i, j) = \frac{\sum_{k=1}^{n} w\left(I_k\left(i, j\right)\right)\left(\frac{f^{-1}(I_k(i,j))}{\Delta t_k}\right)}{\sum_{k=1}^{n} w\left(I_k\left(i, j\right)\right)} \qquad [19.1]$$

where N is the total number of images, $I_k(i, j)$ is the color value of the pixel with coordinates (i, j) in image I_k acquired with an exposure time Δt_k and f^{-1} is the inverse function of the camera response (see Chapter 2). This function may be ignored if RAW data are used directly, in which case the data may be considered to be linear.

Different weighting functions w have been proposed to take under- or overexposed pixels into account. A state of the art of these methods is presented by Granados et al. [GRA 10]; each method is differentiated by the type of formula applied. A graphical representation of the performance of these methods is also given, showing that their method and the method put forward by Mitsunaga and Nayar [MIT 99] produce the best results.

In [AGU 12], the method put forward by Granados et al., based on the maximum likelihood estimation [GRA 10], was also shown to produce the best results. Aguerrebere et al. [AGU 12] proposed a new weighting function, allowing all pixels, including saturated pixels, to be taken into account; according to the authors, these pixels contain useful information for HDR data estimations.

Even when a tripod is used to guarantee acquisition stability, the fact that acquisitions occur at successive instants introduces sensitivity to the presence of moving objects or persons, which (or who) will be in a different position in each image. Several methods have been developed to detect and take this movement into account [JAC 08, GAL 09, GRA 08, GRO 06, SAN 04, WAR 03]. In the same context, Khan et al. [KHA 06] and Pedone et al. [PED 08] have calculated the probability that a pixel will belong to a static part of the image. Only Orozco *et al.* [ORO 12] have obtained an HDR value for all pixels, even those affected by movement, using mutual information or the normalized cross-correlation (NCC).

Instead of multiplying view captures to obtain different exposures, another method consists of acquiring sets of pixels at different exposures in a single operation. Nayar and Mitsunaga [NAY 00] adapted a camera by fixing an optical mask, such as the one shown in Figure 19.4, adjacent to a conventional image detector array. This filter permits the acquisition of four different exposures of the same image, distributed regularly by groups of four pixels. The final HDR image is then constructed either by aggregation or by interpolation. The first method allows calculation of the mean value of four neighboring pixels, a value which is then assigned to the center of the group of pixels. Considering an original image of size $N \times N$, using this method, the final image will be of size $(N - 1) \times (N - 1)$. In the second case, the pixels in the image are divided into two categories: on-grid points (black disks in Figure 19.4) corresponding to the center of pixels, and off-grid points (black squares in Figure 19.4) corresponding to the intersection point of four pixels. This gives us a value for each pixel center, so there is no loss in resolution. For each of the two groups, saturated pixels are distinguished from non-saturated pixels. First, the off-grid points are calculated from the non-saturated on-grid points, then all of the off-grid points are interpolated to obtain on-grid points.

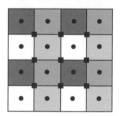

Figure 19.4. *Representation of an optical mask used to acquire four different exposures [NAY 00]: the disks represent on-grid points, and the squares represent off-grid points*

19.2.2. *2D subspace: HDR videos*

Several exposures and several instants are required to obtain HDR video. According to an idea put forward by Kang *et al.* [KAN 03], we may use different acquisition instants to obtain different exposures. In this method, the acquisition procedure alternates long and short exposure times from one image to the next. Reconstructed HDR values for a given image at time t_i are obtained using data from the image at t_{i-1} and the image at t_{i+1}. In this context, pixel shifts may be due to a change in camera viewpoint and to changes in the content of a scene from one instant to the next. Kang *et al.*'s pixel mapping method is based on the use of optical flow to estimate the movement of a pixel from one image to the next, an estimation that is then refined using homography. Once these displacements have been correctly estimated, it becomes possible to combine the values of corresponding pixels to obtain an HDR image. The results may include artifacts when there is rapid movement, as acquisition is limited to 15 images per second because of the alternating exposure times and optical flow is efficient mostly in a near neighborhood. Another limiting factor is the reduced number of exposure times available when reconstructing an image.

HDR video acquisition is also possible by obtaining several exposures for each instant, as with Nayar and Mitsunaga's optical filter [NAY 00] (see section 19.2.1). Tocci *et al.* [TOC 11] have developed another type of camera, using three sensors that receive a different percentage of the incident light by prism diffraction. Three images with different exposures are thus obtained for a single capture, with no shifts between images. Unlike Debevec and Malik's method [DEB 97], which used all pixel values from different acquired images, in this case only the pixels with the highest exposure are taken into account. The pixel at the same position in the image with lower exposure is only taken into account when a pixel is saturated, reducing the quantity of data to manage in lower exposure images, generally affected by different sensor-related noise.

19.2.3. *2D subspace: 3D HDR images*

All HDR image acquisition techniques may be extended to 3D by multiplying viewpoints. In this way, we obtain multiple exposures for each viewpoint, and thus, after estimation, an HDR image for each viewpoint. These are recombined during restitution to obtain a 3D HDR image. Clearly, while this principle is viable, the number of images to acquire makes it costly, except when using the systems developed by Nayar and Mitsunaga [NAY 00] or Tocci *et al.* [TOC 11], which only require a single capture for multiple exposures. For standard capture devices, one way of improving this situation would be to vary exposure at the same time as the viewpoint, thus obtaining one exposure per viewpoint. However, this solution includes problems with luminance matching, as a point in the scene will not be projected onto the same pixel in different images. Mapping therefore needs to be carried out before estimating brightness values. In this section, we consider the matching methods used in HDR reconstruction.

19.2.3.1. *Stereo matching for HDR reconstruction*

Many different methods exist for pixel matching. In this particular context, the input data contain a variety of intensity values. Dark, or saturated, zones have poor or erroneous data that vary across the sequence of considered images. Moreover, if this sequence is captured using several lenses, the data will have a higher degree of variability. We therefore need to establish a procedure for calibrating data to make it consistent (see section 19.2.3.2) and adapt or propose new matching algorithms. In this section, we explore four recent methods for tackling this problem.

Lin and Chang [LIN 09] aimed to match pixels contained in two images acquired from different viewpoints with different exposures, supplied by Middlebury[3]. To do this, they applied Sun *et al.*'s algorithm [SUN 03], based on belief propagation, after modifying the images to obtain a shared exposure time. This algorithm establishes a correspondence between pixels using three Markov random fields, corresponding, respectively, to three important problems that must be addressed during the matching phase: disparities, discontinuities and occlusions in the different images. While Lin and Chang [LIN 09] only used one set of stereoscopic data, Sun *et al.*'s method [SUN 03] has also been tested on multiscopic image sets (5 and 11 viewpoints), where an additional cost function is minimized in order to match pixels with the lowest cost.

Sun *et al.* [SUN 10] also proposed a solution for matching pixels taken from stereoscopic images acquired with two exposure times (Middlebury images[3]). As we saw in Chapter 7, different similarity measurements may be taken into account for matching purposes. In this case, the authors chose to use NCC, which is invariant to exposure changes. Different similarity measurements have been compared for mapping pixels taken from images with different exposures [BLE 08, ORO 12], and the NCC method currently produces the best results. Its invariance to changes in brightness under certain conditions was demonstrated by Troccoli *et al.* [TRO 06], who used it to improve results obtained using Kang and Szeliski's method [KAN 04]. To do this, two matching operations were carried out, the first with NCC and the second with the sum of square differences (SSD) in the luminance space to

---

3 http://vision.middlebury.edu/stereo/data/.

refine initial results. This method used N viewpoints and four different exposures.

Lu *et al.* [LU 11] considered 3D HDR reconstruction and have not directly addressed the production of 3D HDR images. They proposed the use of projection to assist matching, as shown in Figure 19.5, using a dome of 20 cameras to obtain images with two exposures simultaneously. Ten cameras acquired images with one exposure time and the ten others with the second exposure time. If we know pixel P_2 in image I_2, its projection X in the scene to acquire is also known (the point belongs to the optical center/pixel line). By inverse projection onto images acquired by other cameras, it is possible to find the points corresponding to this 3D point in all images in which it features. Points P_1 and P_3 thus correspond to point P_2 in images I_1 and I_3. The zero normalized cross-correlation (ZNCC) is then used to limit correspondences and improve their coherency.

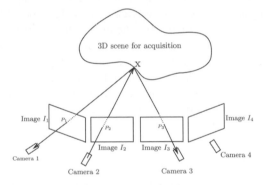

Figure 19.5. *Diagram showing the matching method used by Lu et al. [LU 11] for 4 of the 20 images. Pixel P_2 of image I_2 is known, and we need to find matches in images I_1, I_3 and I_4. Using projection, the 3D point x of the scene is identified, and, using inverse projection, we obtain points P_1 and P_3 in the images*

Bonnard *et al.* [BON 12] proposed an original approach, extending a purely 3D acquisition method to the context of HDR. The camera is presented in Chapter 3 and is built so that the objectives are in decentered parallel optical geometry, thus simplifying matching algorithms. In this case, Niquin *et al.*'s method [NIQ 10] was used for the matching stage in which pixels were matched by color similarity in a neighborhood and on the same line. The acquisition of different exposure times is simulated by applying a neutral density filter to each lens. Three filter pairs are selected: 0.3, 0.6 and 0.9, permitting simultaneous acquisition of eight images with four different exposures. This method is sensitive to data calibration, as the matching technique is based on a color similarity calculation. Calibration is difficult as each lens is independent.

Of the four methods presented above, the best results have been produced by Sun *et al.*'s approach   [SUN 10]. These results are comparable to those obtained by Lin and Chang [LIN 09] due to the use of the same image set. Lu *et al.*'s method   [LU 11] cannot be directly compared with the two methods above, as the authors wished to reconstruct HDR textures. Nevertheless, their results show the possibility of reading the text contained within an image using HDR rendering, something which cannot be done in an LDR context. For Bonnard *et al.*'s method, the artifacts that appear during the matching phase demonstrate reconstruction errors in different HDR images.

19.2.3.2. *Discussion of color data consistency*

The brightness values for each pixel come into play when estimating their value for the final HDR image, but they may also be necessary to bring all different exposures into the same space. This stage is essential in matching methods which have not been designed for images with different exposures.

Two cases are possible: if a single device is used (and moved to obtain different viewpoints), a single response curve may be used to linearize data; if several devices, or several sensors on the same device are used, then a response curve must be estimated for each device. The discussed method works with a single exposure time per device, so the response curve may only be estimated at the pre-processing stage. For this reason, simplifications are often used: for instance, Lu *et al.* [LU 11] considered that the set of camera sensors were of the same type and subject to the same calibration, and therefore reacted to light in the same manner. A single response curve was then calculated for one view, and used to retrieve luminance space values for all of the pixels of the images acquired by the different cameras. This hypothesis was not verified for Bonnard *et al.* [BON 12].

A final point to consider is that the estimation of a response curve requires us to operate on pixels representing the same 3D point in a scene, so preliminary matching may be needed to estimate the response curve. This curve may then be used for a second matching phase used in HDR calculations [LIN 09, SUN 10].

19.2.4. *Extension to the whole space: 3D HDR videos*

To date, no method has been developed to generate a 3D HDR video flow. However, an extension of 3D HDR imaging methods for video may be envisaged. These methods use a minimum of two images acquired with different exposure times. For video, the simultaneous acquisition of images for each filmed frame must be guaranteed, and problems generated by different exposure times (for example non-identical blurring of a fast-moving object) would need to be solved.

Kang *et al.* [KAN 03] encountered this problem in developing their HDR video method, a method that might be extended to produce 3D HDR video using N viewpoints and two exposures, by multiplying the number of cameras and ensuring synchronization. This procedure could also be used to extend Tocci *et al.*'s method [TOC 11] (prism cameras) or the Nayar and Mitsunaga method [NAY 00] (with a modified camera filter). These last two methods pose fewer problems as different exposures are obtained for each image.

The final question to consider concerns the performance of these envisaged methods. Although video postprocessing is generally accepted as a necessary step, the ultimate aim of video is live retransmission. This presents a considerable challenge, as, in addition to matching operations, HDR values need to be estimated for all of the images used in producing a 3D image. For the moment, this goal is out of reach. In addition to the problems discussed above, the restitution phase itself can require specific calculations (see section 19.3).

19.3. 3D HDR restitution

Visualizing 3D HDR content is problematic, as none of the displays currently available are able to present both HDR and 3D content. In this section, we consider a compromise based on available technologies and algorithms. We propose two approaches, which aim to combine the benefits of HDR and 3D display, based particularly on stereoscopic and multiscopic display techniques. HDR data may either be transformed for a non-HDR-dedicated display (section 19.3.1), or displayed directly in stereo on an HDR-dedicated display (section 19.3.2).

19.3.1. *Rendering on a 3D-dedicated display*

Screens allowing 3D content to be displayed with or without glasses have existed for a few years and are discussed in further detail in Chapter 14, but do not allow visualization of HDR content. We therefore need to adapt generated 3D HDR content to show on a standard display. Tone mapping algorithms may be applied [TUM 93] to convert an HDR image into a 24 bit RGB image, enabling perceptual preservation of contrasts in the image. A variety of tone mapping algorithms have been proposed, prioritizing either human perception, the quality of color rendering or computing efficiency; a list of these methods is presented in [BAN 11, DEV 02]. An evaluation method for these tone mapping operators is given in [CAD 08].

A single image may be rendered after applying algorithms such as those proposed by Reinhard et al. [REI 05] or Fattal et al. [FAT 02]. However, in our case, tone mapping algorithms must operate both on HDR and multiview video content. Certain tone mapping operators have already been proposed for HDR video by Drago et al. [DRA 03] and Kang et al. [KAN 03]. In this case, temporal consistency must be maintained for the operator to consider the images as a sequence rather than independently, based on calculations carried out for previous images. Yang et al. [YAN 12] propose an original approach to tone mapping, using the properties of human binocular vision and stereoscopic rendering systems: different tone mapping is applied for the left and right eyes, and the human visual system uses this information to recreate an image with a higher dynamic range.

None of the algorithms developed to date is suited to both multiview and HDR video content. Most tone mapping operators use global data to obtain a perceptual optimization of the values to display. If we simply apply existing algorithms, the chosen operations may be different for distinct viewpoints, leading to visual inconsistency when all views are displayed simultaneously. The point at which the tone mapping operator is applied also requires consideration: before data processing for 3D display (choice of views and/or interleaving) or afterwards.

19.3.2. *Displaying on an HDR-dedicated screen*

The first HDR display was proposed by Heidrich et al. [SEE 04]. This display allowed the contrast relationship to be extended to 50,000:1 (compared to 300:1 for standard screens at the time). This notably involved a maximum brightness of 8,500 cd m^{-2}. A commercial version of this screen was offered by BrightSide, a company bought out by Dolby Canada[4]. The type of screens available has changed considerably; they are now available as

---

4 www.dolby.com/.

light-emitting diode (LED)-based flat screens, commercialized by Sim2[5]. Current image technology consists of storing each color component in 16 bits, with American national standards institute (ANSI) contrast of 20,000:1 and luminance of 4,000 cd m^{-2}. This corresponds to a luminance spectrum over five orders of magnitude, as opposed to three for current liquid-crystal display (LCD) screens. One specificity of the Sim2 screen is the representation of total black. Other LED-based screens also increase perceived brightness, but this remains lower than the values offered by the Sim2 screen.

For a display frequency suitable for stereo, it is possible to send an HDR image flow reduced to the format accepted by the display, alternating right and left views and using active shutter glasses to create depth perception. The influence of the opacity of these glasses on perceived brightness remains to be measured.

19.4. Conclusion

In this chapter, we have presented methods used to extend the interval of color intensities to 3D video. These approaches are based on the reconstruction of HDR values. Although there is currently no stable approach for 3D HDR video generation, we have seen that advances have been made in complementary directions in multiscopic HDR imagery and in HDR video. We have seen that, while HDR data are popular, they cannot yet be rendered with the whole range of intensity used in their creation. Current display procedures may be adapted to provide better data display on 3D or HDR screens, but no procedure has been validated to date.

The storage of 3D HDR videos also needs to be considered. The standard formats used to store HDR images are listed in [REI 10]. The OpenEXR format has been adapted for HDR video, but was not intrinsically designed for this use. An HDR video format based on MPEG-4 has been proposed by Mantiuk *et al.* [MAN 04]. The use of standard formats might lead to faster adoption of HDR data in the industrial domain. The first approach has recently been put forward for compressing stereo and HDR data [SEL 12], which is compatible with standard formats. Given the speed of progress in HDR imaging, display and storage methods for 3D HDR data are likely to emerge in the near future.

This domain is still highly experimental but is expanding rapidly. The remaining issues are mostly technological, with a need for new capture

---

5 www.sim2.com/HDR/.

devices, and algorithmic, requiring better data calibration and reliable matching. HDR reconstruction should tend toward better noise control and the conservation of consistency in reconstructed data in terms of space and time. Finally, live transmission will only be possible if both technical equipment and data processing operate in real time, and when an operational, standardized compression, transmission and display format becomes available.

19.5. Bibliography

[AGG 04] AGGARWAL M., AHUJA N., "Split aperture imaging for high dynamic range", *International Journal of Computer Vision*, vol. 58, pp. 7–17, 2004.

[AGU 12] AGUERREBERE C., DELON J., GOUSSEAU Y., *et al.*, Best algorithms for HDR image generation. A study of performance bounds, Technical report, 2012.

[BAN 11] BANTERLE F., ARTUSI A., DEBATTISTA K., *et al.*, *Advanced High Dynamic Range Imaging: Theory and Practice*, AK Peters (CRC Press), Natick, MA, 2011.

[BLE 08] BLEYER M., CHAMBON S., POPPE U., *et al.*, "Evaluation of different methods for using colour information in global stereo matching approaches", in CHEN J., JIANG J., FÖRSTNER W. (eds), *Congress of the International Society for Photogrammetry and Remote Sensing*, vol. XXXVII, Part B3a, Beijing, China, pp. 415–420, July 2008.

[BON 12] BONNARD J., LOSCOS C., VALETTE G., *et al.*, "High-dynamic range video acquisition with a multiview camera", *Proceedings of SPIE Optics, Photonics, and Digital Technologies for Multimedia Applications II*, SPIC, vol. 8436, no. 1, p. 84360A, 2012.

[CAD 08] CADÍK M., WIMMER M., NEUMANN L., *et al.*, "Evaluation of HDR tone mapping methods using essential perceptual attributes", *Computers & Graphics*, vol. 32, no. 3, pp. 330–349, June 2008.

[DEB 97] DEBEVEC P.E., MALIK J., "Recovering high dynamic range radiance maps from photographs", *Proceedings of SIGGRAPH97, Computer Graphics Proceedings, Annual Conference Series*, pp. 369–378, August 1997.

[DEV 02] DEVLIN K., CHALMERS A., WILKIE A., *et al.*, "STAR: tone reproduction and physically based spectral rendering", in FELLNER D., SCOPIGNIO R. (eds), *State of the Art Reports, Eurographics 2002*, The Eurographics Association, pp. 101–123, 2002.

[DRA 03] DRAGO F., MYSZKOWSKI K., ANNEN T., *et al.*, "Adaptive logarithmic mapping for displaying high contrast scenes", *Computer Graphics Forum*, vol. 22, pp. 419–426, 2003.

[FAT 02] FATTAL R., LISCHINSKI D., WERMAN M., "Gradient domain high dynamic range compression", *ACM Transactions on Graphics*, vol. 21, no. 3, pp. 249–256, July 2002.

[FER 01] FERWERDA J.A., "Elements of early vision for computer graphics", *IEEE Computer Graphics and Applications*, vol. 21, no. 5, pp. 22–33, 2001.

[GAL 09] GALLO O., GELFAND N., CHEN W., et al., "Artifact-free high dynamic range imaging", *IEEE International Conference on Computational Photography (ICCP)*, San Francisco, CA, USA, April 2009.

[GRA 08] GRANADOS M., SEIDEL H.-P., LENSCH H.P.A., "Background estimation from non-time sequence images", *Proceedings of graphics interface 2008*, GI '08, Canadian Information Processing Society, Toronto, Ontario, Canada, pp. 33–40, 2008.

[GRA 10] GRANADOS M., AJDIN B., WAND M., et al., "Optimal HDR reconstruction with linear digital cameras", *2010 IEEE Conference on Computer Vision and Pattern Recognition (CVPR)*, San Francisco, CA, USA, pp. 215–222, 2010.

[GRO 06] GROSCH T., "Fast and robust high dynamic range image generation with camera and object movement", *Vision, Modeling and Visualization, RWTH Aachen*, pp. 277–284, 2006.

[JAC 08] JACOBS K., LOSCOS C., WARD G., "Automatic high-dynamic range generation for dynamic scenes", *IEEE Computer Graphics and Applications*, vol. 28, no. 2, pp. 24–33, March 2008.

[KAN 03] KANG S.B., UYTTENDAELE M., WINDER S., et al., "High dynamic range video", *ACM Transactions on Graphics*, vol. 22, no. 3, pp. 319–325, ACM, 2003.

[KAN 04] KANG S.B., SZELISKI R., "Extracting view-dependent depth maps from a collection of images", *International Journal of Computer Vision*, vol. 58, no. 2, pp. 139–163, 2004.

[KHA 06] KHAN E.A., AKYZ A.O., REINHARD E., "Ghost removal in high dynamic range images", *IEEE International Conference on Image Processing*, Atlanta, GA, USA, pp. 2005–2008, 2006.

[LIN 09] LIN H.-Y., CHANG W.-Z., "High dynamic range imaging for stereoscopic scene representation", *Proceedings of the 16th IEEE International Conference on Image Processing (ICIP)*, Cairo, Egypt, pp. 4305–4308, 2009.

[LOS 10] LOSCOS C., JACOBS K., "High-dynamic range imaging for dynamic scenes", in RATISLAV L. (ed.), *Computational Photography: Methods and Applications*, CRC Press/ Taylor & Francis, pp. 259–281, October 2010.

[LU 11] LU F., JI X., DAI Q., et al., "Multi-view stereo reconstruction with high dynamic range texture", *Proceedings of the Computer Vision ACCV 2010*, Springer, pp. 412–425, 2011.

[MAN 95] MANN S., PICARD R.W., "On being 'undigital' with digital cameras: extending dynamic range by combining differently exposed pictures", *Proceedings of IS&T*, pp. 442–448, 1995.

[MAN 04] MANTIUK R., KRAWCZYK G., MYSZKOWSKI K., *et al.*, "Perception-motivated high dynamic range video encoding", *ACM SIGGRAPH 2004 Papers*, SIGGRAPH '04, ACM, New York, NY, pp. 733–741, 2004.

[MER 07] MERTENS T., KAUTZ J., REETH F.V., "Exposure fusion", *Computer Graphics and Applications, Pacific Conference*, pp. 382–390, 2007.

[MIT 99] MITSUNAGA T., NAYAR S., "Radiometric self calibration", *IEEE Conference on Computer Vision and Pattern Recognition (CVPR)*, vol. 1, pp. 374–380, June 1999.

[NAY 00] NAYAR S., MITSUNAGA T., "High dynamic range imaging: spatially varying pixel exposures", *IEEE Conference on Computer Vision and Pattern Recognition (CVPR)*, vol. 1, pp. 472–479, June 2000.

[NIQ 10] NIQUIN C., PRÉVOST S., REMION Y., "An occlusion approach with consistency constraint for multiscopic depth extraction", *International Journal of Digital Multimedia Broadcasting (IJDMB), special issue Advances in 3DTV: Theory and Practice*, vol. 2010, no. 857160, pp. 1–8, February 2010.

[ORO 12] OROZCO R.R., MARTIN I., LOSCOS C., *et al.*, "Full high-dynamic range images for dynamic scenes", *Proceedings of SPIE*, vol. 8436, pp. 843609–843609-16, 2012.

[PED 08] PEDONE M., HEIKKILÄ J., "Constrain propagation for ghost removal in high dynamic range images", *3rd International Conference on Computer Vision Theory and Applications (VISAPP)*, Funchal, Madeira - Portugal, vol. 1, pp. 36–41, 2008.

[REI 05] REINHARD E., DEVLIN K., "Dynamic range reduction inspired by photoreceptor physiology", *IEEE Transactions on Visualization and Computer Graphics*, vol. 11, no. 1, pp. 13–24, January 2005.

[REI 10] REINHARD E., WARD G., PATTANAIK S., *et al.*, *High Dynamic Range Imaging: Acquisition, Display, and Image-based Lighting*, The Morgan Kaufmann series in Computer Graphics, 2nd ed., Elsevier (Morgan Kaufmann), Burlington, MA, 2010.

[SAN 04] SAND P., TELLER S., "Video matching", *ACM Transactions on Graphics*, vol. 23, no. 3, pp. 592–599, ACM, 2004.

[SEE 04] SEETZEN H., HEIDRICH W., STUERZLINGER W., *et al.*, "High dynamic range display systems", *Proceedings of SIGGRAPH '04 (Special issue of ACM Transactions on Graphics)*, August 2004.

[SEL 12] SELMANOVIC E., DEBATTISTA K., BASHFORD-ROGERS T., *et al.*, "Backwards compatible JPEG stereoscopic high dynamic range imaging", *Theory and Practice of Computer Graphics (TPCG)*, pp. 1–8, 2012.

[SUN 03] Sun J., Zheng N.-N., Shum H.-Y., "Stereo matching using belief propagation", *IEEE Transactions on Pattern Analysis and Machine Intelligence*, vol. 25, no. 7, pp. 787–800, July 2003.

[SUN 10] Sun N., Mansour H., Ward R.K., "HDR image construction from multi-exposed stereo LDR images", *IEEE International Conference on Image Processing (ICIP)*, pp. 2973–2976, 2010.

[TOC 11] Tocci M.D., Kiser C., Tocci N., *et al.*, "A versatile HDR video production system", *ACM SIGGRAPH 2011 papers (SIGGRAPH '11)*, ACM, New York, NY, USA, pp. 41:1–41:10, 2011.

[TRO 06] Troccoli A., Kang S.B., Seitz S., "Multi-view multi-exposure stereo", *Proceedings of the 3rd International Symposium on 3D Data Processing, Visualization, and Transmission (3DPVT'06)*, IEEE Computer Society, Washington, DC, pp. 861–868, 2006.

[TUM 93] Tumblin J., Rushmeier H.E., "Tone reproduction for realistic images", *IEEE Computer Graphics and Applications*, Los Alamitos, CA, USA, vol. 13, no. 6, pp. 42–48, November 1993.

[WAR 03] Ward G., "Fast, robust image registration for compositing high dynamic range photographs from handheld exposures", *Journal of Graphics Tools*, vol. 8, pp. 17–30, 2003.

[YAN 12] Yang X., Zhang L., Wong T.-T., *et al.*, "Binocular tone mapping", *ACM Transactions on Graphics (SIGGRAPH 2012 issue)*, vol. 31, no. 4, pp. 93:1–93:10, July 2012.

Chapter 20

3D Visualization for Life Sciences

20.1. Introduction

Stereoscopic and autostereoscopic techniques are not limited to the entertainment industry, and are widely used in numerous scientific applications.

Imaging has gained considerable importance in the field of medicine, where it is used to study organs or organisms *in vivo* with techniques that allow us to study the inside of the body without intervention: radiography, ultrasound, computerized tomography (CT) and Magnetic Resonance Imaging (MRI), radionuclide imaging, Single Photon Emission Computerized Tomography (SPECT) and Positron Emission Tomography (PET).

Imaging also plays an important role in biochemistry, where it is used for molecular modeling. Using computers, scientists are able to approach the structures they manipulate in an intuitive manner, and 3DTV has increased the ergonomics of the systems in question.

20.2. Scientific visualization

Within the domain of graphical information, visualization is often considered to be the simple action of showing the results of data processing

---

Chapter written by Aassif BENASSAROU, Sylvia PIOTIN, Manuel DAUCHEZ and Dimitri PAPATHANASSIOU.

on screen. However, the first meaning of visualization is "render visible", and visualization consists of constructing graphical representations using data which are not necessarily of a visual nature. Work on visualization often differentiates between information visualization and scientific visualization. In the first case, abstract data are given visual attributes (position, color, size, etc.); the second case concerns the presentation of concrete data with an intrinsic representation.

20.2.1. *3D construction*

The human sensory system is so well suited to the representation of perspective that it opts for a 3D interpretation, even when faced with a 2D representation. This occurs to the point where errors may be induced by ambiguous images, making us see impossible objects.

3D rendering engines generate images which are supposed to represent a 3D scene seen from a given angle, that of the observer. "Realistic" engines calculate images which appear real to the human eye using a number of techniques that imitate natural phenomena, such as lighting, depth of field and fog. The calculation of cast shadows is more costly, but significantly increases the perceptibility of an image.

Finally, other procedures such as environment mapping, Fresnel reflection or BRDF modeling[1] produce indisputable levels of photorealism. However, it is not strictly necessary to use a multitude of perceptual tricks in order for the human visual system to invent a 3D representation; the careful use of certain metaphors is generally sufficient.

20.2.2. *Interactivity*

Interactivity is central to the visualization of complex data [FEK 02], as in virtual reality environments [BOW 99]. Visualization tools are not limited to the production of static representations of data. Users require a certain degree of freedom in order to manipulate the representation and understand the information space. Interactions may take two different forms:

– In the first case, the user is in direct contact with the information, and acts on data by modification through a visualization interface.

---

1 Bidirectional reflectance distribution functions.

– In the second case, the user acts on a representation with the aim of understanding the structure of the information. This interaction involves manipulation of the interface, with no effect on data.

The combination of these two approaches allows users to gain a clear understanding of the visualized data. A user will perceive the volume of a scene more clearly if he or she is able to modify the viewing angle, and will better understand data if he or she is able to intervene in the visualization process.

20.2.3. *3D visualization*

No 2D visual artifice, however perfect, can equal *real* 3D perception. Autostereoscopic screens and glasses kits, such as Nvidia 3D Vision, have democratized "real" 3D solutions. The programming approach varies depending on the target material, and may require two or more views. Glasses-based solutions require two renderings, one per eye; this is known as *quad buffering*, as opposed to the *double buffering* generally used in graphical programming. This technique is supported by the OpenGL and Direct3D graphics libraries. Solutions for autostereoscopic screens, which are less invasive, require a minimum of five viewpoints and involve more "bulky" implementation, based on shaders, due to the absence of standardization (see Chapter 14).

While some applications can natively produce stereoscopic content, almost none of these approaches are designed to calculate the interleaved views needed for 3D display on autostereoscopic screens. One solution consists of modifying and recompiling the original code, but this is time consuming and not always feasible. To avoid rewriting and recompilation, we can use another application to transform the monoscopic output of the host application into multiscopic output.

The OpenGL call interception feature has been used for a variety of purposes, from debugging to code analysis. A number of OpenGL debuggers have been developed based on this principle, including SpyGLass, GLIntercept and BuGLe. Duca *et al.* [DUC 05] intercepted and recorded OpenGL command flows, storing the data taken from the pipeline in a relational database, allowing analysis of the whole rendering chain, including shaders. Mohr and Gleicher [MOH 01] have shown how to change the rendering mode in an OpenGL application without modifying the source code by intercepting OpenGL commands and changing drawing function calls. The

Chromium [HUM 02] system also allows manipulation of graphical command flows. Its flexibility allows users to add multiscopic functions to generic OpenGL applications using configuration scripts written in Python.

20.3. Medical imaging

The use of imaging in medicine is constantly increasing due to progress being made in various techniques. The most spectacular and immediately visible progress in the domain relates to image acquisition techniques, which give access to an increasing variety of information; the benefits of technological advances are also apparent in data visualization. Scientific visualization may thus be applied to this particular field of imaging; medical imaging has clearly benefited from progress in terms of visualization techniques, notably as this imaging increasingly concerns a volumetric image representing the distribution of information in the human body.

Currently, objects studied using 3D imaging techniques are usually presented to users through the representation of information as a series of slices, and the doctor studying these slices has the mental imaging capacity to integrate and understand the presented information. In this way, we are generally able to clearly locate structures in relation to one another; certain tools exist which effectively represent the environment of a structure in an organism, due to rapid image processing methods (e.g. rapid and dynamic display of slices passing along an axis fixed by the operator). It is also possible to represent a set of 3D information as a whole using techniques such as volume rendering, although these are not the most widespread techniques. They require the use of information selection methods before visualizing acquired information in the whole volume, eliminating parasites (technical artifacts) and organs or tissues which are not important for a particular exploration. Threshold methods may be used, but access to the inside of organs, not just their surface, may involve a combination of 3D renderings and slices.

20.3.1. *Volumetric visualization*

Historically, the imperatives of interactivity have been achieved to the detriment of image quality. A number of methods have been developed with the aim of speeding up the rendering process [BRO 00], which is known for its high CPU usage[2]. The main innovations in the field appeared at the same

---

2 Central processing unit.

time as the new generation of graphics maps, with powerful GPUs[3] enabling us to obtain renderings of a quality formerly restricted to costly graphics stations [HAD 09]. These graphics stations took several minutes to calculate images, which can now be created in a fraction of a second by any home, personal computer (PC). The generalization of this type of material is one of the factors that enables high levels of user interaction [VAN 96].

At a time when mass texture use was not a possibility, the first methods for visualizing 3D images often made use of intermediary surfaces. This was the case with the famous marching cubes algorithm [LOR 87, WYV 86], which was able to extract the surface of structures contained in images based on their level of intensity. "Direct" volume rendering techniques, on the other hand Levoy [LEV 88], did not reduce the set of volumetric data to the surface of the observed objects, and allowed users to see the inside of these objects.

Shirley and Tuchman [SHI 90] were among the first to use graphics maps to produce an approximation of volume rendering. Based on a breakdown into simplicial cells, their projected tetrahedra (PT) algorithm involved five stages. First, it projected the tetrahedra (camera + perspective), then sorted projections (painter's algorithm) and categorized them according to the obtained profile, before breaking them down into triangles and, finally, displaying the results. This idea has been reused on a number of occasions, for example by Stein *et al.* [STE 94], who attempted to improve these estimations using a more precise sorting algorithm.

Methods based on the use of textured slices may be considered as an additional approximation, where polygons show whole volume slices. Yagel *et al.* [YAG 96] proposed a technique for the volumetric rendering of non-structured grids by calculating intersections between data and a sweep plane. They showed that their algorithm generated visually comparable images in less time, without even needing to conserve vertex adjacency information, essential for other methods.

Textured slices constitute the most practical and efficient method as the material automatically carries out the bulk of essential calculations.

Nowadays, 2D texture placement is part of the standard range of functions of PCs. Moreover, the technique is perfectly integrated into the production process for slices aligned along object axes. The principle is similar to that applied in Lacroute and Levoy's shear warp [LAC 94]. It requires the use of

---

3 Graphics processing unit.

three variants of the 3D image: one series of slices aligned along x-axis, another along the y-axis and a third along the z-axis. These variants are then applied to the most appropriate pile of rectangles (see Figure 20.1). The choice of a pile to display is made based on the dominant axis, in order to produce slices with the least possible slant. The rendering is applied by placing textured quadrilaterals from back to front and mixing their colors by alpha blending [POR 84]. During this stage, the texture coordinates are bilinearly interpolated on the surface and transcribed in color.

Figure 20.1. *Axis-aligned slices requiring several 2D textures*
(or one 3D texture)

The use of material solutions for 3D texture application has taken over from 2D solutions in the domain of interactive volumetric rendering [CAB 94, WES 98]. The management of 3D textures and the subsequent trilinear interpolation makes it possible to select a slice plane at random in the volume for visualization. This allows us to select optimal slices by imposing new constraints. The flaws observed in 2D also disappear if slices are all calculated parallel to the screen (see Figure 20.2).

Figure 20.2. *View-aligned slicing requiring a single 3D texture*

The slice definition stage may be carried out by specifying englobing rectangles or by expressing exact intersection polygons. The first solution presents the advantage of being direct and requiring little calculation effort. The second is more complex, and requires calculation of all intersection points and the definition of polygons based on these points. However, the second approach is quicker, as it minimizes the calculations carried out by the GPU, which does not need to process fragments located outside of the data volume. Kniss *et al.* [KNI 01] took a pseudo-angle approach; unfortunately, this algorithm is dependent on the sequencing of the vertices of the polygon. Inversely, the method proposed in [BEN 05] provides pre-sorted vertex sequences, which may be used as they are and drawn using OpenGL functions.

Finally, Engel *et al.* [ENG 01] proposed a technique known as *pre-integration*, which notably reduces the artifacts produced during slicing. When a polygon is grafted, it is considered to be volumetric rather than planar, considering the ray traversing each fragment. An input point is calculated along the ray midway from the previous slice, and an output point midway toward the next slice. The texture is then sampled at these two points, and the obtained values act as an index in a pre-calculated 2D color table.

20.3.2. *3D medical imaging*

3D visualization techniques, such as glasses or stereoscopic screens, are currently in their infancy. Their full potential should become clear with experience, refinements and technical progress which will facilitate their use and increase their power. Nevertheless, these techniques have begun to appear at specialist medical imaging conferences.

The passage from traditional imaging, using 2D projections, to 3D slicing techniques was revolutionary; the shift toward volumetric rendering for interpretation should be easier. However, outside of specific contexts, these techniques are not easy to implant into everyday practice, at least for now; the cost of essential software and materials limits their use to certain centers.

The same evolution may occur in the future for stereoscopic 3D visualizations, which, in theory, represent a relatively small step from volume rendering. Again, usage may be limited to centers with the means of purchasing costly equipment which is not strictly necessary in general practice. The advantages and disadvantages of these new techniques also need to be weighed against established techniques. All biological or medical activities which involve observation of all or part of the human body may, potentially, benefit from scientific visualization, whether at microscopic or macroscopic level. The initial field for identification of profitable applications is therefore vast.

In addition to biochemical and pharmacological applications, with 3D visualization of molecules, a deeper understanding of cellular biological processes may be obtained by visualizing cells during different phases of normal or pathological biological phenomena. 3D visualization of a single organism as a whole, potentially over time, should also allow a clearer understanding of certain phenomena in embryology. These examples demonstrate the degree to which different medical professions may be affected by 3D visualization techniques. Other possible applications exist in dentistry, both in surgical preparation and in the creation of implants and prosthetics.

In this section, we consider three domains, within a medical context, in which the advantages of these new techniques should become clear: teaching, diagnosis and certain treatment methods.

20.3.2.1. *Teaching*

The potential utility of 3D visualization methods in teaching is clear, for example in anatomy teaching. The ability to "look" at an organ in space, from different angles and magnifications, based on "real" data obtained from the natural world, would facilitate understanding of the relationships between organs. The teaching of pathology could also benefit from these visualizations, as realistic representations enable better comprehension of diseases. Pathological anatomy is not the only domain that would benefit, as dynamic visualizations would enable us to understand a pathological process over time, or consider the abnormal operation of an organ (such as the heart [HER 08]). As with research, scientific visualization in an educational context may lead to a better understanding of the shape and temporal evolution of a variety of structures, for example in embryology or cell biology (e.g. considering cell division) and for whole organs. The combination of morphological information with functional or metabolic information, an important trend in current medical imaging, should also benefit from 3D anatomical visualization, with, for example, the superposition of a spatial distribution of a metabolism (using a different range of colors), on condition that the inside of organs is represented in a suitable manner (e.g. by modifying transparency or showing slice planes within objects shown in 3D). This type of teaching may be envisaged in the near future for small groups (practical work), before being proposed for small lecture theaters, depending on screen possibilities. Clearly, the concept of collaborative visualization is perfectly suited to this context.

20.3.2.2. *Diagnosis*

In diagnosis, instant global visualization of data can be expected to increase comfort in the analysis of imaging data. This could lead to an increase in the speed of data evaluation for a volume, saving time in the interpretation of images, something which is clearly valuable from a professional perspective. The gains represented by the most advanced techniques in scientific visualization in terms of diagnostic precision (proportion of correct affirmations of the presence or absence of a disease) compared to attentive study of slice information, or volumetric rendering without stereoscopic vision, are not yet clear, and require methodical evaluation. As with all new techniques, this visualization method brings new disadvantages alongside the new possibilities it offers.

3D visualization requires a new mode of mental representation of anomalies, different to that used in slice visualization, for the correct recognition of anomalies for diagnostic purposes. New, or additional, science of diseases presentation and signs may emerge based on new visualization possibilities. A learning period is likely to be necessary, in addition to the time involved in learning to use the tool, due to the loss of normal reference points and the presence of particular artifacts in this mode of visualization. Any technique which moves away from raw data may amputate information and introduce artificial information. The realism of a representation, applied to a parasite signal or incomplete information, may lead to erroneous understanding of the structures in question. Poor knowledge of the artifacts generated by a technique and excessive trust in something which is merely an image of reality constitute potential pitfalls of these new techniques, as with all medical imaging methods.

In teaching anatomy, 3D visualization of the surface of organs is sufficient in a large number of circumstances. However, in diagnostics, the ability to visualize the inside of organs is important for the identification of lesions. The ease of use of software and the relevance of proposed tools in relation to this point are of crucial importance in developing the use of 3D visualization in practice. The limitations of use of 3D visualization for diagnosis are linked to the concrete conditions of use, not to speak of the material and its cost: these methods offer a more immediate understanding of the position of information in a space, but require the use of eye equipment, or limitations in user position in relation to the screen. Prolonged use of the tool may generate ocular fatigue. However, these limitations still need to be evaluated within the framework of everyday use. One potential stumbling block is the fact that different subjects may find it more or less difficult to gain an understanding of relief using these techniques.

While this mode of scientific visualization may not immediately become systematic in diagnosis, for some or all of the reasons mentioned above, it may play a role in certain complex cases where the possibilities offered by collaborative visualization would allow group study; it might also be used in training medical imaging specialists. The contribution made by new scientific visualization options is likely to vary depending on the selected imaging techniques, the organs studied or the diseases considered; this will become clearer as more experiments are carried out over the coming years. For the moment, 3D visualization in the context of medical diagnosis has been used with promising results in studying blood vessels, which have a complex spatial form and which may be isolated from the rest of the volume with relative ease, as we can obtain significant contrast between their contents and neighboring tissue [ABI 10, MAG 12].

20.3.2.3. *Therapy*

In terms of therapeutic applications, scientific visualization tools may be used to improve planning for certain treatments. This is particularly true in surgery, where the visualization of lesions and organs, with the possibility of interacting with these images, would allow "rehearsal" of an intervention within a virtual volume. The 3D visualization of morphological data recorded before the operation might also be used to guide the progression of surgical actions. Image preparation is likely to be necessary in this case, with an impact on the time required from specialists. The limitations of correlation between images, whatever their level of realism, and real objects need to be evaluated, along with the impact this "virtual reality" may have on the mental image of the operation site. In practice, the use of an autostereoscopic screen in operating theaters is easier to imagine than the use of goggles, which would disrupt the visual passage from the operating table to image observation. Radiotherapy is another domain where these new scientific visualization techniques may prove particularly useful. The current aim is to limit the irradiated zone as far as possible around the tumor zone, and new tools are increasingly able to do this. The establishment of contours around the treated zone when planning treatment implies a clear understanding of the shape and position of the tumor. The possible use of 3D visualization needs to be evaluated in this context, both in terms of the precision of delimitation and in the time required.

20.4. Molecular modeling

In chemistry and biochemistry, the understanding of molecular systems as 3D structures is crucial, not only in terms of spatial perception but also in terms of associated functional and mechanical consequences. 3D imaging enables us to "see", understand and interpret chemical and biological phenomena, which are often highly complex and involve multiple aspects, usually in a dynamic manner. The 3D observation of atomic, molecular and macromolecular interactions allows scientists to go further in understanding and interpreting these phenomena and makes it possible to predict new functions. This vision also facilitates the assisted creation of new molecules, which may then be constructed chemically or through molecular biology before experimental testing.

Atomic components (from 10^{-18} to 10^{-10} m, for the electron and the atom, respectively) and molecular components (10^{-10} to 10^{-8} m from the atom to "standard"-sized macromolecules) are too small for direct "visualization" through experimentation to be possible in current conditions, except in the case of considerably larger macromolecular structures such as fibrillar systems, transcription machinery or transmembrane receptors, for which we are currently able to evaluate shapes using an ultrastructural (but non-atomistic) vision using electron or atomic force microscopes. The only remaining possibility is virtual representation, using the power of computers and graphics stations, using data obtained through crystallography or nuclear magnetic resonance.

Since the first structural resolutions of biological macromolecules were obtained by X-ray crystallography in the 1950s, as in the case of Watson and Crick's work on DNA [WAT 53], Sanger's resolution of insulin [SAN 53] and Perutz's resolution of hemoglobin [PER 60], scientists have aimed at producing 3D reconstructions of these objects, either physical, using ball and stick models, or using early visualizations on oscilloscope screens in the 1970s. For 3D representation, the complexity of the objects in question and the fact that no real representation of these macromolecules was available led the scientific community to propose new models for representation and simplification, based on the work by Pauling and Corey [PAU 51a, PAU 51b], allowing easy understanding of associated 3D folding. While the sequence of basic components is generally linear (in proteins and nucleic acids), their spatial arrangement depends on the sequence and their physical properties, determining their folding functionality. Thus, for proteins, several graphical models have been used not only for helix representation of α helix amino acids, but also for β chains that may themselves be grouped into β sheets, and finally the bonds connecting all of these secondary structures. A 3D understanding of the arrangement of these secondary structural elements is essential in understanding the relative positions of domains and for the spatial interpretation of dynamics, which may be correlated with this 3D structure.

20.4.1. *Classic visualization modes*

Macromolecules are made up of several thousands or even tens of thousands of atoms. They are highly complex, and it is difficult to obtain information by simply observing them. For this reason, several different visualization modes are used to simplify the complexity of macromolecules based on the aspects we wish to observe.

To represent the overall structure of a protein, we may use wireframe rendering, a historic visualization mode [LEV 66]. Using this type of rendering, atoms and their connections are represented by lines. Each atom has a color code, which may be represented on the line. In Figure 20.3, we see that while this mode of visualization is in 3D, it does not give an impression of depth, due partly to the fact that all lines, no matter how distant from the viewpoint, have the same thickness, and to the fact that they do not make use of lighting effects.

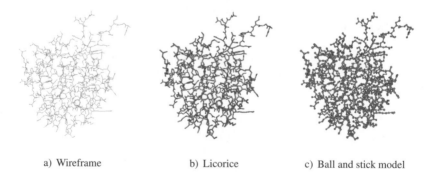

a) Wireframe   b) Licorice   c) Ball and stick model

Figure 20.3. *Basic visualization modes*

Another mode often associated with wireframe rendering is licorice rendering, which uses cylinders instead of lines. Cylinders are surfacic objects, and may thus interact with light. The best-known type of visualization uses the ball and stick method. This method allows us to precisely locate atoms by representing them as spheres.

To represent the overall shape of a protein, we may use methods such as van der Waals visualization, representing van der Waals radii, or modes such as the solvent accessible surface model or the solvent excluded surface model [LEE 71]. The van der Waals surface is represented by the edge of the set of balls formed by spheres centered on an atom and with a radius determined by the van der Waals radius of the atom (see Figure 20.4). The solvent accessible surface is the set of possible positions of the center of a spherical probe representing a solvent, rolling on the van der Waals surface. The solvent-excluded surface takes account of the fact that the probe covers hollows.

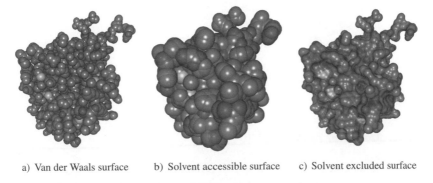

a) Van der Waals surface   b) Solvent accessible surface   c) Solvent excluded surface

Figure 20.4. *Surfacic visualization modes*

In cases where we only wish to present certain structures of a protein, we may use simplified or symbolic renderings.

The most widespread form of symbolic rendering is cartoon rendering, made popular by the 2D tracing possibilities of the Molscript program developed by Kraulis [KRA 91] in the 1990s. Using this type of rendering, we aim to represent secondary structures. The α helices take the form of coils, and β chains, the basic components of β sheets, are shown as arrows.

Other visualization modes exist [O'DO 10], but we have discussed the most important and widely used methods. However, these modes present certain drawbacks. Observing a protein using the licorice mode, for example, we are unable to observe secondary structures due to the complexity of the object. The cartoon mode appears to be the most appropriate, but we can only clearly distinguish the α helices (Figure 20.5), which are easy to identify even in a large-scale structure. The real difficulty comes from the fact that only β chains are represented, and not the β sheets. In Figure 20.5, we are able to see certain β chains, but it is impossible to locate, visualize, enumerate or characterize the corresponding sheets. To date, only Nolin *et al.* [NOL 09] have offered a visualization mode able to represent a β sheet as a whole.

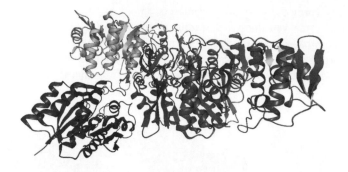

Figure 20.5. *Cartoon representation*

20.4.2. *Molecular modeling in 3D*

The joint emergence of "office" computing, which gives everyone the ability to access structural models developed using graphics stations and computers with specialist software, and high-speed molecular biology, through a number of genome projects generating more and more solved 3D structures, have led users of biological molecules to consider 3D visualization. High-performance computers and 3D screens have been introduced in an increasing number of laboratories, and are consequently used in scientific projects. These tools are now used on a regular basis for teaching purposes, and allow students to rapidly come to grips with complex notions, giving access to invisible nanoscopic objects, which are difficult or impossible to materialize in a laboratory setting.

Stereoscopic vision is highly relevant, for example, in representing the interaction of a ligand in an active macromolecular site. Using 3D techniques, we can provide a symbolic representation of these macro-objects with a sequence of elements of secondary structures, alongside additional elements such as solvent accessible surfaces, and representations of the main elements involved in interactions, such as weak bonds (e.g. hydrogen bonds) between various partners, using different visualization modes. While a 2D representation allows us to locate atoms within a hydrogen bond, 3D vision allows us to consider both the structural location of this hydrogen bond and, for example, the immediate environment of other chemical groupings. It is then possible to evaluate the consequences of modifications to chemical groupings, amino acid mutations and the influence of these modifications. The use of 3D representation requires us to reconsider the objects in question and, where necessary, to propose new mechanisms or approaches, which may then be tested biologically. The recent possibility of augmenting reality by applying additional textures to show different physicochemical properties without losing information on atomic connectivity is also extremely useful, allowing us to add an interpretation of physicochemical phenomena.

Autostereoscopic displays make it easier to work with 3D images. The ability to work without goggles is important when using two screens: the command screen, used to supervise work and for console commands, is a standard screen that does not require goggles, whereas the second screen, used for visualization and as the graphical work zone, previously required specific eye wear. Autostereoscopic displays increase comfort levels and working efficiency, although there is still room for improvement in terms of definition and protocols. Learning times are somewhat longer for these displays, and workspace reorganization is often needed to gain the required distance from the screen, improving ocular relaxation for sessions of several hours. These new screens and visual techniques offer a range of possibilities and future applications which will allow us to progress in molecular graphics and, particularly, interact in real time with 3D movement through the use of augmented reality tools.

20.5. Conclusion

While costly, specialized systems have offered the possibility of 3D visualization for a number of years, these possibilities, with or without glasses, have only recently become available to a wider public. 3D content is clearer and easier to use than 2D images, as they remove the need for inference. In this chapter, we have seen the ways in which stereoscopic capacities may vastly improve biomedical imaging and molecular graphics; hospital practitioners have great hopes of applications for teaching, diagnosis and patient treatment, and biochemists hope to achieve a clearer structural understanding of the macromolecules which they study. Despite certain reservations concerning a prolonged learning period, the scientific community has welcomed 3D technology with open arms; this technology, far from being simply a new gadget, promises real progress in the observation, interpretation and manipulation of data.

20.6. Bibliography

[ABI 10] ABILDGAARD A., WITWIT A.K., KARLSEN J.S., *et al.*, "An autostereoscopic 3D display can improve visualization of 3D models from intracranial MR angiography", *International Journal of Computer Assisted Radiology and Surgery*, vol. 5, pp. 549–554, 2010.

[BEN 05] BENASSAROU A., BITTAR E., JOHN N.W., *et al.*, "MC slicing for volume rendering applications", *Proceeding of the International Conference on Computational Science*, Lecture Notes in Computer Science, Springer, Berlin, Heidelberg, Atlanta, GA, vol. 3515, pp. 314–321, May 2005.

[BOW 99] BOWMAN D.A., JOHNSON D.B., HODGES L.F., "Testbed evaluation of virtual environment interaction techniques", *Proceedings ACM Symposium on Virtual Reality Software and Technology*, London, UK, pp. 26–33, 1999.

[BRO 00] BRODLIE K., WOOD J., "Recent advances in visualization of volumetric data", *Proceedings of Eurographics 2000*, Interlaken, Switzerland, pp. 65–84, 2000.

[CAB 94] CABRAL B., CAM N., FORAN J., "Accelerated volume rendering and tomographic reconstruction using texture mapping hardware", *Proceedings of ACM Symposium on Volume Visualization '94*, Washington, DC, pp. 91–98, 1994.

[DUC 05] DUCA N., NISKI K., BILODEAU J., *et al.*, "A relational debugging engine for the graphics pipeline", *ACM SIGGRAPH 2005 Papers*, SIGGRAPH '05, New York, ACM, pp. 453–463, 2005.

[ENG 01] ENGEL K., KRAUS M., ERTL T., "High-quality pre-integrated volume rendering using hardware-accelerated pixel shading", *Proceedings of the ACM SIGGRAPH/EUROGRAPHICS Workshop on Graphics Hardware*, HWWS '01, New York, ACM, pp. 9–16, 2001.

[FEK 02] FEKETE J., PLAISANT C., "Interactive information visualization of a million items", *Proceedings of IEEE Symposium on Information Visualization*, Boston, MA, pp. 117–124, October 2002.

[HAD 09] HADWIGER M., LJUNG P., REZK-SALAMA C., *et al.*, "Advanced illumination techniques for GPU-based volume raycasting", *ACM SIGGRAPH 2009 Courses*, SIGGRAPH '09, New York, ACM, pp. 1–166, 2009.

[HER 08] HERLAMBANG N., LIAO H., MATSUMIYA K., *et al.*, "Real-time autostereoscopic visualization of registration-generated 4D MR image of beating heart", in DOHI T., SAKUMA I., LIAO H., (eds), *Medical Imaging and Augmented Reality*, vol. 5128 of *Lecture Notes in Computer Science*, pp. 349–358, Springer, Berlin, Heidelberg, 2008.

[HUM 02] HUMPHREYS G., HOUSTON M., NG R., *et al.*, "Chromium: a stream-processing framework for interactive rendering on clusters", *ACM Transactions on Graphics*, vol. 21, no. 3, pp. 693–702, 2002.

[KNI 01] KNISS J., KINDLMANN G., HANSEN C., "Interactive volume rendering using multi-dimensional transfer functions and direct manipulation widgets", *Proceedings Visualization 2001*, pp. 255–262, 2001.

[KRA 91] KRAULIS P.J., "MOLSCRIPT: a program to produce both detailed and schematic plots of protein structures", *Journal of Applied Crystallography*, vol. 24, no. 5, pp. 946–950, 1991.

[LAC 94] LACROUTE P., LEVOY M., "Fast volume rendering using a shear-warp factorization of the viewing transformation", *Computer Graphics (SIGGRAPH'94 Proceedings)*, vol. 28, no. 4, pp. 451–458, 1994.

[LEE 71] LEE B., RICHARDS F., "The interpretation of protein structures: estimation of static accessibility", *Journal of Molecular Biology*, vol. 55, no. 3, pp. 379–400, 1971.

[LEV 66] LEVINTHAL C., "Molecular model-building by computer", *Scientific American*, vol. 214, pp. 42–52, 1966.

[LEV 88] LEVOY M., "Display of Surfaces from Volume Data", *IEEE Computer Graphics and Applications*, vol. 8, no. 3, pp. 29–37, 1988.

[LOR 87] LORENSEN W., CLINE H., "Marching cubes: a high resolution 3D surface construction algorithm", *Computer Graphics (SIGGRAPH'87 Proceedings)*, vol. 21, no. 4, pp. 163–169, 1987.

[MAG 12] MAGALHES D.S.F., RIBEIRO F.H., LIMA F.O., *et al.*, "3D brain MR angiography displayed by a multi-autostereoscopic screen", *Medical Imaging 2012: Image Perception, Observer Performance, and Technology Assessment*, SPIE, pp. 83181P–83181P-6, 2012.

[MOH 01] MOHR A., GLEICHER M., "Non-invasive, interactive, stylized rendering", *Proceedings of the 2001 Symposium on Interactive 3D Graphics*, I3D '01, New York, ACM, pp. 175–178, 2001.

[NOL 09] NOLIN L., BENASSAROU A., DAUCHEZ M., *et al.*, "Beta sheet simulation in molecular modeling", in WEGHORN H., ISA P.T., (eds), *IADIS Applied Computing 2009*, Rome, pp. 259–266, 2009.

[O'DO 10] O'DONOGHUE S.I., GOODSELL D.S., FRANGAKIS A.S., *et al.*, "Visualization of macromolecular structures", *Nature Methods*, vol. 7, pp. S42–55, 2010.

[PAU 51a] PAULING L., COREY R.B., "Atomic coordinates and structure factors for two helical configurations of polypeptide chains", *Proceedings of the National Academy of Sciences of the United States of America*, vol. 37, no. 5, pp. 235–240, 1951.

[PAU 51b] PAULING L., COREY R.B., "The pleated sheet, a new layer configuration of polypeptide chains", *Proceedings of the National Academy of Sciences of the United States of America*, vol. 37, no. 5, pp. 251–256, 1951.

[PER 60] PERUTZ M.F., ROSSMANN M.G., CULLIS A.F., *et al.*, "Structure of haemoglobin: a three-dimensional fourier synthesis at 5.5 angstroms resolution, obtained by X-Ray analysis", *Nature*, vol. 185, no. 4711, pp. 416–422, 1960.

[POR 84] PORTER T., DUFF T., "Compositing digital images", *Computer Graphics (SIGGRAPH'84 Proceedings)*, vol. 18, no. 3, pp. 253–259, 1984.

[SAN 53] SANGER F., THOMPSON E.O.P., "The amino-acid sequence in the glycyl chain of insulin. I. The identification of lower peptides from partial hydrolysates", *Biochemical Journal*, vol. 53, no. 3, pp. 353–366, 1953.

[SHI 90] SHIRLEY P., TUCHMAN A., "A polygonal approximation to direct scalar volume rendering", *Computer Graphics (SIGGRAPH'90 Proceedings)*, vol. 24, no. 5, pp. 63–70, 1990.

[STE 94] STEIN C., BECKER B., MAX N., "Sorting and hardware assisted rendering for volume visualization", *Proceedings of ACM Symposium on Volume Visualization '94*, Washington, DC, pp. 83–90, 1994.

[VAN 96] VAN GELDER A., KIM K., "Direct volume rendering with shading via three-dimensional textures", *Proceedings of ACM Symposium on Volume Visualization '96*, pp. 23–30, 1996.

[WAT 53] WATSON J.D., CRICK F.H., "The structure of DNA", *Cold Spring Harbor Symposia on Quantitative Biology*, vol. 18, pp. 123–131, 1953.

[WES 98] WESTERMANN R., ERTL T., "Efficiently using graphics hardware in volume rendering applications", *Computer Graphics (SIGGRAPH'98 Proceedings)*, vol. 32, no. 4, pp. 169–179, 1998.

[WYV 86] WYVILL B., WYVILL G., MCPHEETERS C., "Data structure for soft objects", *The Visual Computer*, vol. 2, no. 4, pp. 227–234, 1986.

[YAG 96] YAGEL R., REED D., LAW A., *et al.*, "Hardware assisted volume rendering of unstructured grids by incremental slicing", *Proceedings of ACM Symposium on Volume Visualization '96*, pp. 55–63, 1996.

Chapter 21

3D Reconstruction of Sport Scenes

21.1. Introduction

Nowadays, the use of applications involving image analysis, geometric modeling or computer vision is common in transmitting sporting events. These applications allow, for example, annotation of sequences or the automatic selection of highlights of an event for *a posteriori* access [EKI 03, REN 09, YAN 08]. However, applications for three-dimensional (3D) reconstruction of sporting scenes remain extremely rare. This is due to the quantity of scientific issues caused by highly variable capture conditions (lighting, field, player kits, etc.) and the absence of information concerning acquisition parameters, leading to the use of a "raw" video flow (unknown camera parameters). Furthermore, the use of these applications during live events can only be envisaged if reconstruction is possible in real time (or almost real time). Complex systems have been proposed to overcome these difficulties, involving the use of multiple dedicated cameras; these systems represent a considerable outlay, can be difficult to implement and do not produce the desired automatic reconstruction [AMI 13, ORA 13, RED 13]. Certain research projects have focused on the automation of the reconstruction process using uncalibrated cameras and have offered innovative solutions [GRA 07, KIL 07, LE 06, LLU 11, TRI 13].

In this chapter, we will present the key stages of the reconstruction process, focusing on certain difficult points. In section 21.2, we will consider

Chapter written by Sébastien MAVROMATIS and Jean SEQUEIRA.

the analysis of color images with the aim of automatically selecting the playing surface. Section 21.3 concerns the extraction of markings on the playing field using Hough transforms. In section 21.4, we will consider the matching of primitives extracted from images with those used in the model of the scene.

21.2. Automatic selection of a region of interest (ROI)

21.2.1. *ROI characterization*

The ROI of a sporting scene is, naturally, the playing area. This playing area may often be characterized both as a region with a relatively uniform shade and as the largest quasi-connected component of the image under consideration (see Figure 21.1).

Figure 21.1. *ROI of a football scene*

ROI characterization is an important stage in the reconstruction process. Images often include the environment of the playing area, such as stands, athletic tracks, security personnel and advertisements. Knowledge of the ROI, as shown in Figure 21.2, allows us to immediately validate elements present in the playing area.

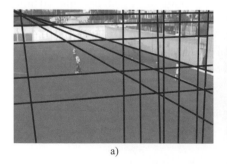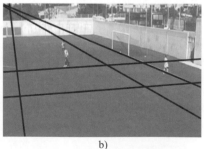

a)   b)

Figure 21.2. *Automatic pitch mark detection without ROI a) and with ROI b)*

21.2.2. *Color coherence*

ROI determination, which is essentially defined by the notion of color coherence, may appear easy at first glance. However, the color is only apparently homogeneous. In fact, the discrete representation of color, in any representation space, uses discrete components that are not sufficiently similar. Various authors have proposed approaches to characterizing regions of the color space which support this homogeneity [DAI 95, LEF 00, LUC 01, VAN 00]. The interest of using a 3D space to segment a color image has been demonstrated on a number of occasions [CAR 95]. In this section, we use the hue lightness saturation (HLS) representation space to illustrate an entirely automated selection method for the color attributes of an ROI. In many respects, the methodology represented here is similar to those using Parzen windows [PAR 62]. It consists of grouping similar colors within the representation spaces and separating clouds of different colors in order to isolate and extract as many "dense" clouds as there are dominant color regions in the image: this is the color coherence criterion. The effectiveness of this approach lies in the continuous representation of the colors of pixels in the image [LE 04]. The use of computers engenders a discrete representation of color spaces, which creates difficulties in analyzing the distribution of colors associated with the pixels in an image.

The first idea consists of calculating a histogram of colors in an image, and retaining only the most common colors. Clearly, the results obtained using this method are not particularly satisfactory. However, based on this histogram, we can construct a continuous and efficient representation. First, we must divide the color space into cells, roughly representing a single color (see Figure 21.3).

During the second stage, the use of a potential function in this rough space allows us to define a continuous color representation space. The purpose of the potential function is to establish links not only between neighboring points, but also between similar regions (e.g. different greens), while separating characteristic clouds of each dominant color in the image. To do this, each point of the color space (corresponding to a pixel in the image) is considered as a source of potential. The potential function may be a simple decreasing linear function of the distance from the point, with a value of 1 at the point, supported by a disk of radius $D_{max} = R * (1 + \epsilon)$, where R represents the radius of the cell in question and ϵ represents an absorption value.

The retained dominant color regions are obtained by thresholding in the HLS space modified by the potential function. The threshold may be obtained automatically by segmenting the cumulated histogram of potential values. In cases where several dominant color regions are present in the image, these may be identified by propagation using neighborhood relationships from the rough partitioning chosen for the HLS space.

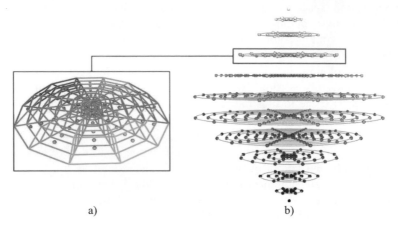

a)   b)

Figure 21.3. *a) A slice of the HLS space; b) partition of this space into 554 cells*

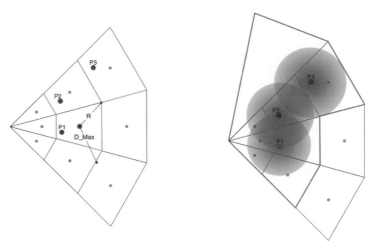

Figure 21.4. *Influence of points P_1, P_2 and P_3 on neighboring cells*

Figure 21.5 shows the results of fully automated selection for a football scene. These results show the desired region, and also show pitch markings, which need to be extracted for the purposes of 3D reconstruction.

Figure 21.5. *On the right, the pixels retained by colorimetric analysis are shown in white*

21.2.3. *Spatial coherence*

While ROI selection is essentially based on the notion of color, we also need to take into account the spatial coherence of this region. At the end of the color-based treatment, the binary mask produced may contain small regions associated with elements of the same color in the image (advertising boards, banners, flags etc.). The ROI is considered to be the largest connected component of the image. In reality, the region is "almost" connected: small channels remain, which must be filled without connecting elements from outside the pitch. Moreover, certain elements of the ROI are not of the same color, such as pitch markings. Elementary mathematical morphology operators [SER 88] may be used to remove these elements. The ROI is selected once the largest region has been filled (see Figure 21.6).

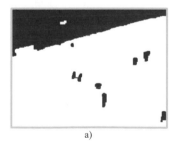 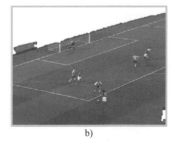

a)   b)

Figure 21.6. *a) Binary image of Figure 21.5 after closing and opening; b) selected ROI*

21.3. The Hough transform

A variety of methods may be used to extract characteristic elements from an image, but our goal in this section is not to provide a complete overview of these approaches. For sporting scenes, pitch markings are often made up of straight and ellipse arc segments, and the Hough transform [ILL 88] is ideally suited to characterizing these simple geometric primitives.

The detection of straight lines using the Hough transform is a textbook case. However, detection becomes more difficult when segments cross over, are affected by noise or are partially obscured. In sporting scenes, lines can be partially thickened or hidden by players. In these cases, pairs must be carefully selected from the retained characteristic points to participate in determining the Hough space. Figure 21.7 shows this type of image and demonstrates the effectiveness of pitch marking detection.

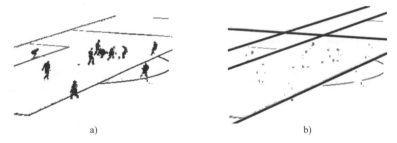

a)   b)

Figure 21.7. *a) the results of colorimetric analysis; b) line detection after player deletion*

21.3.1. *Detection of ellipse arcs*

The detection of ellipses or ellipse arcs is harder than the detection of straight lines. Certain methods use variable ellipses, which adapt to data according to different criteria [KAN 01, MAT 00] with a variable degree of success. In this section, we demonstrate the use of the Hough transform for such cases.

The most general Cartesian equation for an ellipse is expressed using six parameters connected by a proportionality relationship, meaning that an ellipse is characterized by five effective parameters. A Hough transform of 5–1 cannot be used for detection in these cases for reasons of complexity (parameter space of dimension 5), lack of precision and homogeneity in the parameter space.

Several studies have been carried out based on the division of the Hough space into multiple sub-spaces [AGU 95, HO 95, PAO 93, YUE 89].

A characteristic property of the tangents of an ellipse is: "let M_1 and M_2 be two points of the ellipse, T_1 and T_2 the tangents at these points, I the point of intersection of T_1 and T_2, and J the middle of $M_1 M_2$; the straight line $(I J)$ passes through the center O of the ellipse". Using this property, it is possible to break the search down into two searches, one in dimension 2 and the other in dimension 3. Based on this breakdown, the main stages of the detection process are as follows:

– Characterization of relevant pairs (point, tangent): for each point, we use neighbors to determine whether or not we can associate an estimated tangent.

– Search for the center of ellipse: using the retained set of pairs (point, tangent) and the property described above, it is possible to detect the center of ellipses with arcs appearing in the image.

– Calculation of parameters a, b, θ: once the ellipse has been recentered, an expression exists as a function of the angle θ made by its long axis with axis Ox and the parameters a and b of its normalized expression (where a and b are half-lengths of the long and short axes).

– Calculation of the ellipse arc: once an ellipse has been detected, the corresponding ellipse arcs can be extracted by studying the distribution of points associated with the shape.

Figures 21.8 and 21.9 show the obtained results. In Figure 21.8, the detected ellipse is formed by four ellipse arcs.

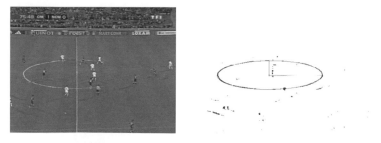

Figure 21.8. *Detection of the center circle*

Figure 21.9. *Detection of the 18 m arc*

21.4. Matching image features to the geometric model

The matching of characteristic elements between an image and a geometric model is a particularly difficult process, as it requires the comparison of elements of different natures: pixels from the image and elements of the model. In reconstructing sporting scenes, work has naturally been carried out on matching pitch markings, often lines and ellipses as discussed in section 21.3, with the same elements defined in the model of the scene to reconstruct [FAR 05, HAY 05, OKU 04, THO 07]. The images for treatment are a partial representation of the model. Thus, detected elements form a very limited subset of the set of primitives defined in the model. Matching these elements with elements of the model may be guided by similarity measures based on knowledge of the model: relative line positions, distances and dimensions. This type of solution is too strongly linked to the characteristics of the model to be used in a more general context. However, the use of parallel lines as markers is common to many sports pitches; it is useful to identify this information within the detected elements in order to introduce it into the mapping stage. The following section offers a geometric method for selecting converging lines in an image.

21.4.1. *Converging lines*

The detection of converging lines has often been compared to the detection of receding points. There are intrinsic problems in the detection of receding points in an image, linked to the search space. This problem is often resolved by using methods based on the use of a Gaussian sphere for the search space [BAR 83, MAG 84]. These methods use accumulation-type approaches, which create certain other difficulties (discretization, maxima searches, etc.). In order to be robust, these methods require heavy calculations, additional knowledge or the use of uncertainty models. Furthermore, accumulation-type methods involve the implicit hypothesis that the number of occurrences of an observed phenomenon (converging lines) is high, something that is not guaranteed in images of sporting scenes. In reality, these images are projections of a model (the field), which is often made up of two sets of parallel lines (groups formed by small numbers of lines).

The use of certain geometric properties allows us to detect sets of converging lines in an image without needing to use an accumulation space. First, we must identify the sets formed by exactly three lines in the image. Second, these sets are combined by considering their shared lines.

Using a Thales circle (a semicircle whose diameter is the hypotenuse of a right-angled triangle), we may give the following corollary:

COROLLARY 21.1 (THALES).– Take three lines D_1, D_2 and D_3. Let A be any point not belonging to D_1, D_2 or D_3. Let A_1, A_2 and A_3 be the orthogonal projections of A onto D_1, D_2 and D_3, respectively. If the circle formed by A_1, A_2 and A_3 passes through A, then we have a Thales circle. Thus, lines D_1, D_2 and D_3 intersect at the same point.

Figure 21.10 illustrates this corollary. It is important to note that, by construction, lines A_1, A_2 and A_3 form a set of lines that intersects the set defined by D_1, D_2 and D_3 at a right angle. Moreover, the set formed by D_1, D_2 and D_3 and the set formed by A_1, A_2 and A_3 are in homographic correspondence. Thus, we obtain an application of the Chasles–Steiner theorem, which shows that a homography between two sets of lines defines a conic, in the Euclidean space. As the sets intersect at right angles, the conic is a circle.

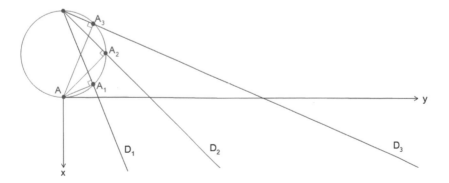

Figure 21.10. *Point A and its projections form a circle*

It is not difficult to verify that point A and its projections are cocyclic, but this may be a source of numerical problems. To avoid explicit calculation of the parameters of the circle, we may use Ptolemy's theorem, which simply requires calculation of the points (orthogonal projections of A onto the different lines in the image) located in the image (or in close vicinity).

THEOREM 21.1 (PTOLEMY'S THEOREM).– A convex quadrilateral is inscribable if, and only if, the product of the lengths of the diagonals is equal to the sum of the products of the lengths of opposite sides.

It is interesting to note that infinite receding points are treated "naturally". The Thales circle formed by a set of straight lines (participating in an infinite receding point) corresponds to a "flat circle", and the lines therefore verify Ptolemy's theorem.

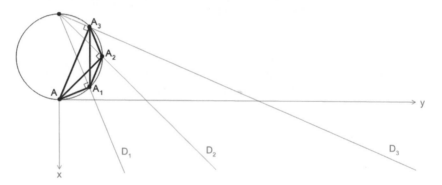

Figure 21.11. *The quadrilateral defined by points A, A_1, A_2, A_3 is inscribable*

A similarity criterion is needed to identify line triplets belonging to the same set. An intuitive idea would be to compare the Thales circles defined by these line triplets, but this requires us to calculate the parameters of these circles, something which we are able to avoid. Another possibility is the use of an accumulation space to highlight equivalent Thales circles, but this is meaningless given the limited number of circles shown in the images. By considering the lines shared between triplets, we are able to identify triplets with potential sets. If a triplet of competing lines shares none of its lines with another triplet, then it corresponds to a set of three lines. If, on the other hand, a triplet of competing lines shares two lines with other triplets, then each line of these triplets belongs to the same set. As the number of triplets is limited, we may simply combine triplets to verify these two properties and, where possible, to fuse them.

Figure 21.12 illustrates the results of detection of two sets of converging lines in the presented binary image. In this way, a line resulting from a detection error is eliminated. Moreover, in addition to providing additional information for the lines, this step allows us to validate the set of elements that are candidates for matching.

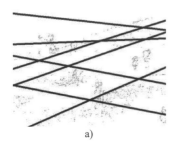
a)

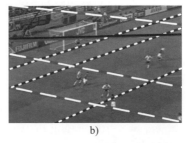
b)

Figure 21.12. *a) Set of detected lines; b) two groups are retained, and the line shown in black is excluded*

The information associated with converging lines generally allows us to class lines that are candidates for matching into two groups. This greatly facilitates the matching stage, in that the problem is reduced to a "which is which?" search with a reduced set of parallel lines.

21.5. Conclusion

The segmentation of sequences in team sport scenes constitutes a major hindrance to the establishment of fully automatic reconstruction mechanisms. These sequences include variable bases and changing lighting, motion blurring and camera distortion, factors that explain the difficulties involved. The identification and tracking of multiple players on the pitch also remains difficult, both as a result of segmentation problems and because of multiple occultations in the course of a sequence. In terms of televised diffusion, the added value produced by player tracking does not justify the significant cost involved in installing specific cameras and the use of dedicated operators, although information such as player positioning and the distances covered in the course of a match constitutes a useful contribution to a post-match analysis.

In this chapter, we have demonstrated the possibility of automating certain stages of the reconstruction process. The proposed methods are robust, automatic and operate in real time. They have been implemented in the form of a software prototype that has been used by several football training centers [PET 08, POP 08, VAL 13]. Nevertheless, it is difficult to envisage fully automatic reconstruction without a material infrastructure offering several views of the same playing surface in order to resolve problems of occultation between players, for example. While these systems do already exist, the algorithms they use do not offer sufficiently good performances, and they only offer supervised reconstruction of short image sequences.

Figure 21.13 shows a circular movement at a constant height produced from the position of the camera: the first image (a) is that produced from the camera, image (b) is a combination of image (a) with a synthesized image (c), i.e. the 3D reconstruction of the image. Images (d), (e) and (f) are synthesized. These images are obtained by sequential rotation of the virtual camera by $-10°$ around the vertical axis passing through the projection on the pitch in the center of the image.

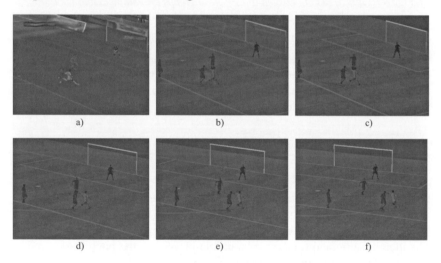

Figure 21.13. *Reconstruction of a real scene*

Figure 21.14 is an illustration of a real situation processed by the SimulFoot training program [MAV 13]: an image is selected from a video (a), and then reconstructed in a realistic (b) and symbolic manner (c). This virtual representation may then be viewed from different perspectives, including those of players (d). It may also be replayed, either identically or with a different outcome.

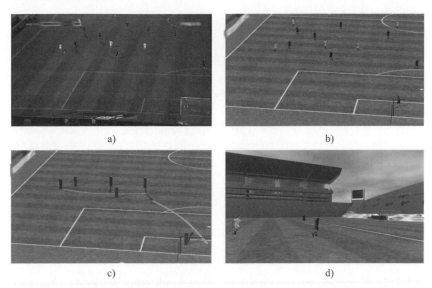

a)   b)

c)   d)

Figure 21.14. *Different views of a reconstruction*

21.6. Bibliography

[AGU 95] AGUADO A., MONTIEL M., NIXON M., "Ellipse extraction via gradient direction in the hough transform", *Proceedings of IEE 5th International Conference on Image Processing and its Applications*, IEE, Edinburgh, UK, pp. 375–378, July 1995.

[AMI 13] AMISCO, "Sport universal", available at http://www.sport-universal.com, 2013.

[BAR 83] BARNARD S.T., "Interpreting perspective images", *Artificial Intelligence*, vol. 21, pp. 435–462, 1983.

[CAR 95] CARRON T., Segmentations d'images couleur dans la base teinte-luminance-saturation: approche numérique et symbolique, PhD Thesis, University of Savoie, France, 1995.

[DAI 95] DAILIANAS A., ALLEN R.B., PAUL E., "Comparison of automatic video segmentation algorithms", *Proceedings of SPIE Photonics West*, IOP Publishing Ltd., pp. 2–16, 1995.

[EKI 03] EKIN A., TEKALP A.M., "Automatic soccer video analysis and summarization", *IEEE Transactions on Image Processing*, vol. 12, pp. 796–807, 2003.

[FAR 05] FARIN D., HAN J., DE P.H.N., "Fast camera-calibration for the analysis of sports sequences", *Proceedings of IEEE International Conference on Multimedia and Expo (ICME)*, Amsterdam, The Netherlands, pp. 482–485, 2005.

[GRA 07] GRAU O., HILTON A., KILNER J., *et al.*, "A free-viewpoint video system for visualisation of sport scenes", *Motion Imaging*, vol. 116, no. 5–6, pp. 213–219, May 2007.

[HAY 05] HAYET J.-B, PIATER J.H., VERLY J.G., "Fast 2D model-to-image registration using vanishing points for sports video analysis", *Proceedings of IEEE International Conference on Image Processing, ICIP 2005*, Genoa, Italy, pp. 417–420, 2005.

[HO 95] HO C.-T., CHEN L.-H., "A fast ellipse/circle detector using geometric symmetry", *Pattern Recognition*, vol. 28, no. 1, pp. 117–124, 1995.

[ILL 88] ILLINGWORTH J., KITTLER J., "A survey of the Hough transform", *Computer Vision, Graphics and Image Processing*, vol. 44, no. 1, pp. 87–116, August 1988.

[KAN 01] KANATANI K.-I., OHTA N., "Automatic detection of circular objects by ellipse growing", *Memoirs of the Faculty of Enginieering*, vol. 36, no. 1, pp. 107–116, 2001.

[KIL 07] KILNER J., STARCK J., HILTON A., *et al.*, "Dual-mode deformable models for free-viewpoint video of sports events", *Proceedings of the 6th International Conference on 3-D Digital Imaging and Modeling, 3DIM '07*, IEEE Computer Society, Washington, DC, pp. 177–184, 2007.

[LEF 00] LEFÈVRE S., HOLLER J., VINCENT N., "Real time temporal segmentation of compressed and uncompressed dynamic colour image sequences", *Proceedings of International Workshop on Real Time Image Sequence Analysis*, Oulu, Finland, pp. 56–62, 2000.

[LE 04] LE TROTER A., MAVROMATIS S., SEQUEIRA J., "Soccer field detection in video images using color and spatial coherence", *Proceedings of ICIAR - International Conference on Image Analysis and Recognition*, vol. 2, ICIAR, Porto, Portugal, pp. 265–272, 2004.

[LE 06] LE TROTER A., Mise en correspondance d'images et de modèles : application à la reconstruction 3D de scènes sportives, PhD Thesis, University of Aix-Marseilles, 2006.

[LLU 11] LLUCIA L., Suivi d'objets à partir d'images issues de caméras mobiles non calibrées, PhD Thesis, Université Aix-Marseille II, 2011.

[LUC 01] LUCCHESE L., MITRA S.K., "Color image segmentation: a state-of-the-art survey", *Proceedings of the of the Indian National Science Academy (INSA-A)*, vol. 67, pp. 207–211, 2001.

[MAG 84] MAGEE M.J., AGGARWAL J.K., "Determining vanishing points from perspective images", *Computer Vision, Graphics, and Image Processing*, vol. 26, no. 2, pp. 256–267, 1984.

[MAT 00] MATEI B., MEER P., "Reduction of bias in maximum likelihood ellipse fitting", *Proceedings of the 15th International Conference Pattern Recognition*, vol. 3, Barcelona, pp. 794–798, September 2000.

[MAV 13] MAVROMATIS S., "SimulFoot", available at http://sebastien.mavromatis. perso.luminy.univ-amu.fr/SimulFoot/docs/assets/SimulFoot_08.pdf, 2013.

[OKU 04] OKUMA K., LITTLE J.J., LOWE D.G., "Automatic rectification of long image sequences", *Asian Conference on Computer Vision*, Jeju Island, Korea, January 2004.

[ORA 13] "ORAD – 3D replay", available at http://www.orad.tv/products/ 3dreplay, 2013.

[PAO 93] PAO D.C.W., LI H.F., JAYAKUMAR R., "A decomposable parameter space for the detection of ellipses", *Pattern Recognition Letters*, vol. 14, no. 12, pp. 951–958, 1993.

[PAR 62] PARZEN E., "On estimation of a probability density function and mode", *The Annals of Mathematical Statistics*, vol. 33, no. 3, pp. 1065–1076, 1962.

[PET 08] PETIT J.-P., RIPOLL H., "Scene perception and decision making in sport simulation: A masked priming investigation", *International Journal of Sport Psychology*, vol. 39, no. 1, pp. 1–19, 2008.

[POP 08] POPLU G., RIPOLL H., MAVROMATIS S.J.B., "How do expert soccer players encode visual information in order to make decisions in simulated game situations?", *Research Quarterly for Exercise and Sport*, vol. 79, no. 3, pp. 392–398, 2008.

[RED 13] "Red Bee Media – Piero", available at http://www.redbeemedia.com/ piero/piero, 2013.

[REN 09] REN R., JOSE J., "General highlight detection in sport videos", in HUET B., SMEATON A., MAYER-PATEL K., AVRITHIS Y., (eds), *Advances in Multimedia Modeling, Lecture Notes in Computer Science*, vol. 5371, Springer, Berlin, Heidelberg, pp. 27–38, 2009.

[SER 88] SERRA J., (ed.), *Image Analysis and Mathematical Morphology*, vol. 2, Theoretical Advances, Academic Press, London, San Diego, New York, 1988.

[THO 07] THOMAS G.A., "Real-time camera tracking using sports pitch markings", *Journal of Real-Time Image Processing*, vol. 2, no. 2-3, pp. 117–132, 2007.

[TRI 13] TRIC TRAC, "TricTrac", available at http://www.multitel.be/trictrac/ ?mod=1, 2013.

[VAL 13] VALORPACA, "SimulFoot", available at http://valorpaca.univmed.fr/IMG/fiches marketing/fiche marketing_simulfoot.pdf, 2013.

[VAN 00] VANDENBROUCKE N., Segmentation d'images couleur par classification de pixels dans des espaces d'attributs colorimétriques adaptés. Application à l'analyse d'images de football, PhD Thesis, University of Lille, 2000.

[YAN 08] YAN F., CHRISTMAS W., KITTLER J., "Layered data association using graph-theoretic formulation with application to tennis ball tracking in monocular sequences", *IEEE Transactions on Pattern Analysis and Machine Intelligence*, vol. 30, no. 10, pp. 1814–1830, October 2008.

[YUE 89] YUEN H.K., ILLINGWORTH J., KITTLER J., "Detecting partially occluded ellipses using the Hough transform", *Image and Vision Computing*, vol. 7, no. 1, pp. 31–37, February 1989.

Chapter 22

Experiments in Live Capture and Transmission of Stereoscopic 3D Video Images

22.1. Introduction

From 2009 to 2011, the INTELSIG Laboratory of the University of Liège (ULg)[1] produced 12 live "retransmissions" in stereoscopic 3D (S-3D). For each of them, INTELSIG acted as organizer and technical coordinator for the whole retransmission chain (capture, transmission, and visualization). These retransmissions were carried out for purely experimental, illustrative, and educational purposes, with the aim of pushing the boundaries of the state of the art. Each experiment involved a large number of partners, bringing to bear their expertise and technical means.

The idea behind the experiments initially arose in the course of the first 3D Stereo MEDIA (3DSM) event, in 2009. This international event was the brainchild of one of the authors of this chapter (Jacques G. Verly). It is devoted to all aspects and all applications of all forms of 3D. It aims to provide a meeting point for the scientific, technological, artistic, and business aspects of 3D.

Chapter written by David GROGNA and Jacques G. VERLY.

1 The INTELSIG Laboratory is part of the Department of Electrical Engineering and Computer Science of the School of Engineering of the University of Liège in Belgium.

These 12 retransmissions included four shows of varying types (2009), two surgical operations (2010), three steadicam interviews (2010), a transatlantic video presentation (2010), and two bicycle races (2011).

22.2. Retransmissions of various shows

Our first experiments with live S-3D retransmissions were carried out in the course of the first 3DSM event, which took place in the *Palais des Congrès* of Liège in Belgium, in 2009.

22.2.1. *Capture*

For each of the four retransmissions, the capture took place in a part of the hall used as an exhibition space (see Figure 22.1(a)), just outside of the conference rooms (where the S-3D images were projected). The first three experiments consisted in the retransmission of three different shows, each on a different day. The fourth – the highlight of the closing session – consisted of capturing a real person on a *green key* background (see Figure 22.1(b)), integrating him into a virtual set, and simultaneously introducing a virtual character conversing with the real person, all – of course – in live S-3D. The virtual set, the virtual person, its animation, and the insertion were produced by NeuroTV.

Figure 22.1. *Capture of a show in live S-3D in a hall of the Palais des Congrès of Liège. a) The artists evolved on a special surface laid onto the carpeting. To the left, one sees exhibition stands. b) The green key used to insert a real person into a virtual set. The wooden floor and the tracks show the relation to the previous image*

22.2.2. *Transmission*

The high-definition (HD) images from the various S-3D rigs were transmitted at full resolution along coaxial cables (HD-SDI signals) to an outside broadcast (OB) van from Outside Broadcast located outside the Palais des Congrès. In the van, the S-3D images were examined, adjusted, and mixed into two synchronized signals, which were then sent down via two optical fiber cables to the projection booth of the 500-seat conference room, where the optical signals were reconverted into HD-SDI signals sent to the S-3D projector.

22.2.3. *Visualization*

For each edition of 3DSM, we have to choose, obtain, install, and select the various S-3D projectors that we need. Most projectors intended for cinemas are designed to project films using contents called digital cinema packages (DCPs). Any other use of these projectors requires specific technical knowledge; and this is the case for live retransmissions.

During the various editions of 3DSM held at the Palais des Congrès of Liège, we used two BARCO projectors of type DP2000 or equivalent. After conversion, the left and right images were brought into the projector via its HD-SDI inputs. Figure 22.4(b) shows an example of a DP2000 projector of the type we used.

22.3. Retransmissions of surgical operations

Our retransmissions of surgical operations [AUD 10, VER 10d, VER 11] took place in 2010 as part of the ninth edition of the ImagéSanté Festival, organized once every two years by the ULg under the direction of Dr. Philippe Kolh, a professor at the ULg and a heart surgeon at the University Hospital of Liège. This hospital is known as the "Centre hospitalier universitaire" (CHU) of Liège. The festival is popular among students in medicine and in other associated domains, professionals in the same sectors, and the general public. The live retransmissions of surgical operations are particularly popular.

The INTELSIG Laboratory had already participated in the festival in 2008, with the addition of a scientific conference called "Medical Imaging". Jacques G. Verly had then proposed to launch another event, independent from ImagéSanté, on all aspects of "3D". This event was called 3D Stereo MEDIA (3DSM), and its first edition took place in 2009. Given that 3DSM had been launched shortly after the Medical Imaging conference, Philippe Kolh wished to examine the possibility of a live S-3D retransmission of a surgical operation as part of the ImagéSanté Festival of 2010.

The aim of the retransmission was to film in S-3D a surgical operation in one of the operating rooms of the CHU of Liège (on the Sart-Tilman site, in the suburbs of Liège), transmit the live images to the Cinéma Sauvenière in the center of Liège (almost 12 km away as the crow flies), and to project the S-3D images in a 300-seat movie theater. A bidirectional vocal link was also required, allowing the surgeon to comment on the operation, the audience to ask questions, and the surgeon to provide answers. It should be noted that, at the time, the cinema was not equipped with a S-3D projector. We thus had a lot of things to do...

We envisioned various types of surgical operations. In the end, we selected a non-endoscopic operation and, specifically, a neurosurgical operation. One reason for this choice was the collaboration between Dr. Didier Martin, a professor at the ULg and a neurosurgeon at the CHU, and Jacques G. Verly in the domain of image-assisted neurosurgical navigation. Another reason is that the zone concerned by the operation was relatively limited, which would allow us to optimize the S-3D capture and the depth budgets.

In addition to the operation planned for live transmission, the decision was made to film another operation beforehand in order to identify potential problems, optimize S-3D capture, and provide a backup recording in case of problems with the live retransmission.

The first surgical operation took place on 25 February 2010. It consisted of treating an aneurysm by installing a clip. It lasted about five hours. The operation was captured and recorded, but was not retransmitted. However, a variety of transmission tests were carried out between the CHU and our laboratories during the month of March, to check that we had correctly identified the access points at the hospital for the appropriate optical fibers, and that the desired data rate of 500 Mbits/s could be maintained over several hours without experiencing any contention. The tests showed that the data rate could reach almost 1 Gbits/s in a continuous manner.

The recordings were then used to create a montage, made up of a variety of clips; this represented a challenge in itself as few local companies had the necessary facilities for S-3D montage and S-3D visualization. The clips were also shown during the live surgery during periods judged to be of limited interest for the audience.

The second surgical operation took place on 18 March 2010. It was recorded and retransmitted live. A relatively short surgical procedure was selected to fit into the scheduled 2.5 h session in the cinema. The operation consisted of the extraction of a meninx tumor (meningioma). As a result of a bidirectional vocal link, the members of the audience were able to ask questions and the surgeon was able to answer them or to comment on the surgery. Figure 22.2 shows the architecture of the whole retransmission chain, including capture, transmission, and projection, which we describe below.

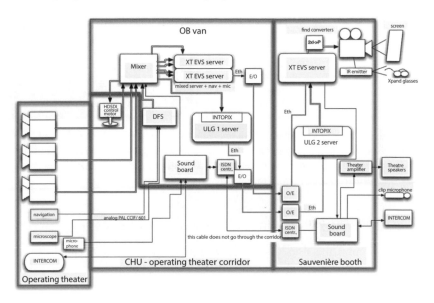

Figure 22.2. *Block diagram of the full retransmission chain, including capture, transmission, and projection*

22.3.1. *Capture*

The two operations took place in the operating room 16 at the CHU (see Figure 22.3(a)). As the operation was going to be projected onto a large screen of a 300-seat movie theater, the left and right images of each S-3D pair were both transmitted in lightly compressed full HD, meaning a total data rate of about 500 Mbits/s, which should be compared to the traditional data rates of 12–15 Mbits/s used for the festival.

The capture and stereography were carried out by Binocle, under the direction of Yves Pupulin. We decided to install three S-3D rigs (see

Figure 22.3.(b)): the static rig at the patient's feet, the second next to the first and opposite to the surgeon, and the third mounted on a crane to film over the shoulder of the surgeon's assistant. One challenge was to place all of the required equipment in the restricted space of an operating room, while allowing the personnel to easily move around the patient. Another challenge was to meet the sterilization requirements.

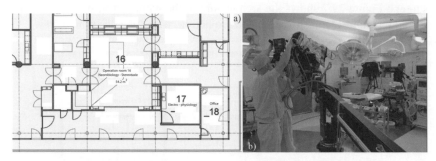

Figure 22.3. *a) Plan of part of the operating room of interest at the CHU of Liège, with, at bottom right, the balcony and its wall that various cables were drawn over. b) Arrangement of the three S-3D rigs around the operating table. The capture operators wore standard operating room clothing*

22.3.2. *Transmission*

To minimize the risks associated with transmission, we decided that we could not tolerate the slightest competition on our communication line, which notably excluded the use of Internet. The required data rate of 500 Mbits/s over the 2 h of the operation excluded the possibility of a satellite link due to bandwidth restrictions and the cost of renting a satellite channel for 2 h, alongside the near impossibility of advance testing. The only possible solution was to use dedicated optical fibers between the CHU and the cinema.

SOFICO indicated that they had fibers between the CHU and the city center, but not up to the cinema. Given the size of the CHU, it was difficult to find the point of entry/exit of the optical fibers. Once the point had been located, technical personnel from the CHU and the ULg installed a cable from this point to the technical room 17, close to the operating room 16 (see Figure 22.3(a)). On the cinema side, SOFICO opened up almost 300 m of sidewalk in the city center and installed the required optical fibers, which were then extended to the sixth floor of the cinema. The required electrical/optical converters were installed at both ends. For added security, the fibers, which were initially butted, were fused together to produce a monolithic "glass" medium between the converters, i.e. over a distance of almost 16 km. Furthermore, two separate communication "tubes" were laid over completely different geographical pathways in case of an accident occurring along one of them (e.g. caused by diggers, etc.). Each cable contained dozens of individual fibers, but only one was used.

During the planning stage, we realized that a specially equipped OB van would be needed at the CHU for the manipulation and the mixing of the video streams from the three S-3D rigs. The van, provided by Notélé, was installed in the parking lot at the foot of the relevant building (see Figure 22.4(a)). Coaxial cables from the three rigs (HD-SDI signals) were run over the balcony wall outside of the operating room to the van (see Figures 22.3(a) and 22.4(a)). In the van, the S-3D images were corrected, adjusted, and mixed to create the image stream to be ultimately shown on the cinema screen. Before transmission, the images were compressed into JPEG 2000 in real time using a PRISTINE electronic board from INTOPIX. The sound was also integrated into the data stream. After electrical/optical conversion in the van, the final signal for transmission was sent to the technical room 17 (see Figure 22.3(a)), where the access point for the 16 km of fibers to the cinema was located. The routing of these cables between the operating room and the car, their protection against mechanical damage, and the sterility requirements presented major logistical challenges, not to mention the wild boars roaming on the CHU grounds!

Figure 22.4. *a) OB van in the parking at the base of the building (far right) where the operating room was located (on the top floor). b) View from the projection booth at the Cinéma Sauvenière, showing the DP2000 projector that we had installed for the occasion and a partial view of an image projected onto the screen (showing the operating field), through the fireproof window of the booth*

A backup plan was developed in case of problems with the transmission strategy: images, with a higher compression rate, were sent block by block over the Internet from an XT server from EVS located in the van to a similar server located in the cinema.

22.3.3. *Visualization*

As the Cinéma Sauvenière was not equipped with S-3D projection equipment at the time, we arranged for the temporary installation of a BARCO DP2000 projector, which we checked out intensively in the test center of the XDC company (now dcinex) in Liège.

In the projection booth, optical signals were converted into electrical signals, and then decompressed by a second PRISTINE board before being sent to the HD-SDI inputs of the BARCO projector (see Figure 22.4(b)). The projection booth also included a sound mixing table that gathered the ambient sounds from the video, the surgeon's microphone signal transmitted using voice-over-IP (VOIP), and the cinema's microphones, also transmitted to the operating room using VOIP.

When we selected the projector, we did not consider the losses caused by the passage of the light beam through the fireproof window of the projection booth (see Figure 22.4(b)). In spite of a significant loss of light, the S-3D images retained a sufficient level of brightness.

22.4. Retransmissions of "steadicam" interviews

For the retransmissions at 3DSM 2009, the capture and visualization locations were in close proximity. For 3DSM 2010, the decision was made to increase the distance between these locations. Visualization was maintained at the Palais des Congrès of Liège, but capture was moved to various sites in Liège [VER 10a, VER 10b]: the "Marché de Noël" (Christmas Market) on 8 December 2010 (see Figure 22.5(a), the "SOS Planet" exhibition at the Gare des Guillemins on 9 December, and the Pôle Image de Liège (PIL) on 10 December for its inauguration (see Figure 22.5(b)). In each case, the scenario consisted in an exploration of the location and in interviews of people. The exploration stage required the use of a *steadicam*; the transmission distance and the chosen locations led us to select a satellite link. A conventional 2D video camera was also used to allow the spectators watching the retransmission to see the S-3D steadicam in operation.

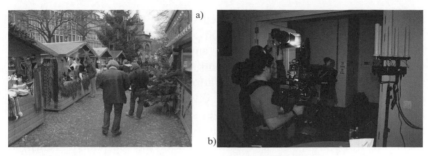

Figure 22.5. *a) First steadicam capture location in Liège: the "Marché de Noël". b) Third steadicam capture location in Liège: the Pôle Image de Liège (PIL) (courtesy of C. Iland from the PIL)*

22.4.1. *Capture*

The left side of Figure 22.6 shows the architecture of the capture system. To be concrete, we will now describe the implementation of this architecture in the case of the "Marché de Noël".

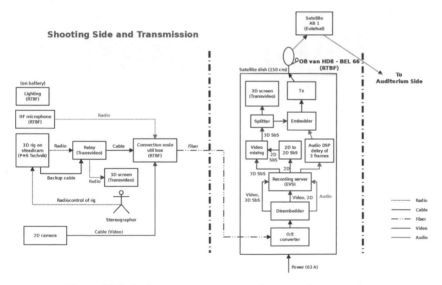

Figure 22.6. *Architecture of capture and of part of transmission*

The main sensor was a P+S Technik S-3D mirror rig mounted on a steadicam supplied by Transvideo. The images acquired by the rig were converted into a side-by-side (SbS) format by the rig and transmitted by radio from a transmitter on the rig to a fixed receiver, positioned in an optimal location. The images were directly converted to SbS as this was required before transmission to the satellite in any case. Figure 22.7(a) shows the steadicam on its base (with the rig, control screen, transmitter, etc.) and the reception/transmission system on the platform. The "candles" on the platform are radio antennas, providing spatial diversity to attenuate the multipath effect. The reception part receives images sent from the rig. The transmission part sends these images to the control monitor of the stereographer. Figure 22.7(b) shows a more detailed view of the reception/transmission system, supplied by Transvideo. The S-3D SbS images were then transmitted via a coaxial cable to a Radio-Tèlèvision Belge Francophone (RTBF) connection unit, which was out of sight, secured, and protected from weather conditions (it was snowing slightly) in one of the "chalets" of the village. From there, the S-3D images were sent via an optical fiber to the RTBF HD6 OB van. The 2D video camera was also linked to the above connection unit, and its images were sent to the OB van at the same time as the S-3D images. The purpose of this 2D camera was to allow spectators (at the Palais des Congrès) to watch the S-3D capture system at work. As this camera was linked to the connection unit via a cable, its field of action was limited, but satisfactory for our needs.

Figure 22.7. *a) View of the steadicam on its base and of the fixed radio reception/transmission system (with "candle" antennas). b) Transvideo reception/transmission system. The system allowed one to receive images from the S-3D rig and to send them directly to the control unit of the stereographer. The system was also able to send images via coaxial cables, in this case to the RTBF connection unit*

The stereographer (Erwan Davigano) had access to a control unit, allowing him to adjust the S-3D rig using a radio remote control, and to observe the obtained images on a Transvideo monitor; these images were transmitted by radio from the control unit to the Transvideo reception/transmission system. Unsurprisingly, the most delicate part of the capture system up to the optical fibers was the radio transmission of images. The frequency of 5 GHz essentially requires a direct line of sight between the transmission antenna and the reception antenna. We thus experimented to find the best position for the fixed radio system (the equipment with "candles"), that would allow us both to control the rig and to receive high quality S-3D images during the live retransmission. Sound was captured separately during retransmission, and inserted into the RTBF connection unit. Figures 22.8(a) and (b) show the capture system in operation. Note that, as a 2D camera was also used to film the S-3D rig in action, spectators could see images similar to the photos shown here.

22.4.2. *Transmission*

As we indicated earlier, we chose (somewhat arbitrarily) the optical fiber arriving at the transmission RTBF HD8 OB van as the transition point between the capture part and the transmission part (see Figure 22.9(a)). Similarly, we choose as the transition between the transmission part and the visualization part the optical fiber leaving the receive RTBF BEL69 satellite van (see Figure 22.9(b)) at the Palais des Congrès. The right side of Figure 22.6 shows the architecture of the *uplink* part of transmission, and the left side of Figure 22.10 shows the *downlink* part.

Figure 22.8. *a) The S-3D steadicam in operation. The steadicammer is followed by the stereographer, who controls the rig by radio remote control. b) One of the exhibitors interviewed during the retransmission, who quickly realized that brandishing a sausage at the client was an excellent way of demonstrating the "jumping out of the sceen" effect*

Upon entering the transmission OB van, the 2D and S-3D images were recorded for future analysis (before possible degradation during satellite transmission). For ease of mixing and projection, it was decided to convert 2D images into SbS, an unusual and somewhat paradoxical decision. An electronic system was even constructed at the last minute to carry out this conversion. The producer thus mixed the 2D and S-3D images during the transmission. The image stream sent to the satellite constituted the content that was shown.

 a)

 b)

Figure 22.9. *a) RTBF HD8 OB van used to mix images and to send signals to the satellite (uplink). b) RTBF BEL69 satellite van used to receive signals from the satellite (downlink)*

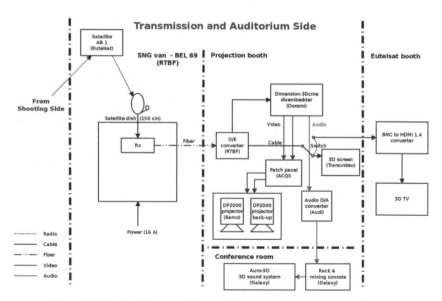

Figure 22.10. *Architecture of part of the transmission system and of the visualization system*

Eutelsat allowed us to use the Eutelsat AB1 satellite, located at 12.5° west. For each retransmission, Eutelsat allowed a period of 40 min of use, with 10 min for testing and 30 min for live retransmission. Several weeks before the retransmissions, Eutelsat also allowed satellite time for transmission testing from the RTBF site at Rhisnes. At the Palais des Congrès, the signals from the satellite were received by the RTBF BEL69 satellite van, and then sent over an optical fiber to the projection booth.

22.4.3. *Visualization*

The right side of Figure 22.10 shows the architecture of the visualization system. In the projection room of the 500-seat room, a Doremi Dimension-3Dcine converter separated the video and audio signals and converted the SbS (2D or 3D) video stream into two distinct synchronized streams, feeding the HD-SDI ports of a BARCO DP2000 projector. The interest of transforming 2D images into 2D SbS resides in the fact that it is then not necessary to activate/deactivate the side-by-side or double-stream functions of the projector with each change of camera.

22.5. Retransmission of a transatlantic video presentation

We invited Mark Schubin from Shubin Café, based in New York City (NYC), to deliver an invited presentation at 3DSM 2010. Due to professional obligations, he was unable to travel to Liège, but he offered to make his presentation via video conference from NYC. We then asked whether he was interested in making the presentation in S-3D, a challenge he was willing to take up with us.

The scenario that was ultimately retained is the following: there were two distinct sets of audiovisual content to project: a classic PowerPoint style presentation and a live S-3D video retransmission of Mark Schubin giving his presentation from NYC. The projection could therefore take different forms: the presentation alone, the presenter alone, or a combination of the two, for example showing the presentation at the base of the screen with a smaller window showing the presenter higher up. These various options were tested during live transmission, and the arrangement of the elements of the composite view was also adjusted live, as only 10 min of testing time was available immediately before the presentation. The solution that was ultimately chosen was to load the presentation onto a PC located in the projection booth and connected to the S-3D projector, and to let Mark Schubin take control of the presentation using the LogMeIn software.

22.5.1. *Capture*

S-3D capture was carried out by All-Mobile Video in NYC, using a *3ality* S-3D rig. All stereographic adjustments were carried out in NYC, taking into account the screen size (9 m wide) and the dimensions of the room.

22.5.2. *Transmission*

Transmission was carried out using the same satellite used for the steadicam interviews, the Eutelsat AB1. It was this transmission which dictated the choice of satellite; as the satellite is within "view" of Liège, it was also suited for retransmissions with uplink and downlink points in Liège. In NYC, All-Mobile Video routed the signals to a satellite uplink point, probably using optical fibers. In Liège, the signals were captured by the RTBF BEL69 satellite van, already used for the retransmission of steadicam interviews. The images were then transmitted in SbS format, allowing them to be inserted into classic transmission schemes. One then talks about *in-frame* transmission.

There was, however, one major difference between the retransmissions from NYC and from Liège. In the case of the retransmissions between two points in Liège, which involved two RTBF vans, the choice of modulation to be used was easy and almost automatic. In the case of the transmission from NYC to Liège, the modulation used for the uplink was different to that generally used by the RTBF. This problem was rapidly identified during preliminary testing. Once the signals had been received and correctly demodulated, they were sent to the projection booth in exactly the same way used for the steadicam interviews.

22.5.3. *Visualization*

The visualization system was very similar to that used for the steadicam retransmissions. For a very short period at the beginning of the live session, we encountered color problems and an absence of depth. These problems were rapidly solved because of the perspicacity and reactiveness of the technical team. When Mark Schubin pressed a key on his computer in NYC, the audience in Liège saw the corresponding movement in the S-3D video, and the appropriate 2D slide appeared on screen.

22.6. Retransmissions of bicycle races

In early 2011, we provided technical assistance to the RTBF (French-speaking Belgian Radio and Television organization) for experimental live S-3D transmissions of two major bicycle races in Belgium [VER 10c]: the Flèche Wallonne (20 April 2011) and Liège-Bastogne-Liège (24 April 2011). In each case, tests were carried out the day before over part of the course and notably at the finish line.

One constraint lay in the fact that these retransmissions could not disrupt the regular 2D retransmissions. Therefore, in addition to the four motorcycles used for standard 2D capture, the RTBF used a fifth motorcycle carrying specific S-3D capture equipment (see figure 22.11(a)). This motorcycle was also required to stay behind the racers. The chosen transmission channel corresponded to the backup channel for regular 2D transmission.

As the retransmissions were purely experimental, the images were visualized live in S-3D by the ULg and RTBF teams and a group of VIPs at the finish line, and they were not broadcast to the general public. They were also recorded for later analysis and for use in developing stereography algorithms. As far as we know, these live S-3D retransmissions over a whole race and using almost standard race infrastructure constituted a first.

22.6.1. *Capture*

As for classic 2D race retransmissions, the cameraman was positioned on the back of the S-3D capture motorcycle. Panasonic placed an AG-3DA1 camera at the disposal of the RTBF. The cameraman generally carried it on his shoulder. No image stabilization equipment (mechanical or electronic) was used. As the interaxial distance of the camera was fixed, the cameraman was only able to modify the convergence and the zoom. As none of the RTBF's motorcycle cameraman had experience in S-3D capture, we provided basic stereography training to the chosen cameraman. In addition, we were in permanent radio contact with him: based on the images visualized at the finish line, we were able to give him precise real-time instructions for adjusting the camera controls, improving image quality, and testing a variety of shooting strategies, showing the advantages and disadvantages of each.

The "language" of S-3D filming for bicycle races remains to be defined, and it can only be developed and refined through field experimentation. With this in mind, we tested a variety of view capture methods: in the direction of movement of the motorcycle, toward the front (close-up and far); toward the back (close-up and far); while overtaking, or being overtaken by cyclists, using a panning to follow them; toward the back with the camera close to the ground (as sometimes seen in 2D retransmissions); and with both a stationary motorcycle and a fixed camera direction.

22.6.2. *Transmission*

To enable in-frame transmission over existing 2D channels, the RTBF added a Davio converter to the classic motorcycle transmission equipment. The converter was used to convert the left and right HD streams produced by the camera into a side-by-side (SbS) HD stream suited for transmission over standard radio channels. The raw images produced by the camera were also recorded on a pair of SD memory maps located in the camera itself. This allowed us to preserve the raw images, avoiding losses of resolution and degradations due to SbS conversion and to transmission. The addition of the Davio converter obliged us to rig up a specific electrical power supply (feeding the motorcycle battery). While the available current was sufficient in continuous conversion mode, it proved to be insufficient to read the memory maps used to program the Davio, something which compromised the transmission tests carried out the day before the first race.

The SbS images were then transmitted in the conventional way to the helicopter and airplane following the race and from these air vehicles to the OB van at the finish line. As the S-3D images were not broadcast to the general public, they were not subject to mixing and transmission over a longer distance, for example via an RTBF tower such as the one located at the Bol d'Air outside of Liège. The helicopter is generally used to "retrieve" images from deep valley areas.

22.6.3. *Visualization*

The images received by the airplane or helicopter were then transmitted to the radio antennas and receivers on the finish line (see Figure 22.11(b)). During testing at the Flèche Wallonne race, the images were examined on a S-3D monitor in the back of a vehicle. For the Liège-Bastogne-Liège race, the images were shown on a large Panasonic 3D-screen (with active glasses) in a tent close to the finish line, accessible to VIP visitors. The line and the tent were located close to the equipment shown in Figure 22.11(b).

Viewers of the S-3D retransmission of the Liège-Bastogne-Liège race on the large S-3D screen were unanimous in recognizing the interest of S-3D for the retransmission of bicycle races. Evidently, image stabilization would have been desirable in many cases.

Figure 22.11. *a) View of the traditional radio transmission equipment on the back of the S-3D capture motorcycle. The side-by-side Davio converter is not visible. b) A variety of radio reception equipment on the finish line at Ans during the 2011 Liège-Bastogne-Liège race*

A variety of interesting phenomena were observed during retransmission. One of the most striking examples involved the rear-view mirror of the motorcycle, which was in the field of vision of the camera: the scene observed in the mirror was also in S-3D! A video montage with highlights of each race was produced at the CRIG in Liège and was shown by the authors during a presentation at 3DSM in December 2011. The presentation was followed by a debate during which experienced stereographers were able to draw lessons from these pioneering, experimental retransmissions.

22.7. Conclusion

In this chapter, we have described the live stereoscopic (S-3D) transmissions in which we were involved from 2009 to 2011 as architects, technical advisers, and/or instigators. At the time when we began planning these retransmissions (in 2008), live S-3D retransmissions were a rarity and constituted highly visible technical exploits. Moreover, our work took place at the very beginning of 3D-TV and S-3D broadcast channels. Sky 3D (Korea) was the first 3D broadcast channel in continuous operation (1 January 2010). In Europe, the British company BSkyB began offering a limited service on 3 April 2010, with retransmission of a match to more than 1,000 bars and sporting clubs in England. Our retransmission of a surgical operation using full HD, rather than side-by-side technology, took place on 25 March 2010, just before this first BSkyB broadcast. All of our experiments may thus be considered to be pioneering.

Live S-3D retransmission involves all of the difficulties of S-3D, i.e. capture, transmission, and visualization, simultaneously. Each domain is a challenge in itself and presents a rich terrain for scientific and technical development. Moreover, while stereography for live action (as opposed to computer animation) is already difficult in the best possible capture conditions, such as on a film set, it must be carried out almost instantly in the case of live transmission, with no opportunity for later corrections. While significant progress has been made in computer-controlled stereography, there is currently no fully automatic solution to the problem, despite the claims made by some companies.

In our experiments, we have often pushed the boundaries of technical possibilities, whether by adding virtual and real characters into virtual surroundings, transmitting S-3D streams with each image in full HD (in contrast to traditional in-frame methods, such as those resulting from SbS conversion), or capturing mobile subjects (cyclists) from a mobile platform (motorcycle) with constantly changing backgrounds, for which we do not have detailed information in advance.

Because of the new challenges encountered, the retransmissions we performed were generally not feasible (at a reasonable cost) for traditional broadcasters, mostly because they operate in highly limited contexts, with a constant and justified focus on reliability. Because of the technical resources required, often not fully known beforehand, these retransmissions were not feasible for researchers and academics. Our retransmissions were therefore a particularly suitable subject for collaboration between research and broadcasting teams, groups that do not generally meet. In addition to technical questions, some "soft skills" were clearly needed to allow the teams to communicate effectively, and to collaborate and come up with solutions to problems, in a context where the most minor technical difficulty can be a show-stopper. Examples include the conversion of 2D streams into SbS streams, something no one thought would ever be useful, and the need to maintain synchronization between left and right video streams in full HD (plus sound) during transmission. The challenges presented by these retransmissions naturally contributed to the development of close ties between participants, who were all simultaneously confronted with technical difficulties and the stress of live broadcasting.

Above and beyond the close collaboration between S-3D researchers and broadcasters, retransmission projects also constitute a unique occasion for encounters between scientists, engineers, artists or doctors, broadcast technicians and directors, and logistics personnel. During experimentation at the CHU of Liège, the 20 to 30 people involved in the project had their meals together in the hospital cafeteria. A live S-3D retransmission exercise thus lies at the convergence of science, technology, art, and business. While no equations have been presented in this chapter, many calculations are concealed in the various technologies used, from stereographic calculations to radio link budget calculations. The retransmission projects we have described also present a unique opportunity for engineers to take a "system" approach. From this perspective, one of our most useful tools was the precise block diagram of the whole retransmission chain. The creation and updating of diagrams of this type require a strong discipline, and it was often difficult to convince all of our partners to submit to this delicate art. In spite of having followed such a discipline, numerous modifications were required up until the last minute. At the end of each retransmission, the reality barely corresponded to the final block diagram drawn; and the speed with which retransmission chains were dismantled and the teams dispersed often precluded bringing the block diagram up to date for future use.

One striking observation for the architects that we were was that, often, at the moment of retransmission, no one had a clear idea of the precise nature (in terms of content or format) of the signals traveling through the multiple cables in use! While the near-perfect operation of all of our retransmissions always seemed miraculous, their success was due to excellent preparation and teamwork, where each person demonstrated (almost) perfect mastery of his/her equipment!

We hope that the various technical solutions described in this chapter will be of assistance to those who would wish to undertake live S-3D transmissions, and will help them to save time. At the time of this writing (early 2013), commercial S-3D broadcasting channels are still few and far between. Only time will tell what the future holds in terms of live S-3D retransmissions and broadcasting.

22.8. Bibliography

[AUD 10] AUDRIT P., LEENS J., VERLY J., "La S-3D et Réalité Augmentée au service de la médecine et de la science", *Dimension 3*, Paris, June 2010.

[VER 10a] VERLY J., EVRARD M., GROGNA D., "Steadicam Demo 1: the live 3D transmission from a steadicam at the Holiday Season Village in Liège to the Convention Center via terrestrial and satellite radio links", *3D Stereo MEDIA*, Liège, Belgium, December 2010.

[VER 10b] VERLY J., EVRARD M., GROGNA D., "Steadicam Demo 2: the live 3D transmission from a steadicam at the SOS Planet Exhibit in Liège to the Convention Center via terrestrial and satellite radio links", *3D Stereo MEDIA*, Liège, Belgium, December 2010.

[VER 10c] VERLY J., GROGNA D., EVRARD M., *et al.*,, "Experimental live 3D transmissions of two classical bicycle races in Belgium: screening of S-3D clips, analysis of images, lessons learned, and the dos and don'ts", *3D Stereo MEDIA*, Liège, Belgium, December 2010.

[VER 10d] VERLY J., LEENS J., GROGNA D., *et al.*,, "Live, interactive, 3D-stereo, full-HD, high-bandwidth capture, transmission, and projection of a neurosurgical operation", *3D Stereo MEDIA*, Liège, Belgium, December 2010.

[VER 11] VERLY J., "Live 3D-stereo full-HD retransmission of a surgical operation", *Workshop on 3D Imaging*, Stanford University, CA, January 2011.

Conclusion

The development of technologies linked to 3D video has considerably transformed our relation with images over the last decades. The democratization of augmented reality, for example, now allows the whole cinematography industry to mix real and virtual images, using quality criteria that make it more and more difficult to identify computer-generated items in a natural image. The same applies to 3D television (3DTV); the emergence of 3DTV technologies has brought out a number of changes that are certain to revolutionize the whole image production chain, from capture to diffusion.

In this book, we aimed to cover a broad range of subjects connected with 3D video. In five main parts, we have successively considered issues linked to:

– 3D capture;

– 3D representation, structuring and reconstruction;

– 3D coding and compression;

– 3D rendering and display.

In the final part, we considered a number of applications in relation to healthcare and/or audiovisual production.

As a whole, this book is the result of considerable effort in knowledge accumulation. In publishing this book, the editors and the authors had two

---

Conclusion written by Laurent LUCAS, Céline LOSCOS and Yannick REMION.

aims in terms of 3D video: to provide students with a basis and motivation to successfully tackle this domain, and to allow researchers in the field to reinforce their knowledge and add to their understanding of sectors outside of their respective specialisms, observing trends and discovering new perspectives.

List of Authors

Marc ANTONINI
I3S
University of Nice-Sophia Antipolis
France

Marcus BARKOWSKY
IRCCyN
LUNAM University
University of Nantes
France

Benjamin BATTIN
OPEXMedia
Clichy
France

Aassif BENASSAROU
CReSTIC
University of Reims
Champagne-Ardenne
France

Marie-Odile BERGER
LORIA
INRIA Nancy Grand Est
University of Lorraine
France

Venceslas BIRI
LIGM
University Paris-Est Marne-la-Vallée
France

Ludovic BLACHE
CReSTIC
University of Reims
Champagne-Ardenne
France

Jennifer BONNARD
CReSTIC
University of Reims
Champagne-Ardenne
France

Marco CAGNAZZO
LTCI
Telecom ParisTech
France

Sylvie CHAMBON
IRIT
ENSEEIHT
University of Toulouse
France

Afef CHAMMEM
ARTEMIS
Telecom SudParis
France

Mohamed DAOUDI
LIFL
Telecom Lille 1
France

Manuel DAUCHEZ
MeDYC
University of Reims
Champagne-Ardenne
France

Michel DESVIGNES
GIPSA-Lab
University of Grenoble
France

Frédéric DEVERNAY
PRIMA
INRIA Grenoble
France

Frédéric DUFAUX
LTCI
Telecom ParisTech
France

Florent DUPONT
LIRIS
University Claude Bernard-Lyon 1
France

Guillaume GALES
National University of Ireland
Maynooth
Republic of Ireland

Josselin GAUTIER
IRISA
University of Rennes 1
France

David GROGNA
Institut Montefiore
University of Liège
Belgium

Christine GUILLEMOT
SIROCCO
INRIA Rennes
France

Nicolas HAUTIÈRE
LEPSiS
IFSTTAR
Marne la Vallée
France

Muhannad ISMAËL
CReSTIC
University of Reims
Champagne-Ardenne
France

Vincent JANTET
Artefacto
Rennes
France

Joël JUNG
Orange Labs
France

Min H. KIM
VCLAB
KAIST
Daejeon
South Korea

Guillaume LAVOUÉ
LIRIS
INSA de Lyon
France

Patrick LE CALLET
IRCCyN
LUNAM University
University of Nantes
France

Antoine LEJEUNE
Institut Montefiore
University of Liège
Belgium

Olivier LE MEUR
IRISA
University of Rennes 1
France

Jing LI
IRCCyN
LUNAM University
University of Nantes
France

Céline LOSCOS
CReSTIC
University of Reims
Champagne-Ardenne
France

Laurent LUCAS
CReSTIC
University of Reims
Champagne-Ardenne
France

Sébastien MAVROMATIS
LSIS
University of Aix-Marseille
France

Benoît MICHEL
Université catholique de Louvain
Louvain-La-Neuve
Belgium

Mihai MITREA
ARTEMIS
Telecom SudParis
France

Elie Gabriel MORA
LTCI
Telecom ParisTech and Orange Labs
France

Luce MORIN
IETR
INSA Rennes
France

Cédric NIQUIN
XD Productions
Issy Les Moulineaux
France

Jean-Michel NOURRIT
CReSTIC
University of Reims
Champagne-Ardenne
France

Vincent NOZICK
LIGM
University Paris-Est Marne-la-Vallée
France

Dimitri PAPATHANASSIOU
Institut Jean Godinot
Reims
France

Béatrice PESQUET-POPESCU
LTCI
Telecom ParisTech
France

Sylvia PIOTIN
CReSTIC
University of Reims
Champagne-Ardenne
France

Françoise PRÊTEUX
MINES ParisTech
France

Stéphanie PRÉVOST
CReSTIC
University of Reims
Champagne-Ardenne
France

Jessica PRÉVOTEAU
CReSTIC
University of Reims
Champagne-Ardenne
France

Yves PUPULIN
Binocle 3D
Bry-sur-Marne
France

Yannick REMION
CReSTIC
University of Reims
Champagne-Ardenne
France

Barbara ROMANIUK
CReSTIC
University of Reims
Champagne-Ardenne
France

Jean SEQUEIRA
LSIS
University of Aix-Marseille
France

Gilles SIMON
LORIA
University of Lorraine
France

Philippe SOUCHET
XD Productions
Issy Les Moulineaux
France

Hedi TABIA
ETIS
ENSEA
University of Cergy-Pontoise
France

Jean-Baptiste THOMAS
Le2i
University of Burgundy
France

Matthieu URVOY
IRCCyN
University of Nantes
France

Gilles VALETTE
CReSTIC
University of Reims
Champagne-Ardenne
France

Jean-Philippe VANDEBORRE
LIFL
Telecom Lille 1
France

Philippe VAUTROT
CReSTIC
University of Reims
Champagne-Ardenne
France

Jacques G. VERLY
Institut Montefiore
University of Liège
Belgium

Lara YOUNES
CReSTIC
University of Reims
Champagne-Ardenne
France

Index